Soil Biology

Series Editor: Ajit Varma

11

Volumes published in the series

Volume 1
Applied Bioremediation and Phytoremediation (2004)
A. Singh, O.P. Ward (Editors)

Volume 2
Biodegradation and Bioremediation (2004)
A. Singh, O.P. Ward (Editors)

Volume 3
Microorganisms in Soils: Roles in Genesis and Functions (2005)
F. Buscot, A. Varma (Editors)

Volume 4
In Vitro Culture of Mycorrhizas (2005)
S. Declerck, D.-G. Strullu, J.A. Fortin (Editors)

Volume 5
Manual for Soil Analysis – Monitoring and Assessing Soil Bioremediation (2005)
R. Margesin, F. Schinner (Editors)

Volume 6
Intestinal Microorganisms of Termites and Other Invertebrates (2006)
H. König, A. Varma (Editors)

Volume 7
Microbial Activity in the Rhizosphere (2006)
K.G. Mukerji, C. Manoharachary, J. Singh (Editors)

Volume 8
Nucleic Acids and Proteins in Soil (2006)
P. Nannipieri, K. Smalla (Editors)

Volume 9
Microbial Root Endophytes (2006)
B.J.E. Schulz, C.J.C. Boyle, T.N. Sieber (Editors)

Volume 10
Nutrient Cycling in Terrestrial Ecosystems (2007)
P. Marschner, Z. Rengel (Editors)

Ajit Varma
Ralf Oelmüller (Eds.)

Advanced Techniques in Soil Microbiology

With 94 Figures, 2 in Color

Springer

Prof. Dr. Ajit Varma
Director, Amity Institute of Microbial Sciences
and
Vice Chairman (International), Amity Sciences Technology Foundation
Amity University Uttar Pradesh
Sector-125, Noida 201303
India
E-mail: ajitvarma@aihmr.amity.edu

Prof. Dr. Ralf Oelmüller
Friedrich-Schiller-University Jena
Institute of General Botany and Plant Physiology
Dornburger Str. 159
07743 Jena
Germany
E-mail: b7oera@rz.uni-jena.de

Library of Congress Control Number: 2007921585

ISSN 1613-3382
ISBN-978-3-540-70864-3 Springer-Verlag Berlin Heidelberg New York

Springer-Verlag is a part of Springer Science+Business Media
springer.com

Editor: Dr. Dieter Czeschlik, Heidelberg, Germany
Desk Editor: Dr. Jutta Lindenborn, Heidelberg, Germany
Cover design: WMXDesign GmhH, Heidelberg, Germany
Typesetting and production: LE-TEX Jelonek, Schmidt & Vöckler GbR, Leipzig, Germany
Printed on acid-free paper SPIN 11543862 31/3100 YL 5 4 3 2 1 0

Preface

There is general belief and admission that important, innovative and novel ideas emerge over a cup of 'Indian Darjeeling tea' or a glass of 'German beer'. The editors of this book were sipping a cup of tea on the lush green garden lawns of North Maharastra University, Jalgaon, India. The weather was congenial and most suitable for materializations of original ideas. The genesis of this book underlines the concept developed in 2006.

The field of microbiology began concurrently with the discovery of microorganisms by two Fellows of The Royal Society, Robert Hooke and Antony van Leeuwenhoek, during the period 1665–1683. Later, during the golden era of microbiology, noted scientists Louis Pasteur and Robert Koch laid a sound foundation for the modern microbiology. The study of microorganisms has became a valuable science in the last 100 years as it has provided both the means to control a number of infectious diseases and the experimental systems for the development of molecular biology. New developments in biotechnology and environmental microbiology signify that microbiology will continue to be an exciting field of study in the future. Various modern tools and techniques are required for a proper understanding of the roles of microbes in the causation of infectious diseases and the recycling of chemical elements in the biosphere. Assorted laboratory experiments not only motivate researchers and students by stimulating interest and enjoyment but also enhance the acquisition of scientific knowledge along with the development of 'scientific attitudes', such as open-mindedness and objectivity.

There are numerous textbooks and review papers dealing with state-of-the-art of various aspects of molecular biology of microorganisms. However, the readers get lost in initiating the experiments due to lack of suitable and easy protocols. They have to search for diverse methods and techniques in a variety of literature and journals and still do not obtain the complete information dealing with the protocols in a concise manner. This book is an attempt to overcome the inherent cumbersome search process. Every effort was made to present the protocols in a very simple manner for easy understanding of undergraduate, graduates, postgraduates, post doctorates, active scientists and researchers.

Soil, the main contributor to plant nourishment, is the top layer of the Earth's surface and consists of rock and mineral particles mixed with organic matter. Soil microbiology is the study of the microorganisms in soil, their functions, and the consequences of their activities on the nature of the soil and the effect on the growth and health of plant life. Just a few grams of soil, less than a teaspoonful, may contain hundreds of millions to billions of microbes. Not only is the total number of microorganisms in fertile soil quite high, but also, together, they weigh a lot. Soil microbial biomass can range from several hundred to thousands of pounds per acre.

The most plentiful microbes in soil are one-celled bacteria and fungi, which produce long, slender strings of cells called filaments or hyphae. The actinomycetes come between these two organisms. It is the actinomycetes that give soil its characteristic earthy smell. In this volume, the editors have accumulated various advanced molecular approaches for studying the different soil microorganisms for the benefit of humankind. Different techniques for measuring microbial biomass and activity in soil have been developed. Primers in Random Amplified Polymorphic DNA (RAPD) techniques for species identification and other forgotten tools like quantitative histochemistry are discussed in details in this book with the hope that this would promote the understanding of microbes by students and advanced researchers alike.

The editors have brought together the diverse topics related to various aspects of molecular approaches to the detection of soil microbes, namely assessing and detecting soil micro-fungal diversity and providing insight into their feasibility. Various problems associated with the dilution plating technique, importance of the rDNA gene in fungal systematics, the reliability of other molecular approaches (especially Denaturing Gradient Gel Electrophoresis) and their drawbacks are discussed. Various modern tools and techniques like automated fluorescent DNA sequencing strategy, mRNA quantitation using real time PCR, RNAi technology, transcriptome analysis and immuno-techniques are handled by subject experts of these specific fields for clear and easy understanding for all. Different widely used methods like fatty acid methylester (FAME), phospholipid fatty acid (PLFA) analyses and denaturing gradient gel electrophoresis (DGGE) are elucidated with their advantages and limitations outlined. DGGE and RISA protocols for microbial community analysis in soil are also one of the highlights of this book.

The soil zone located in and around the active roots is called the rhizosphere. This zone has high microbial activity. Materials released from roots, called exudates, create a food-rich environment for the growth of microorganisms. Rhizosphere microorganisms in turn help plants by fixing nitrogen from the soil air, dissolving soil minerals and decomposing organic matter, all of which allow roots to obtain essential nutrients. Plant-Growth-Promoting Rhizobacteria (PGPRs) generate a variety of chemicals that stimulate plant growth. The bacteria grow and persist in the rhizosphere of non-woody roots. Various screening methods for PGPRs are described in this book.

A special kind of fungus called mycorrhizae also associates with higher plants. By colonizing large areas of roots and reaching out into the soil, mycorrhizae assist in transport of soil nutrients and water into the plant. The latest methods for conducting experiments and research in mycorrhiza have been described.

Cultivation of a group of mycorrhiza-like fungi belonging to family Sebacinales is enumerated. One of the members of Sebacinales which provides stress tolerance activity against heavy metals and induced pathogen resistance in cereals is discussed.

Authors have brought forth diverse approaches and methods to study the mechanisms behind the observed pathogen resistance induced by *Piriformospora indica*.

Model organism *A. thaliana* was used as the plant partner to understand the molecular basis for beneficial plant/microbe interactions and this is also discussed in this edition. Several other techniques like ion cyclotron resonance Fourier transform mass spectrometry (ICR-FT/MS) for non-targeted metabolomics of molecular interactions in the rhizosphere are presented. Immunotechnology for the localization of acid phosphatase using native gel bands in *P. indica* and other soil microorganism are elaborated in this volume of the Soil Biology series.

We are grateful to the many people who helped to bring this volume to light. We wish to thank Dr. Dieter Czeschlik and Dr. Jutta Lindenborn, Springer Heidelberg, for generous assistance and patience in finalizing the volume. Finally, specific thanks go to our families, immediate, and extended, not forgetting those who have passed away, for their support or their incentives in putting everything together. Ajit Varma in particular is very thankful to Dr. Ashok K. Chauhan, Founder President of the Ritnand Balved Education Foundation (an umbrella organization of Amity Institutions), New Delhi, for the kind support and constant encouragement received. Special thanks are due to my esteemed friend and well-wisher Professor Dr. Sunil Saran, Director General, Amity Institute of Biotechnology and Adviser to Founder President, Amity Universe, all faculty colleagues Drs. Amit C. Kharkwal, Harsha Kharkwal, Shwet Kamal, Neeraj Verma, Atimanav Gaur and Debkumari Sharma and my Ph.D. students Ms. Aparajita Das, Mr. Ram Prasad, Ms. Manisha Sharma, Ms. Sreelekha Chatterjee, Ms. Swati Tripathi, Mr. Vipin Mohan Dan and Ms. Geetanjali Chauhan. The technical support received from Mr. Anil Chandra Bahukhandi is highly appreciated.

New Delhi, India
Jena, Germany
March 2007

Ajit Varma
Ralf Oelmüller

Foreword

There is no doubt that biotechnology is one of the leading disciplines in modern biology. Concerning its ever-growing impact on the development of new products, its importance cannot be overestimated; in terms of generating new jobs and industries it is certainly that section of biology which is responsible for the largest financial volume and the highest degree of application of biological knowledge. Interestingly, biotechnology is also the most interdisciplinary science as it uses efficiently the various biological disciplines which were often separated in the past and which even kept their own characteristics at the expense of neighboring disciplines. In biotechnology the product counts more than the origin, and the frontiers between animal, plant and bacterial cells are of minor importance. Today, the central role of the new genes dominates a good part of biotechnology; it creates new products; however, the cellular environment must obey the laws of efficiency, practicability and production costs.

For the above reasons it is important to assemble the ever-improving methods of modern biotechnology in a book under these new guidelines, i.e. practical aspects and immediate use in the laboratory and beyond. These methods involve all the essential methods of molecular biology, immunology, microbiology and structural biology; the complexity of the systems involved ranges from individual molecules to the eukaryotic organisms themselves, with a focus on bacteria, fungi and higher plants. As it is extremely difficult to cover even the most important state-of-the-art methods from the whole field, a comprehensive book with selected authors and methods such as this is extremely useful: it encourages students to look at biology in a different focus, assembling methods with a clear aim at a product, and it tells the experienced researcher about the leading laboratories and the most promising strategies.

The 26 chapters of this book are indeed an excellent and outstanding contribution towards this end.

Mattthias Rögner — Ruhr Universität Bochum, Germany
Govindjee — University of Illinois at Urbana-Champaign, USA

Contents

Contributors

Achatz, Beate
Institute of Phytopathology and Applied Zoology (IPAZ), Justus-Liebig-Universität Gießen, Heinrich-Buff-Ring 26–32, D-35392 Gießen, Germany
and
Institute for Vegetables and Ornamental Crops, D-14979 Großbeeren, Germany

Bamezai, R. N. K.
National Centre of Applied Human Genetics, School of Life Sciences, Jawaharlal Nehru University (JNU), New Delhi-110067, India

Bazzicalupo, Marco
Dipartimento di Biologia Animale e Genetica 'Leo Pardi', Via Romana 17, I-50125, Firenze, Italy

Bhaskar, Sangeeta
National Institute of Immunology, Aruna Asaf Ali Marg, New Delhi-110067, India

Bhatnagar, Kamya
Amity Institute of Microbial Sciences, Amity University Uttar Pradesh, Sector 125, Noida, 201303, India

Bukhari, S. I.
National Centre of Applied Human Genetics, School of Life Sciences, Jawaharlal Nehru University (JNU), New Delhi-110067, India

Dangre, Devanand
Department of Biotechnology, SGB Amravati University, Amravati-444602, Maharashtra, India

Das, Aparajita
Amity Institute of Microbial Sciences, Amity University Uttar Pradesh, Sector 125, Noida, 201303, India

DasGupta, Sangeeta Mehta
Amity Institute of Microbial Sciences, Amity University Uttar Pradesh, Sector 125, Noida, 201303, India

Englmann, M.
GSF – Research Center for Environment and Health, Institute of Ecological Chemistry, Molecular BioGeoanalysis / BioGeomics, Ingolstädter Landstraße 1, D-85764 Neuherberg, Germany

Fekete, A.
GSF – Research Center for Environment and Health, Institute of Ecological Chemistry, Molecular BioGeoanalysis / BioGeomics, Ingolstädter Landstraße 1, D-85764 Neuherberg, Germany

Frommberger, M.
GSF – Research Center for Environment and Health, Institute of Ecological Chemistry, Molecular BioGeoanalysis / BioGeomics, Ingolstädter Landstraße 1, D-85764 Neuherberg, Germany

Gaur, A.
Amity Institute of Microbial Sciences, Amity University Uttar Pradesh, Sector 125, Noida, 201303, India

Gebefugi, I.
GSF – Research Center for Environment and Health, Institute of Ecological Chemistry, Molecular BioGeoanalysis / BioGeomics, Ingolstädter Landstraße 1, D-85764 Neuherberg, Germany

Giuntini, E.
Dipartimento di Biologia Animale e Genetica 'Leo Pardi', Via Romana 17, I-50125, Firenze, Italy

Gochhait, S.
National Centre of Applied Human Genetics, School of Life Sciences, Jawaharlal Nehru University (JNU), New Delhi-110067, India

Haag, S.
University of Tübingen, Physiological Ecology of Plants, Auf der Morgenstelle 1, 72076 Tübingen, Germany

Hahn, Nadin
Institute of General Botany, Department of Environmental Sciences, University of Jena, Dornburger Straße 159, D-07743 Jena, Germany

Hampp, Rüdiger
University of Tübingen, Physiological Ecology of Plants, Auf der Morgenstelle 1, 72076 Tübingen, Germany

Hertkorn, N.
GSF – Research Center for Environment and Health, Institute of Ecological Chemistry, Molecular BioGeoanalysis / BioGeomics, Ingolstädter Landstraße 1, D-85764 Neuherberg, Germany

Hyde, K. D.
Department of Ecology and Biodiversity, University of Hong Kong, Pok Fu Lam Road, Hong Kong

Jeewon, R.
Department of Ecology and Biodiversity, University of Hong Kong, Pok Fu Lam Road, Hong Kong

Karan, Ratna
School of Life Sciences, Jawaharlal Nehru University, New Delhi, India

Kashem, Abul
Proteomics Laboratory of Pathology, School of Medical Science, Blackburn Building (D06), University of Sydney, NSW 2006, Australia

Kharkwal, Amit C.
Amity Institute of Microbial Sciences, Amity University Uttar Pradesh, Sector 125, Noida, 201303, India

Kharkwal, Harsha
Amity Institute of Microbial Sciences, Amity University Uttar Pradesh, Sector 125, Noida, 201303, India

Khurana, Satish
National Institute of Immunology, Aruna Asaf Ali Marg, New Delhi-110067, India

Kogel, Karl-Heinz
Institute of Phytopathology and Applied Zoology (IPAZ), Justus-Liebig-Universität Gießen, Heinrich-Buff-Ring 26-32, D-35392 Gießen, Germany

Kumar, P.
Virology Group, International Centre for Genetic Engineering & Biotechnology (ICGEB), New Delhi, India

Kumari, Sumita
School of Life Sciences, Jawaharlal Nehru University, New Delhi, India

Lal, S. K.
Virology Group, International Centre for Genetic Engineering & Biotechnology (ICGEB), New Delhi, India

Lucio, M.
GSF – Research Center for Environment and Health, Institute of Ecological Chemistry, Molecular BioGeoanalysis / BioGeomics, Ingolstädter Landstraße 1, D-85764 Neuherberg, Germany

Malhotra, D.
National Centre of Applied Human Genetics, School of Life Sciences, Jawaharlal Nehru University (JNU), New Delhi-110067, India

Marschner, Petra
School of Earth and Environmental Sciences, The University of Adelaide, DP 636, SA 5005, Australia

Mengoni, A.
Dipartimento di Biologia Animale e Genetica 'Leo Pardi', Via Romana 17, I-50125, Firenze, Italy

Mukhopadhyay, Asok
National Institute of Immunology, Aruna Asaf Ali Marg, New Delhi-110067, India

Nautiyal, C. Shekhar
Plant–Microbe Interaction Laboratory, National Botanical Research Institute, Rana Pratap Marg, Lucknow, 226001

Oberoi, R.
Virology Group, International Centre for Genetic Engineering & Biotechnology (ICGEB), New Delhi, India

Oelmüller, Ralf
Institute of General Botany and Plant Physiology, University of Jena, Dornburger Straße 159, D07743, Jena, Germany

Ortaş, Ibrahim
University of Çukurova, Faculty of Agriculture, Department of Soil Science, 01330 Balcali, Adana, Turkey

Pandey, A.
Department of Biological Sciences, The University of Alabama in Huntsville, Huntsville, AL 35899, USA

Pareek, Ashwani
Stress Physiology and Molecular Biology Laboratory, School of Life Science, Jawaharlal Nehru University, Aruna Asaf Ali Marg, New Delhi-110067, India

Podila, G. K.
Department of Biological Sciences, The University of Alabama in Huntsville, Huntsville, AL 35899, USA

Pokharel, Utprekshya
Department of Microbiology, Punjab University, Chandigarh, India

Prasad, Ram
Amity Institute of Microbial Sciences, Amity University Uttar Pradesh, Sector 125, Noida, 201303, India

Rai, E.
National Centre of Applied Human Genetics, School of Life Sciences, Jawaharlal Nehru University (JNU), New Delhi-110067, India

Rai, Mahendra
Department of Biotechnology, SGB Amravati University, Amravati-444602, Maharashtra, India

Rajani Malla
Department of Microbiology, Tri-Chandra Campus, Tribhuvan University, Kathmandu, Nepal

Schmitt-Kopplin, Philippe
GSF – Research Center for Environment and Health, Institute of Ecological Chemistry, Molecular BioGeoanalysis / BioGeomics, Ingolstädter Landstraße 1, D-85764 Neuherberg, Germany

Shahollari, Bationa
Institute of General Botany and Plant Physiology, University of Jena, Dornburger Straße 159, D-07743 Jena, Germany

Sherameti, Irena
Institute of General Botany and Plant Physiology, University of Jena, Dornburger Straße 159, D-07743 Jena, Germany

Singla-Pareek, Sneh L.
Plant Molecular Biology, International Centre for Genetic Engineering and Biotechnology, Aruna Asaf Ali Marg New Delhi-110067, India

Solaiman, Zakaria
School of Earth and Geographical Sciences, The University of Western Australia, Crawley, WA 6009, Australia

Strasser, Reto J.
Laboratory of Bioenergetics, University of Geneva, Chemin des Embrouchis 10, CH-1254 Jussy-Geneva, Switzerland

Tsimilli-Michael, Merope
Laboratory of Bioenergetics, University of Geneva, Chemin des Embrouchis 10, CH-1254 Jussy-Geneva, Switzerland

Varma, Ajit
Amity Institute of Microbial Sciences, Amity University Uttar Pradesh, Sector 125, Noida, 201303, India

Waller, Frank
Institute of Phytopathology and Applied Zoology (IPAZ), Justus-Liebig-Universität Gießen, Heinrich-Buff-Ring 26-32, D-35392 Gießen, Germany

White, H.
Department of Biological Sciences, The University of Alabama in Huntsville, Huntsville, AL 35899, USA

Yadav, Sudesh Kumar
Biotechnology Division, Institute of Himalayan Bioresource Technology, CSIR, Palampur-176061 (HP), India

1 Detection and Diversity of Fungi from Environmental Samples: Traditional Versus Molecular Approaches

R. Jeewon and K.D. Hyde

1.1 Introduction

Microbial life within the soil ecosystem is a fascinating aspect of soil biology, and has recently caught the attention of microbiologists. Many fungi grow in the soil and some have evolved to thrive in harsh conditions, such as those found in acidic or alkaline soils. These microorganisms can be considered as "highly developed" as they flourish and reproduce in these ecological niches and unusual habitats and have successfully made use of soil and its nutrients for their energy sources. Fungi are an important component of the soil microbiota, they mediate important ecological processes such as nutrient recycling, and they maintain important symbiotic relationships with plants and bacteria (Garrett 1981; Parkinson 1983; Yu et al. 2005). Many fungi are pathogenic (e.g Jaworski et al. 1978; Cahill and Mohr 2004) and some may be useful in bio-exploitation (e.g Vinokurova et al. 2003). The realms of soil mycota are possibly the largest on the planet.

A diverse range of fungi are present in soil ecosystems and include ascomycetes, basidiomycetes, some being ectomycorrhizal fungi, anamorphic fungi and arbuscular mycorrhizal fungi (AMF). At present, there is no clear morphological, phylogenetic or ecological definition of soil fungi. Any definitions based on these concepts are very difficult to implement because the soil ecosystem harbours a plethora of fungi with great morphological, genetic and functional diversity and lacks geographic boundaries. Perhaps the best definition of soil fungi should be encapsulated in the word itself (fungi from soil!). Most of our current knowledge of soil mycota is based on traditional systematics, which does not reflect any real sense of evolutionary relationships. The interaction between

Department of Ecology and Biodiversity, University of Hong Kong,
Pok Fu Lam Road, Hong Kong, email: rjeewon@hku.hk; rjeewon@graduate.hku.hk

Soil Biology, Volume 11
Advanced Techniques in Soil Microbiology
A. Varma, R. Oelmüller (Eds.)

these fungi with plant roots and other biotic or abiotic factors within the soil constitutes a challenge to soil microbiologists. Obviously there must have been a long evolutionary history of adaptation and competition that permitted fungi to evolve in diverse forms and interact with other organisms.

In this chapter we explore the limits of conventional and molecular techniques used to assess and detect soil microfungal diversity and provide insights into their feasibility. In particular we address the problems associated with the dilution plating technique, importance of the rDNA gene in fungal systematics, the reliability of other molecular approaches (especially denaturing gradient gel electrophoresis; DGGE) and their drawbacks.

1.2 Microscopy and Culture-Based Methods

Traditional methods to assess fungal diversity in soil environment rely mainly on the dilution-plating technique (coupled with the use of selective media) and microscopy to identify sporulating fungal bodies. Davet and Rouxel (1997) have already detailed all the experimental procedures commonly used in the dilution plate method and direct comparison. Both methods are direct isolation techniques; and the dilution-plating method involves a combination of gentle dispersion, soil dilution and serial dilution, small amounts of which are ultimately plated on artificial media and incubated. The direct comparison method involves sprinkling of a known amount of soil onto a medium, which is then incubated (Davet and Rouxel 1997). Both methods provide a reasonably sensitive recognition of soil fungi and have been widely used in diversity studies in different habitats (e.g. Elmholt et al. 1999; Cho et al. 2001; Cabello and Arambarri 2002). Cultural methods, coupled with morphological details from microscopy, are among the earliest techniques used and allow one to detect exactly which taxon is present (identification). They have also commonly been used because of their simplicity, low cost and the fact that they are easy to conduct. Williams et al. (1965) has already detailed the efficiency of the soil washing technique, its applicability and potential for studying soil microhabitats and these are not detailed here. While these methodologies are easy, fast and reliable in finding the dominant culturable fungal taxa, they have a number of limitations which impede a proper diversity assessment.

Davet and Rouxel (1997) mentioned that the traditional methods outlined above tend to overestimate species that sporulate in soil, while those in mycelial state or those that have slow growth in culture are largely overlooked. In addition, most of these methods result in isolation of only the most common and abundant fungi (often referred to as "generalists"), such as the asexual ascomycetes *Fusarium, Penicillium* and *Trichoderma* and oomycetes (*Pythium*). These cultivated organisms are those that can utilise the energy source under the physical and chemical limitations of the growth medium. The continuous isolation of similar fungi following these traditional approaches clearly indicates that many others do not respond readily to cultural techniques. Therefore, the diversity data cannot be considered as accurate (Bridge and Spooner 2003). Although these unculturable fungi play a vital role in the soil ecosystem, they were not previously thought to be central part of any biological processes in soil. Altered and optimised growth medium, coupled with 16S rRNA gene comparative analysis, has demonstrated that a larger proportion of uncultured bacteria (above the 5% level postulated) and belonging to novel bacterial lineages could be isolated and identified (Janssen et al. 2002). Similar strategies are required for fungi. However, there is insufficient knowledge on the nutritional and environmental demands of soil fungi and these present methodological drawbacks in providing a clear assessment of fungal communities associated with soil.

Another major complication with cultural studies is that a large number of other fungi existing as mycelial (vegetative) propagules or dormant spores can be numerically dominant populations in their natural environment but never grow in culture. These organisms will escape normal isolation-based detection procedures and therefore provide bias data regarding fungal diversity. Even for fungi that sporulate and can be cultured, it is not always easy to correctly identify them with certainty. Our knowledge regarding the taxonomy and classification of these fungi are still limited. In addition, there are being many species that appear to be similar under cultural conditions and exhibit similar morphology, but are in fact different species. It is thought that only a small fraction (0.1% to 10.0%) of microorganisms existing in the nature can be cultured artificially (e.g. Muyzer et al. 1993; Torsvik and Øvreås 2002). Hawskworth and Rossman (1997) suggested that commonly used methods have probably only recovered 17% of known fungal population and the majority of them await discovery. Even if morphological assessment of some taxa is possible, nothing conclusive regarding the viability, percentage occurrence, physiologic and phylogenetic information can be accrued.

Processing of cultures can be time-consuming and laborious when a large number of isolates has to be handled. During these processes, the risk of culture contamination is always high and in most cases the fast-growing fungi will overgrow others and occupy the whole medium (even when Rose Bengal solution is used). Many fungi assume different life forms (e.g. existence as vegetative hyphae or dormant spores) depending upon environmental or seasonal factors. Therefore it is highly probably that many fungi are only either collected in forms that: (1) do not allow them grow in artificial media or (2) preclude their identification via microscopy. Given that fungal diversity may be quite high in soil and each population or species may occupy a specific niche, there is no single method that is appropriate to target all of them efficiently.

Garbeva et al. (2004) and Buckley and Schmidt (2002) have reviewed the effects of factors, such as plant type, soil type, soil management regime, microenvironment and disturbance, on soil microbial diversity, from single soil aggregates to entire landscapes. These are not detailed here. Generally it appears that both cultural and direct morphological methods have specific bias, as data generated is largely dependent upon the methodologies involved.

1.3 Molecular-Based Methods

The drawbacks associated with culture-dependent methods for the detection and identification of fungi in soil samples prompted the development of alternative methods which largely circumvent cultivation of target organisms. Molecular techniques have been employed, basically involving the application of hybridisation probes, PCR amplification of rDNA genes and other DNA fingerprinting techniques. These include terminal restriction fragment length polymorphism (T-RFLP), amplified rDNA restriction analysis (ARDRA), amplified random intergeneric spacer analysis (ARISA), denaturing gradient gel electrophoresis (DGGE), temperature gradient gel electrophoresis (TGGE), oligonucleotide fingerprinting of rRNA genes or single-stranded conformation polymorphism (SSCP) and have been used frequently in combination with traditional techniques to analyse fungal community composition (e.g. Egger 1995; van Elsas et al. 2000; Lowell and Klein 2001; Maarit-Niemi et al. 2001; Ranjard et al. 2001; Kirk et al. 2004). Several freshwater fungi have successfully been identified with fluorescence in situ oligonucleotide hybridisation (FISH) (Baschien et al. 2001). Another important PCR-based fingerprinting technique recently applied to assess fungal diversity is oligonucleotide fingerprinting of ribosomal RNA genes (ORFG), a new method which sorts arrayed ribosomal RNA gene clones into taxonomic clusters through a series of hybridisation experiments (Valinsky et al. 2002). These DNA-based techniques can provide a comprehensive measure of the diversity and composition of fungal communities, since they survey both the cultured and often-predominant non-culturable members of a community (Muyzer et al. 1993; van Elsas et al. 2000; Borneman and Hartin 2000; Landeweert et al. 2001; May et al. 2001; Kirk et al. 2004).

The implications of PCR-based methodologies have altered our views about the way we used to think about soil fungal diversity. For instance, Baek and Kenerley (1998) assessed the feasibility of quantitative competitive PCR in the detection and quantification of a genetically modified strain of *Trichoderma virens*. They found that the detection limit of PCR was 10–1000 times lower when compared with traditional dilution plating. By using a combination of culture-dependent and culture-independent approaches (PCR-RFLP), Viaud et al. (2000) found that the latter was an efficient molecular tool for ecological studies and for assessing unexplored fungal diversity. These methods have also been extremely useful in assessing the diversity of fungi that are difficult to isolate from soil, such as basidiomycete and arbuscular mycorrhizal fungi (AMF: Bougoure and Cairney 2005; Kouichi et al. 2005). Other methods relevant to these aspects are outlined by Akkermans et al. (1995).

1.4 The Nuclear-Encoded Ribosomal DNA Gene: Phylogenetic and Systematic Value

Morphological characters provide the basis of current fungal systematics. They provide a wealth of information to distinguish taxa and have been used extensively at different hierarchies. In some cases, however, morphological criteria present some problems and fail to resolve taxonomic relationships. This is true in cases where morphological characters are inadequate, convergent, reduced, missing or overlapping. As a consequence, many taxonomists have combined available morphological characters with biochemical or molecular characters to clarify taxonomic relationships, as well as to infer phylogenies among fungal species. Various molecular techniques that have been applied successfully in fungal systematics and the application of DNA sequencing coupled with phylogenetic analysis have greatly expanded, owing to the ever-increasing amount of sequence data available from a myriad of organisms. Molecular characters offer considerable potential, as they not only close the gap between the traditional and molecular methods, but also may determine relationships between uncultured and cultured fungi.

For several decades, the nuclear-encoded ribosomal DNA (rDNA) gene has been the gene of choice to assess phylogenetic relationships and resolve taxonomic questions at different taxonomic levels (Gouy and Li 1989; Bruns et al. 1991; Spatafora 1995; Liew et al. 2000; Jeewon et al. 2002, 2003a, b, 2004; Duong et al. 2004; Cai et al. 2005). Genes of eukaryotic rDNA are organised in a cluster that includes a small subunit gene (18S), a large subunit gene (28S) and the 5.8S gene that lies in between two internal transcribed spacers (ITS; White et al. 1990). The region that separates the cluster of three genes along the chromosome is called the non-transcribed spacer (NTS) and prior to where the 18S gene is transcribed, there is another small spacer region called the externally transcribed spacer (ETS). Together the ETS and NTS regions comprise the intergeneric spacer region (IGS; Fig. 1.1). These components are repeated in a tandem array but they evolve as a single unit and vary in length around 3000–4500 base pairs (Mitchell et al. 1995).

The ribosomal DNA has attracted increased attention among fungal systematists, especially those interested in applying DNA sequencing analysis to study taxonomic relationships and genetic variation in fungi. The most remarkable feature of the rDNA is the overall sequence homogeneity among repeat units of the gene family (Hillis and Dixon 1991, Dixon and Hillis 1993). This gene shares the same function in all organisms and evolves at approximately the same

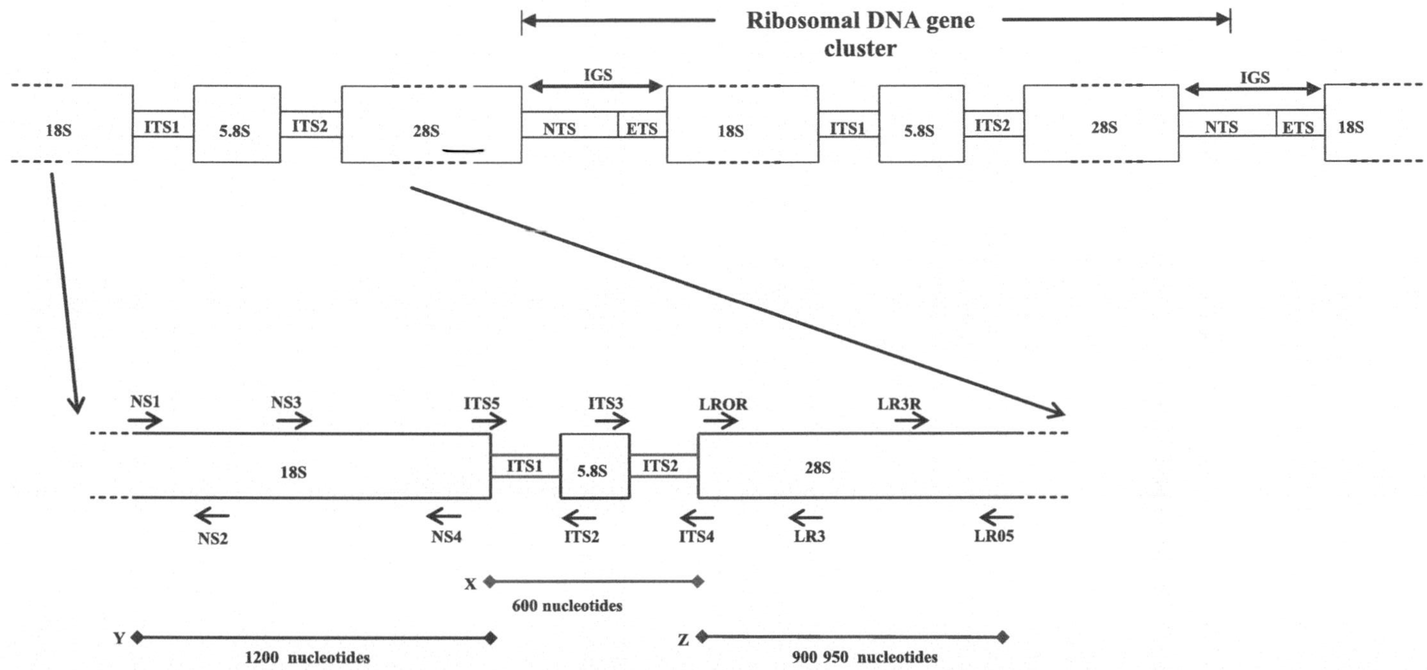

Fig. 1.1 Diagrammatic representation of the rDNA gene cluster, showing the positions of the PCR and sequencing primers. The gene is split into coding regions (*18S*, *5.8S*, *28S* genes) and non-coding regions (*IGS*, *ITS*). Positions of the primers and their direction of replication are indicated by *arrows*. *X* Product amplified with primers ITS4 and ITS5, *Y* product amplified with primers NS1 and NS4, *Z* product amplified with primers LROR and LR05. Sizes of products are approximate. Modified from Mitchell et al. (1995)

rate. However, the three different regions (structural genes, transcribed spacers, NTS) evolve at different rates, thus yielding informative data to reconstruct the phylogeny at different taxonomic levels. The 18S rDNA (small subunit; SSU), which evolves relatively slowly and is quite conserved, has been used to provide insights into the phylogeny of distantly related organisms, particularly at the ordinal and family level. The 28S (large subunit; LSU) is moderately conserved but provides sufficient variation to study relationships at the generic as well as species level. The ITS and IGS regions evolve faster and are highly variable and therefore valuable for comparing fungal species at the intraspecific level. Sequence comparisons of selected regions within the rDNA have been useful for inferring phylogenetic relationships among fungi for several reasons. Universal single primers that are complementary to several regions within this gene are ready available (Vilgalys and Hester 1990; White et al. 1990). The region is short and its multicopy nature makes it easy to amplify. It is easily accessible and a large number of sequences are available for comparison. It has a high nucleotide variability, which makes it feasible to estimate genetic distances as well as investigating systematics.

1.5 Denaturing Gradient Gel Electrophoresis: Applicability, Usefulness and Bias

While rDNA has been the most widely used gene for systematics studies, DGGE has been the most useful genetic fingerprinting technique to investigate complex microbial communities from a diversity of environmental samples. Basically this method involves separation of individual sequences (with different base composition and melting properties) from a mixture. DNA extracted from environmental samples is amplified with a primer pair (specific to the groups of organisms under investigation and one of them attached to a GC clamp) and then purified PCR samples are separated electrophoretically through a gradient of increasing chemical gradient (urea: formamide). Based on the melting behaviour, different sequences migrate at different positions, producing different banding patterns where each presumably represents a microbial taxon. The bands can then be excised from the gel and processed (either by construction of clone libraries and screening clones, or reamplified and sequenced) to obtain phylogenetic sequence information on individual microbial members of the microbial community. DGGE has been used to profile fungal microbial communities from many diverse environments (Kowalchuk et al. 1997; Smit et al. 1999; Omar and Ampe 2000; Gurtner et al. 2001; May et al. 2001; Möhlenhoff et al. 2001; Nikolcheva et al. 2003; Nikolcheva and Bärlocher 2005).

In view of the fact that so little is known about the distribution and abundance of fungi in soil environments, DGGE coupled with phylogenetics has been successfully applied to assess fungal diversity in soil samples and, in most cases, it has been reported that soil possibly consists of a much more diverse micromycota than that observed. van Elsas et al. (2000) assessed the efficiency of two DNA extraction protocols from soil microcosms, the applicability of the NS2f/Fung5r primer pair, and the persistence of *Trichoderma harzianum* and *Arthrobotrys oligospora* in response to petrol treatment. DGGE fingerprints of total DNA from tropical soil and rhizosphere revealed that there was a relationship between fungal community composition and rhizosphere development (Gomes et al. 2003). In the same study, phylogenies revealed that fungal taxa from the order Pleosporales (Ascomycetes) and basidiomycetons yeast were the most dominant phylotypes. Fungal community diversity from organic soil was investigated by PCR-DGGE followed by sequence analyses of ITS fragments (Anderson et al. 2003a). DGGE profiles revealed a clear shift in fungal community composition along a moorland pine forest environment gradient. In addition, phylogenies indicated that the majority of phylotypes (sequence types) were ascomycetes, especially Helotiales, and that the fungal communities were different from those derived using cultural methods.

DGGE is the preferred environmental fingerprinting approach as it: (1) enables large and multiple samples to be analysed simultaneously, (2) overcomes diversity bias from traditional approaches (e.g. cultural methods), (3) can successfully monitor community shifts and succession over time, (4) allows the profiling of communities under different environmental conditions (especially in degraded/polluted ecosystems), (5) makes it possible to acquire taxonomic information via phylogenetic analyses, and (6) gives an indication about the possible biological role of specific microorganisms in the sample (e.g. those that can be involved in the decomposition of organic matter or degradation of pollutants).

Nevertheless there are limitations. The lysis of cells to release DNA in the external environment is the most crucial step. Given that soil is a heterogeneous environment, there can be abundant fungi that are free-living and not localised and are therefore easily extracted. In contrast, those that are less abundant and localised in microhabitats (e.g. inside soil particles, in water-filled spaces) are difficult to extract (van Elsas and van Overbeck 1993). There is always a possibility that fungi that do not release their DNA will not contribute to diversity or that vigorous extraction procedures can result in highly fragmented DNA, producing chimeric PCR products (Wintzingerode et al. 1997). In addition, different fungal structures (spores, mycelia) have different lysing efficiency; and an inappropriate extraction method can potentially give a biased estimate of diversity (Prosser 2002). There are no specific protocols for soil fungi, although there has been considerable improvement in the procedures involved, for instance the addition of PVPP to precipitate PCR inhibitors (Wintzingerode et al. 1997; Prosser 2002; Anderson and Cairney 2004; Kirk et al. 2004). Caution is required because, in bacterial diversity studies, it has been shown that different DNA protocols and purification methods yield different DGGE profiles (Maarit-Niemi et al. 2001). The efficiency of different DNA extraction protocols and the effect of different soil types have partially been dealt with (Laurent et al. 2001; Ranjard et al. 2001; Anderson and Cairney 2004).

PCR is the basis of most molecular methods involved in diversity estimates. However, DNA from environmental samples contains PCR inhibitors and contaminants that interfere with PCR reactions (e.g. humic acid from soil). In many cases, there can be differential amplification, loss of DNA following purification, production of PCR artefacts, and contamination (Wintzingerode et al. 1997). PCR amplification of chimeric sequences is not uncommon. Sequence analyses of these usually indicate that they are not phylogenetically related to other known fungi, as they occupy unique position in the phylogenetic tree. In these cases, one will erroneously assume that these sequences represent novel taxa that escape microscopic or cultural detection. Most of the gene regions targeted in community analyses are from the conserved 18S rDNA gene and are less than 600 base pairs, so that a reasonable DGGE resolution can be achieved. This is, however, to the detriment of accurate systematics and phylogeny. In many cases, the primer pairs used are specific to a group of fungi, while some at the same time can amplify DNA from totally unrelated organisms. Our laboratory has undertaken diversity studies on leaves of *Magnolia Liliifera* (Duong et al. 2006) and pine needles using NS1 and GCfung primers as described by May et al. (2001). In both studies based on DGGE, we recovered only ascomycetous fungi, especially those from Dothideales, Helotiales, Hypocreales, Pleosporales, Rhystimatales and Xylariales, but no basidiomycetous taxa. Anderson et al. (2003b) and Anderson and Cairney (2004) have already demonstrated the potential bias of rDNA in estimating fungal diversity in soil and aspects pertaining to primer design and these are not discussed here.

Although DGGE is a promising tool, it can still underestimate fungal diversity (Nikolcheva et al. 2003, 2005). The number of bands depends on the resolution of the gels; this takes time to optimise and is difficult to reproduce (Fromin et al. 2002). The quality of sequence data recovered can be highly variable due to contaminating background sequences. We have repeatedly encountered this phenomenon when sequencing purified PCR-DGGE bands. It is not necessarily true that one "phylotype" or "operational taxonomic unit" or "sequence type" generated from an environmental sample is representative of an individual organism. As the amount of nucleic acid extracted does not necessarily reflect all the species/populations within one sample, interpretation of bands can be difficult. Often, dominant bands might mask more than one species, resulting in an underestimation of diversity. Another ambiguity we have noticed with leaf and pine needle samples is that co-migrating bands (similar melting behaviour) can actually represent taxa that are phylogenetically unrelated. The reverse also holds true. This is not surprising as it has already been demonstrated in previous studies that phylogenetically distant taxa can have co-migrating bands and that one band does not necessarily mean one unique phylotype (Rosado et al. 1998; MacNaughton et al. 1999; Sekiguchi et al. 2001). Therefore careful interpretation is essential.

Sequences obtained from DGGE bands are quite difficult to analyse as they are usually from different orders and classes. Our taxonomic knowledge is still poor and, phylogenetically, most of the sequence types do not fit clearly within any known family/genera or species, although their ordinal classification seems to be reliable. Definitive species identification is very difficult unless a large number of representatives are available from databases and a sufficiently variable gene region is analysed. Another important question is: which genes and what features of that genetic sequence are crucial, useful and reliable to identify uncultured fungi? Most of the available sequences and phylogenies are derived from the rDNA gene, but classification and taxonomic schemes based on this gene alone are inadequate, subject to debate and need to be re-evaluated. Although rDNA provides sufficient variability for evolutionary and phylogenetic inferences, should more genes be sampled?

The degree of similarities/differences of sequence types obtained from environmental samples also poses a problem. It is commonly assumed that, for bacteria, >97% sequence identity can be regarded as different species (Stackebrandt and Goebel 1994). However, there is no report for such concepts in fungal taxonomy. Another important concern is that the number of novel phylogenetic lineages and new phylotypes is on the rise. In a recent paper published in *Science*, a combination of microbiological and molecular techniques revealed three novel phylogenetic clades that constitute three major new groups of fungi (Schadt et al. 2003). As mentioned before, many sequence types cannot be confidently assigned to any particular genus or family and these have been referred to as novel taxa or lineages. Berney et al. (2004) analysed 484 environmental 18S rRNA gene sequences, including 81 new sequences, to test the potential technical and analytical pitfalls and limitations of eukaryotic environmental DNA surveys. Based on phylogenetic analyses, they suggested that the number of novel higher-level taxa revealed by previously published environmental DNA surveys was overestimated possibly due to: (1) the presence of undetected chimeric sequences, (2) the misplacement of several fast-evolving sequences, and (3) the incomplete sampling of described, but yet unsequenced eukaryotes. It is highly possible that a similar situation exist in fungal studies.

In addition, a number of studies involving the use of DNA fingerprinting techniques did not address the evolutionary history and affinities of fungal taxa based on phylogenetic analyses. This is partly because DNA fingerprinting techniques do not provide any real quantitative data regarding community function; it is time-consuming and requires expertise. It is also far easier to generate a putative uncultured sequence than to understand its biological significance from a practical standpoint. Most of the molecular techniques involved do not discriminate between active and inactive stages. This hampers a proper interpretation of the genetic/phylogenetic diversity with respect to ecology and function. For instance, DGGE analyses from pine needles in our laboratory revealed several dominant phylotypes associated with decay stages, but it is still speculative which ones are actively involved in decomposition.

1.6 Conclusions and Future Directions

Current knowledge pertaining to the *diversity*, *detection* and *distribution* of soil fungi and the *dynamics* of soil ecosystem is still rudimentary. Obviously improvement in traditional approaches combined with other biochemical/serological methods and incorporation of various molecular techniques (DNA-based) has provided new data on these aspects but, for a clearer picture and a better understanding, a combination of all approaches (polyphasic) is essential. There is a need to unravel the taxonomic diversity of speciose groups. Diversity of nematode-trapping fungi from soil (either terrestrial, estuarine or marine) is purely based on morphology and cultural studies and the most common species isolated are from *Arthrobotrys, Dactylaria* and *Monacrosporium*. To date, there are no reports on the feasibility of specific primers targeting other nematode-trapping fungi (most importantly those that are possibly unculturable). Given their relative pathological and biotechnological importance, molecular tools should be employed to assess their genotypic diversity in soil. Fungal diversity studies in soil have previously been carried out mainly in terrestrial habitats, especially those around plant roots. Future studies should target different habitats such as freshwater, estuarine or marine environments.

Our knowledge is extremely limited and we are a long way from realising the components of the soil mycota.

References

Akkermans ADL, Elsas van JD, Bruijn de FJ (1995) Molecular microbial ecology manual. Kluwer, Dordrecht, pp 1–488

Anderson IC, Cairney JWG (2004) Diversity and ecology of soil fungal communities: increased understanding through the application of molecular techniques. Environ Microbiol 6:796–779

Anderson IC, Campbell CD, Prosser JI (2003a) Potential bias of fungal 18S rDNA and internal transcribed spacer polymerase chain reaction primers for estimating fungal biodiversity in soil. Environ Microbiol 5:36–47

Anderson IC, Campbell CD, Prosser JI (2003b) Diversity of fungi in organic soils under a moorland–Scots pine (*Pinus sylvestris* L) gradient. Environ Microbiol 5:1121–1132

Baek JM, Kenerley CM (1998) Detection and enumeration of a genetically modified fungus in soil environments by quantitative competitive polymerase chain reaction. FEMS Microbiol Ecol 25:419–428

Baschien C, Manz W, Neu RN, Szewzyk U (2001) Fluorescence in situ hybridization of freshwater fungi. Int Rev Hydrobiol 86:371–381

Berney C, Fahrni J, Pawlowski J (2004) How many novel eukaryotic kingdoms? Pitfalls and limitations of environmental DNA surveys. BMC Biol 2:13

Borneman J, Hartin RJ (2000) PCR primers that amplify fungal rRNA genes from environmental samples. Appl Environ Microbiol 66:4356–4360

Bougoure DS, Cairney JWG (2005) Fungi associated with hair roots of *Rhododendron lochiae* (Ericaceae) in an Australian tropical cloud forest revealed by culturing and culture-independent molecular methods. Environ Microbiol 7:1743–1754

Bridge P, Spooner B (2003) Soil fungi: diversity and detection. Plant Soil 232:147–154

Bruns TD, White TJ, Taylor JW (1991) Fungal molecular systematics. Annu Rev Ecol Syst 22:525–564

Buckley DH, Schmidt TM (2002) Exploring the diversity of soil – a microbial rain forest. In: Staley JT, Reysenbach Al (eds) Biodiversity of microbial life: foundation of Earth's biosphere. Wiley, New York, pp 183–208

Cabello M, Arambarri A (2002) Diversity in soil fungi from undisturbed and disturbed *Celtis tala* and *Scutia buxifolia* forests in the eastern Buenos Aires province. Argentina, Microbiol Res 157:115–125

Cahill D, Mohr P (2004) Plant–pathogen interactions (Plant biology 2004: abstracts of the annual meeting of the American Society of Plant Biologists). American Society of Plant Biologists, New York, p 134

Cai L, Jeewon R, Hyde KD (2005) Phylogenetic evaluation and taxonomic revision of *Schizothecium* based on ribosomal DNA and protein coding genes. Fungal Divers 19:1–17

Cho JH, Rupe JC, Cummings MS, Gubur EE (2001) Isolation and identification of *F solani* f sp From soil and modified Nash and Syder's medium. Plant Dis 85:256–260

Davet P, Rouxel F (1997) Détection et isolement des champignons du sol. INRA, Paris, p 203

Dixon MT, Hillis DM (1993) Ribosomal RNA secondary structure: compensatory mutations and implications for phylogenetic analysis. Mol Biol Evol 10:256–267

Duong LM, Lumyong S, Hyde KD, Jeewon R (2004) *Emarcea castanopsicola* gen. et sp. nov. from Thailand, a new xylariaceous taxon based on morphology and DNA sequences. Stud Mycol 50:253–260

Duong LM, Lumyong S, Jeewon R, Hyde KD (2006) DGGE coupled with ribosomal DNA phylogenies reveal uncharacterized fungal phylotypes on living leaves of *Magnolia liliifera*. Fungal Divers 23:121–138

Egger KN (1995) Molecular analysis of ectomycorrhizal fungal communities. Can J Bot 73:1415–1422

Elmholt S, Labouriau R, Hestbjerg H, Nielsen LM (1999) Detection and estimation of conidial abundance of *Penicillium verrucosum* in soil by dilution plating on a selective and diagnostic agar medium (DYSG). Mycol Res 103:887–895

Elsas van JD, Overbeck van LS (1993) Bacterial responses to soil stimuli. In: Kjelleberg S (ed) Starvation in bacteria. Plenum, New York, pp 55–79

Elsas van JD, Duarte GF, Keijzer-Wolters A, Smit E (2000) Analysis of the dynamics of fungal communities in soil via fungal-specific PCR of soil DNA followed by denaturing gradient gel electrophoresis. J Microbiol Methods 43:133–151

Fromin N, Hamelin J, Tarnawski S, Roesti D, Miserez KJ, Forestier N (2002) Statistical analysis of denaturing gel electrophoresis (DGGE) fingerprinting patterns. Environ Microbiol 4:634–643

Garbeva P, Veen van JA, Elsas van JD (2004) Microbial diversity in soil: selection of microbial populations by plant and soil type and implications for disease suppressiveness. *Annu Rev Phytopathol* 42:243–270

Garrett, SD (1981) Soil fungi and soil fertility: an introduction to soil mycology. Pergamon, Oxford, pp 141–145

Gomes NCM, Fagbola O, Costa R, Rumjanek NG, Buchner A, Mendona-Hagler L, Smalla K (2003) Dynamics of fungal communities in bulk and maize rhizosphere soil in the tropics. Appl Environ Microbiol 69:3758–3766

Gouy M, Li WH (1989) Molecular phylogeny of the kingdoms Animalia, Plantae and Fungi. Mol Biol Evol 6:109–122

Gurtner SC, Pinar G, Lubitz W, Rolleke S (2001) Analysis of fungal communities on historical church window glass by denaturing gradient gel electrophoresis and phylogenetic 18S rDNA sequence analysis. J Microbiol Methods 47:345–354

Hawskworth DL, Rossman AY (1997) Where are all the undescribed fungi? Phytopathology 87:888–891

Hillis DM, Dixon MT (1991) Ribosomal DNA: molecular evolution and phylogenetic inference. Q Rev Biol 66:411–453

Janssen PH, Yates PS, Grinton BE, Taylor PM, Sait M (2002) Improved culturability of soil bacteria and isolation in pure culture of novel members of the divisions Acidobacteria, Actinobacteria, Proteobacteria, and Verrucomicrobia. Appl Environ Microbiol 68:2391–2396

Jaworski CA, McCarter SM, Johnson AW, Williamson RE (1978) Response of onions grown for transplants to soil fumigation. J Am Soc Hortic Sci 103:385–388

Jeewon R, Liew ECY, Hyde KD (2002) Phylogenetic relationships of *Pestalotiopsis* and allied genera inferred from ribosomal DNA sequences and morphological characters. Mol Phylogenet Evol 25:378–392

Jeewon R, Liew ECY, Simpson JA, Hodgkiss IJ, Hyde KD (2003a) Phylogenetic significance of morphological characters in the taxonomy of *Pestalotiopsis* species. Molecular Phylogenet Evol 27:372–383

Jeewon R, Liew ECY, Hyde KD (2003b) Molecular systematics of the Amphisphaeriaceae based on cladistic analyses of partial LSU rDNA sequences. Mycol Res 107:1392–1402

Jeewon R, Liew ECY, Hyde KD (2004) Phylogenetic evaluation of species nomenclature of *Pestalotiopsis* in relation to host association. Fungal Divers 17:39–55

Kirk JL, Beaudette LA, Hart M, Moutoglis P (2004) Methods of studying soil microbial diversity. J Microbiol Methods 58:169–188

Kouichi S, Yoshihisa S, Masanori S, Kazuo S (2005) A new primer for discrimination of arbuscular mycorrhizal fungi with polymerase chain reaction-denature gradient gel electrophoresis. Grassl Sci 51:179–181

Kowalchuk GA, Gerards S, Woldendorp JW (1997) Detection and characterization of fungal infections of *Ammophila arenaria* (Marram grass) roots by denaturing gradient gel electrophoresis of specifically amplified 18s rDNA. Appl Environ Microbiol 63:3858–3865

Landeweert R, Hoffland E, Finlay RD, Kuyper TW, Breemen van N (2001) Linking plants to rocks: ectomycorrhizal fungi mobilise nutrients from minerals. Trends Ecol Evol 16:248–254

Laurent FM, Philippot L, Hallet S, Chaussod R, Germon JC, Soulas G, Catroux G (2001) DNA extraction from soils: old bias for new microbial diversity analysis methods. Appl Environ Microbiol 67:2354–2359

Liew ECY, Aptroot A, Hyde KD (2000) Phylogenetic significance of the pseudoparaphyses in loculoascomycete taxonomy. Mol Phylogenet Evol 16:392–402

Lowell JL, Klein DA (2001) Comparative single-strand conformation polymorphism (SSCP) and microscopy-based analysis of nitrogen cultivation interactive effects on the fungal community of a semiarid steppe soil. FEMS Microbiol Ecol 1239:1–8

Maarit-Niemi R, Heiskanen I, Wallenius K, Lindstrom K (2001) Extraction and purification of DNA in rhizosphere soil samplesfor PCR-DGGE analysis of bacterial consortia. J Microbiol Methods 45:155–165

MacNaughton SJ, Stephen JR, Venosa AD, Davis GA, Chang YJ, White DC (1999) Microbial population changes during bioremediation of an experimental oil spill. Appl Environ Microbiol 65:3566–3574

May LA, Smiley B, Schmidt MG (2001) Comparative denaturing gradient gel electrophoresis analyses of fungal communities associated with whole plant corn silage. Can J Bot 47:829–841

Mitchell JI, Roberts PJ, Moss ST (1995) Sequence or structure? A short review on the application of nucleic acid sequence information to fungal taxonomy. Mycologist 9:67–75

Möhlenhoff P, Muller L, Gorbushina AA, Petersen K (2001) Molecular approach to the characterisation of fungal communities: methods for DNA extraction, PCR amplification and DGGE analysis of painted art objects. FEMS Microbiol Lett 195:169–173

Muyzer G, Waal de EC, Uitterlinden AG (1993) Profiling of complex microbial populations by denaturing gradient gel electrophoresis analyses of polymerase chain reaction-amplified genes coding for 16S rDNA. Appl Environ Microbiol 59:695–700

Nikolcheva LG, Bärlocher F (2005) Molecular approaches to estimate fungal diversity. II. Denaturing gradient gel electrophoresis (DGGE). In: Graça MAS, Bärlocher F, Gessner MO (eds) Methods to study litter decomposition. Springer, Berlin Heidelberg New York, pp 177–183

Nikolcheva LG, Cockshutt AM, Bärlocher F (2003) Determining diversity of freshwater fungi on decaying leaves: comparison of traditional and molecular approaches. Appl Environ Microbiol 69:2548–2554

Nikolcheva LG, Bourque T, Bärlocher F (2005) Fungal diversity during initial stages of leaf decomposition in a stream. Mycol Res109:246–253

Omar B, Ampe NF (2000) Microbial community dynamics during production of the Mexican fermented maize dough pozol. Appl Environ Microbiol 66:3664–3673

Parkinson D (1983) Functional relationships between soil organisms. Proc Int Colloq Soil Zool 8:153–165

Prosser JI (2002) Molecular and functional diversity in soil microorganisms. Plant Sci 244:9–17

Ranjard L, Poly F, Lata JC, Mougel C, Thioulouse J, Nazaret S (2001) Characterization of bacterial and fungal soil communities by automated ribosomal intergenic spacer analyses fingerprints: biological and methodological variability. Appl Environ Microbiol 67:4479–4487

Rosado AS, Duarte GF, Seldin L, Elsas van JD (1998) Genetic diversity of nifH gene sequences in *Paenibacillus azotofixans* strains and soil samples analyzed by denaturing gradient gel electrophoresis of PCR-amplified gene fragments. Appl Environ Microbiol 64:2770–2779

Schadt CW, Martin AP, Lipson DA, Steven K (2003) Seasonal dynamics of previously unknown fungal lineages in Tundra soils. Science 301:1359–1361

Sekiguchi H, Tomioka N, Nakahara T, Uchiyama H (2001) A single band does not always represent single bacterial strains in denaturing gradient gel electrophoresis. Biotechnol Lett 23:1205–1208

Smit E, Leeflang P, Glandorf B, Elsas van JD, Wernars K (1999) Analyses of fungal diversity in the wheat rhizosphere by sequencing of cloned PCR-amplified genes encoding 18S rRNA and temperature gradient gel electrophoresis. Appl Environ Microbiol 65:2614–2621

Spatafora JW (1995) Ascomal evolution of filamentous ascomycetes: evidence from molecular data. Can J Bot 73:S811–S815

Stackebrandt E, Goebel BM (1994) A place for DNA-DNA reassociation and 16S ribosomal-RNA sequence analyses in the present species definition in bacteriology. Int J Syst Bacteriol 44:846–849

Torsvik V, Øvreås L (2002) Microbial diversity and function in soil: from genes to ecosystems. Curr Opin Microbiol 5:240–245

Valinsky L, Vedova GD, Jiang T, Borneman J (2002) Oligonucleotide fingerprinting of rRNA genes for analyses of fungal community composition. Appl Environ Microbiol 68:5999–6004

Viaud M, Pasquier A, Brygoo Y (2000) Diversity of soil fungi studied by PCR–RFLP of ITS. Mycol Res 104:1027–1032

Vilgalys R, Hester M (1990) Rapid genetic identification and mapping of enzymatically amplified ribosomal DNA from several *Cryptococcus* species. J Bacteriol 172:4238–4246

Vinokurova NG, Khmel'nitskaya II, Baskunov BP, Arinbasarov MU (2003) Occurrence of indole alkaloids among secondary metabolites of soil *Aspergillus* spp. Appl Biochem Microbiol 39:192–196

White TJ, Bruns TL, Lee S, Taylor JW (1990) Amplication and direct sequencing of fungal ribosomal RNA genes for phylogenetics. In: Innis M (ed) PCR protocols, a guide to methods and applications, Academic, San Diego, pp 315–322

Williams ST, Parkinson D, Burges NB (1965) An examination of the soil washing technique by its application to several soils. Plant Soil 22:167–186

Wintzingerode FV, Göbel UB, Stackebrandt E (1997) Determination of microbial diversity in environmental samples: pitfalls of PCR-based rRNA analysis. FEMS Microbiol Rev 21:213–229

Yu X, Cheng J, Wong MH (2005) Earthworm–mycorrhiza interaction on Cd uptake and growth of ryegrass. Soil Biol Biochem 37:195–201

2 Functional Genomic Approaches for Mycorrhizal Research

A. K. Pandey, H. White, and G.K. Podila

2.1 Introduction

Mycorrhizal fungi are important and significant biological components of the rhizosphere. These fungi interact with the roots of more than 80% of land plants and form symbiotic associations called mycorrhizas or mycorrhizae (Smith and Read 1997). On the basis of the colonization pattern of host cells, two major types of mycorrhizas can be identified: ectomycorrhizas and arbuscular mycorrhizas. In the ectomycorrhizas the fungus does not penetrate the host cells, but forms a sheath around the roots and only traverses the cortical layers of the roots in the intercellular spaces, forming an interface called the "Hartig Net". However, in endomycorrhizas the fungal hyphae penetrate cells and form intracellular structures like coils or arbuscules (Smith and Read 1997). Mycorrhizal fungi provide improved access to limited soil resources such as minerals and nitrogen to the host plant. In contrast, mycorrhizal fungi receive carbon compounds from host plants to sustain their metabolism and complete the life cycle and also receive protection from other microbes in the rhizosphere.

While the ecology and physiology of mycorrhizal fungi and their uses is well studied, knowledge about cellular and molecular aspects leading to the growth and the development of a mycorrhizal fungi as well as the establishment of a functioning symbiosis is still limited (Harrison 1999; Martin et al. 2001; Podila et al. 2002; Duplessis et al. 2005; Wright et al. 2005). The development of molecular techniques and the recent progress made in the first sequencing of mycorrhizal genomes (Martin et al. 2004) has made it possible to begin to ask important biological questions on the development of symbiotic interactions and the formation of mycorrhizae.

Department of Biological Sciences, The University of Alabama in Huntsville, Huntsville, AL 35899, USA, email: podilag@uah.edu

Soil Biology, Volume 11
Advanced Techniques in Soil Microbiology
A. Varma, R. Oelmüller (Eds.)

An appropriate approach to the study of mycorrhizal fungi is to understand the molecular process leading to the host recognition, development and functioning of mycorrhizae through the analysis of expressed genes. With the advent of many high throughput techniques that have been successfully applied to the functional analysis of genes from many organisms, it is now possible to apply similar strategies to study the various aspects of the mycorrhizal symbiosis. In this chapter, we describe protocols applied to study of ectomycorrhizal symbiosis leading to: (1) Yeast two-hybrid methods to determine the interactions of signaling proteins and signaling cascades and (2) transformation system for gene replacement and functional genomic analysis of ectomycorrhizal symbiosis.

2.2 Yeast Two Hybrid: An Approach for Understanding Signaling Pathways

Ectomycorrhiza encompasses a series of complex and overlapping ontogenic process in symbionts, which includes the switching-off of fungal growth mode, initiation of lateral roots, aggregation of hyphae, arrest of cell division in ensheathed roots, and radial elongation of epidermal cells (Feugey et al. 1999). Early events in the interactions are crucial and result in the activation of a cascade of molecular events in each partner. Understanding the process involved during the interaction of plant root with fungal mycelium might provide an insight into the highly complex developmental process of mycorrhiza formation. One of the key points is to study the early induction of signaling genes, which are activated when they perceive signals from each other, leading to their recognition. This process further determines the symbiotic compatibility of fungus towards plant host.

There have been considerable reports of signaling genes and gene response events between the two partners during ectomycorrhizal symbiosis (Barker et al. 1998; Kim et al. 1998; Martin and Tagu 1999; Tagu et al. 2000; Martin et al. 2001; Sundaram et al. 2001). One of the ways to understand the signaling process is to study interactions between the early-induced signaling genes. Novel interacting genes coding for proteins can be screened using signaling genes as bait. For example, cloning and studying the regulation of G protein-coupled receptors (GPCRs) and RAS, which are early induced during the interaction phase, might help in providing some insight into ectomycorrhizal symbiosis. Utilizing yeast two hybrid in many other systems led to the identification of many genes coding for interacting proteins (Fields and Song 1989; Chein et al. 1991; Bartel et al. 1993; Fields 1993; Bendixen et al. 1994; Fields and Strenglanz 1994; Hao et al. 1999) and such a technique has proven to be quite useful. In the yeast two-hybrid assay two fusion proteins are created: the protein of interest "X" which is constructed to have a DNA binding domain attached to its N-terminus, and its potential binding partner "Y" which is fused to an activation domain. If protein

X interacts with protein Y, the binding of these two forms an intact and functional transcriptional activator (Fields and Song 1989). This newly formed transcriptional activator then goes on to transcribe a reporter gene, which is simply a gene whose protein product can be easily detected and measured. In this way, the amount of the reporter produced can be used as a measure of interaction between our protein of interest and its potential partner (Fig. 2.1).

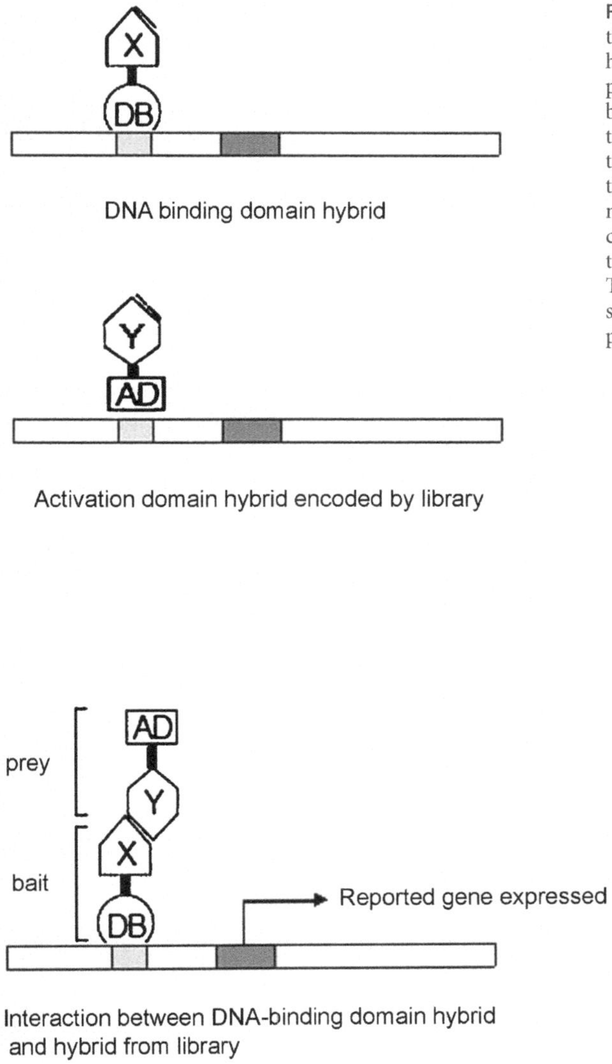

Fig. 2.1 Yeast two-hybrid transcription. The yeast two-hybrid technique measures protein–protein interactions by measuring the transcription of a reporter gene. If protein X and protein Y interact, then their DNA binding domain and activation domain combine to form a functional transcriptional activator (TA). The TA then proceeds to transcribe the reporter gene that is paired with its promoter

The *LbRas* gene cloned from *Laccaria bicolor* has been shown to be regulated during the early stage of the fungal ectomycorrhizal interaction with *Pinus resinosa* (Sundaram et al. 2001). The RAS gene is also expressed in mycorrhizal tissue when compared with free-living fungal mycelium. Such differential expression clearly suggests that *LbRas* plays a key role during ectomycorrhiza formation. Using *LbRas* as a bait and performing yeast two-hybrid interactions with tissue from early stages of *L. bicolor–P. resinosa* led to the isolation of a novel line of Ras-interacting yeast two-hybrid ectomycorrhizal clones (*Rhythm*; Sundaram et al. 2004). One of the important Ras interacting clones showed considerable sequence homology to other eukaryotic clones coding for an AP180-like protein (RhythmA; ~50%). The predicted amino acid sequence of RhythmA shows the presence of an Asn-Pro-Phe (NPF) motif, which is characteristic of all known AP180 proteins (De Camilli et al. 1996; Paoluzi et al. 1998). NPF motifs have been shown to be involved in protein–protein interactions (Paoluzi et al. 1998). AP180 proteins have been shown to play roles in the assembly of clathrin-coated vesicles through protein–protein interactions (Hao et al. 1999). Previously it has been shown the AP180 is also involved in cargo sorting in coated vesicles through its interaction with GTPase (De Camilli et al. 1996). Since establishment of mycorrhizal association is also related to exchange of signals, ligands, and nutrient, one could observe such turnover of vesicles and vesicular trafficking during ectomycorrhizal phenomena. Such regulations were previously reported during the interaction of *L. bicolor–P. resinosa* (Kim et al. 1999).

Similarly performing yeast two-hybrid analysis of pre-infection stage interactions of *L. bicolor* with aspen (*Populus tremuloides*) seedlings led to the isolation of other RAS interacting clones (Table 2.1). Screening the cDNA library prepared from the interaction of *L. bicolor* with *P. tremuloides* led to the isolation of some important RAS interacting clones, including a GPCR, cyclophilin, vacuolar protein sorting (VPR), and biogenesis related protein.

GPCR proteins are a very important group of genes and are one of the largest protein families in human and other animal genomes (Pin et al. 2005). However, in most fungal genomes, the number of GPCRs identified is very low (Kulkarni

Table 2.1 Interacting partners of Lbras. Yeast two-hybrid interactions were performed with tissue from early stages of *L. bicolor–P. tremuloides* using *LbRas* as bait. The table describes some of the interacting clones and their potential function

Description	Putative role
Cyclophilin (peptidylprolyl isomerase-like 2 isoform)	Phosphorylation and dephosphorylation process
Vacuolar protein sorting and organization and biogenesis related protein	Transport, signaling
Cullin-1	Protein targeting, processing, and degradation
GPCR – STE3 like	Receptor and signaling

et al. 2005). GPCRs are involved in a variety of signaling mechanisms of various organisms and are known to be associated with critical biological functions. In fungi, GPCRs have been shown to be involved in signaling (Riquelme et al. 2005), including pheromone signaling, receptors involved in host recognition (Marsh and Herskowitz 1988), and pathogenicity (Kulkarni et al. 2005).

Cyclophilin have peptidyl-prolyl *cis-trans* isomerase (PPIase) and are chaperones and folding catalysts with the ability to catalyze the *cis-trans* isomerization of prolyl bonds, a rate-limiting step in protein folding. Cyclophilin were shown to be involved in different phosphorylation processes of RAF/RAS, thus regulating the signaling events (Dougherty et al. 2005). Northern analysis of cyclophilin showed they are induced at the very early stages of interaction and are present until the time of interaction (Fig. 2.2).

Another Lbras interacting clone showed a high match with VPS. Vacuolar protein sorting genes encode proteins that are involved in protein sorting to the vacuole. A region of vacuolar protein sorting Vps9p from yeast is related to human proteins (Rin1, JC265) that were shown to negatively regulate Ras-mediating signaling in *S. cerevisiae* (Han and Colicelli 1995). In fact Rin1 has been shown to bind directly to RAS in a manner that competes with the binding of RAF (a downstream effector of RAS). This suggests that an effector domain of RAS is a principal binding site for Rin1 and that Rin1 may act as a downstream effector of RAS. There have been previous reports of vesicular traffic protein regulation during ectomycorrhizal interaction and vesicular turnover in ectomycorrhiza (Cole et al. 1998; Kim et al. 1999). Though such functional evidence is lacking to show their role during ectomycorrhiza formation, the involvement of vacuolar sorting proteins in signaling and transport of signals/information cannot be ruled out.

Cloning of the signaling genes like GPCRs, Raf, and Rho and testing their intensity of interaction with each other to build a protein–protein interaction network model could provide us evidence during the upstream and downstream signaling process in ectomycorrhiza formation.

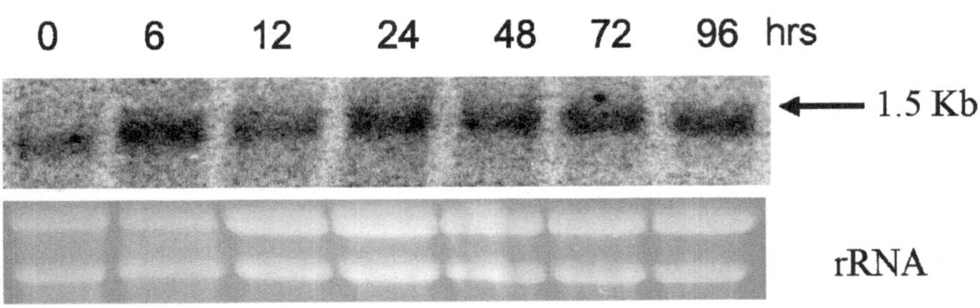

Fig. 2.2 Temporal regulation of mRNA expression for cyclophilin-like protein during the early stages of interaction between *L. bicolor* and *P. tremuloides*. Northern analysis with 10 µg of total RNA samples from *L. bicolor* subjected to interaction with *P. tremuloides* for 6, 12, 24, 48, 72, 96 h was carried out as described in the Materials and Methods. Ethidium bromide staining of rRNA was used to verify the loaded amount of total RNA

2.3 *Agrobacterium*-Mediated Transformation in *Laccaria bicolor*

Agrobacterium tumefaciens is a well known bacterium that causes crown galls on plants by transferring part of a tumor-inducing plasmid into their genomes. Such *Agrobacterium*-mediated transformation has been well developed for gene transfer in plant systems (Tinland 1996) and also in variety of other organisms like yeast, filamentous fungi, and human cells (Bundock et al. 1995; de Groot et al. 1998; Kunik et al. 2001). To date *Agrobacterium*-mediated transformation has been reported in many fungal species, like *Botrytis cinerea, Aspergillus awamori, Magnaporthe grisea, Fusarium oxysporum* (Gouka et al. 1999; Rolland et al. 2003; Khang et al. 2005). Methods of molecular and genetic analysis have progressed more slowly for ectomycorrhizal fungi than for higher fungi. Still, successful transformation has been reported in *Suillus bovinus, Agaricus bisporus, Paxillus involutus, Hebeloma cylindrosporum*, and *Laccaria bicolor* (Pardo et al. 2002, 2005; Hanif et al. 2002; Combier et al. 2003). But the functional aspects and utilization of such a technique for insertional mutagenesis or genetic transformation are largely missing.

Recently, *Agrobacterium*-mediated transformation was used for insertional mutagenesis in the symbiotic ectomycorrhizal fungus *Hebeloma cylindrosporum* (Combier et al. 2003). Though there can be single or multiple insertions, the frequency of single insertions can be controlled by treating bacteria with acetosyringone (AS) prior to co-cultivation, an experimental condition which slightly reduces transformation efficiency. In fact, a higher percentage (60%) of single insertion was obtained in *H. cylindrosporum* using AS-treated bacteria when compared with protoplast-based transformation, which generally led to an unpredictable number of plasmid integration per genome (Marmeisse et al. 1992; Amey et al. 2002). Kemppainen et al. (2005) obtained successful transformation of *L. bicolor* S238N with an efficiency of 55% using the AGL-1 strain.

Using this transformation procedure, a non-mycorrhizal mutant of *H. cylindrosporum* was obtained (Combier et al. 2003). Further, it will be interesting to study the key regulatory steps of mycorrhizal functioning by directing precise silencing of symbiosis-regulated genes using siRNA/RNA interference technology. Though such techniques are well developed in phytopathogens (Fitzgerald et al. 2004), they are still lacking in mycorrhizal fungi.

We have used *Agrobacterium*-mediated transformation of *L. bicolor* strain S238N with a vector for the selection and expression of green fluorescent protein (GFP) reporter gene (Fig. 2.3). In addition, we have also obtained gene replacement for the PF6.2 gene earlier found to be induced very early in interaction between *L. bicolor* and red pine (Kim et al. 1998). This is the first instance of gene replacement in mycorrhizal fungi. Figure 2.4 shows the replacement of the PF6.2 gene in *L. bicolor* transformants. One of these transformants has been tested in its ability to form mycorrhizae on *P. tremuloides* seedlings. The trans-

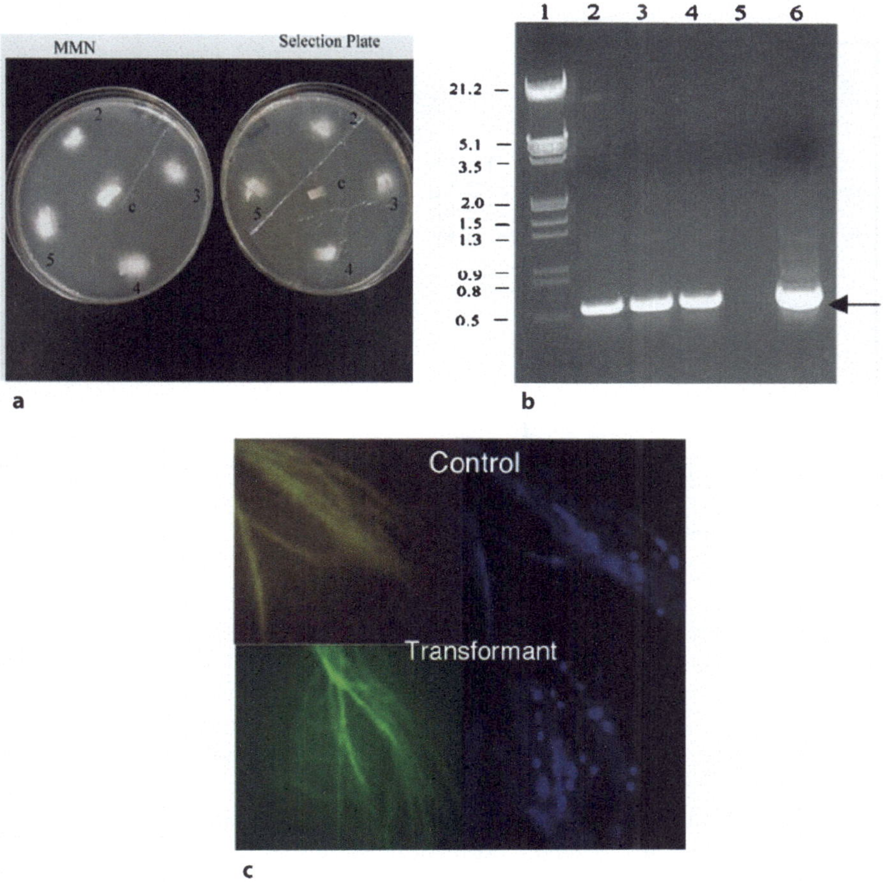

Fig. 2.3 *Agrobacterium*-mediated transformation of *L. bicolor* and expression of selection marker and reporter gene GFP. Panel **a** shows *L. bicolor* wild type (*C*) and transformants grown on non-selective MMN medium (*left dish*). The *right dish* shows the selection of transformants on 300 μg/ml hygromycin. Panel **b** shows colony PCR of transformants selected on hygromycin. The *arrow* points to the GFP PCR product. *Lane 1* DNA molecular weight marker, *lanes 2, 3, 4* transformant DNA samples, *lane 5* blank, *lane 6* positive control from pBGgHg plasmid. Panel **c** shows the expression of GFP protein in the transformants and the DAPI staining of nuclei, compared with a non-transformed control *L. bicolor*

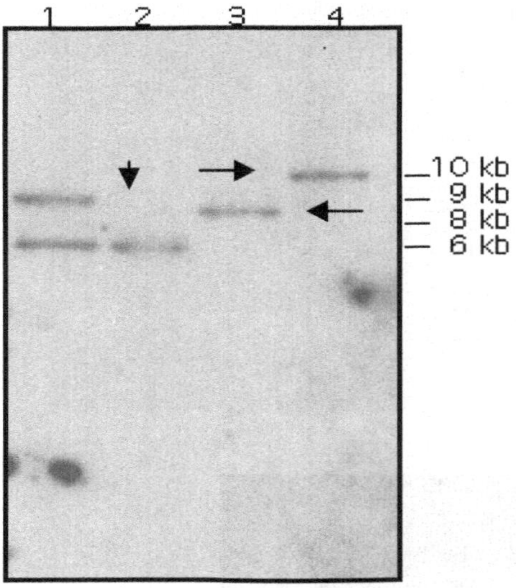

Fig. 2.4 Southern analysis of PF6.2 gene displacement in *L. bicolor*. Genomic DNA samples (10 μg each) from *L. bicolor* were digested with *Bam*H1. Hybridization was done with a ^{32}P-labeled PF6.2 cDNA fragment as described by Kim et al. (1998). Molecular weight markers are indicated in kilobasepairs. *Lane 1* Wild-type *L. bicolor*, *lanes 2–4 L. bicolor* transformants 1, 2, 3, respectively. *Arrows* point to the deletion of one copy of PF6.2 in transformant 1 and displacement in transformants 3 and 4

formant was able to form mycorrhizal roots, but was defective in stopping the formation of root hairs on mycorrhizal roots (Fig. 2.5), which is a common feature under normal conditions. This suggests that the displacement in PF6.2 in the *L. bicolor* genome and reduction in its copy number impacted the symbiosis process. These results also corroborate the earlier hypothesis that PF6.2 from *L. bicolor* may be involved in a signaling process (Kim et al. 1998). Thus, the *Agrobacterium*-mediated gene transformation methods open up the possibility of using gene silencing or ectopic expression techniques in mycorrhizal fungi to study the process of symbiosis.

2.4 Materials and Methods

2.4.1 Interaction Studies of *Laccaria bicolor* with Aspen (*Populus tremuloides*) Seedlings

To construct the cDNA library of *L. bicolor* undergoing interaction with *P. tremuloides*, aspen seeds were incubated overnight at 4 °C and were surface-sterilized using 10% hydrogen peroxide. The sterilized seeds were transferred to Petri dishes (diam. 75 mm) containing woody plant medium (WPM) agar (Sigma, Mo., USA) and was incubated for 1 week at 25 °C. Five seedlings of aspen were

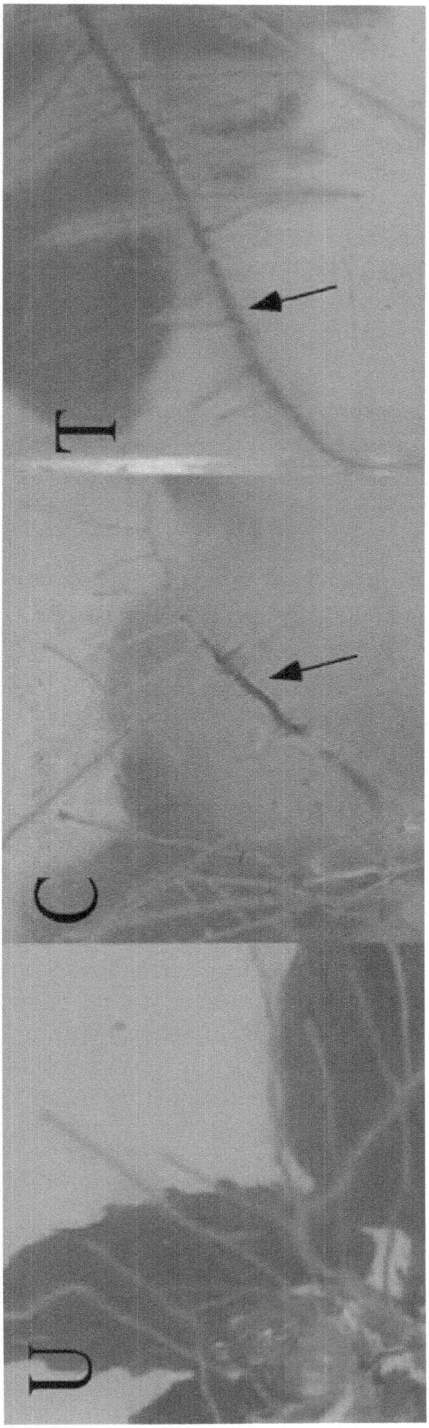

Fig. 2.5 Phenotypic changes in the mycorrhizae formed by *L. bicolor* PF6.2 transformant. **U** Un-inoculated roots. **C** Roots inoculated with control *L. bicolor*. **T** Roots inoculated with transformant 3 showing mantle formation but no loss of root hairs. The loss of root hairs is a hallmark of ectomycorrhizal development

transferred to each magenta box containing WMP media overlaid with a cellophane sheet. The seedlings were incubated in a growth chamber with a cycle of 16 h light and 8 h dark at 25 °C for 4–5 weeks. The *L. bicolor* culture was maintained on MMN medium (Podila et al. 2002) in MMN-filled Petri dishes (diam. 150 mm) at 22 °C. The fungal culture for inoculation of aspen roots was grown from agar plugs of mycelium placed on cellophane-covered MMN medium. Mycelial strips of approximately 15×5 mm were excised from the edge of the culture. Strips were then placed on the cellophane just above the root and were grown for different time intervals (viz. 0, 6, 12, 24, 48, 72, 96 h). This allowed for diffusion of root signals, but prevented physical contact between the roots and the mycelium to study the gene expression before physical contact was made between the fungus and the roots.

2.4.2
Yeast Two-Hybrid Protocol

The yeast two-hybrid experiment was performed using BD Matchmaker library construction and screening kits and protocols (BD Biosciences, Clontech, Calif., USA).

2.4.2.1
cDNA Synthesis and Bait Construction

All RNA samples were treated with RNase-free DNase at 37 °C for 30 min using the DNA-free kit (Ambion, Austin, Tex.) prior to cDNA synthesis, to ensure that the amplicon template originated from RNA and not DNA. Two micrograms of DNA-free RNA was used for first-strand cDNA synthesis for all samples belonging to interaction time points, carried out simultaneously using the BD Smart cDNA synthesis kit. *Lbras* was used to construct a DNA-BD fusion vector using a BD-cloning vector (pGBKT7). The GAL4AD fusion library was constructed using vector pGADT7-Rec and the constructed interaction library. The GAL4AD fusion library samples along with the bait (BD vector) were co-transformed in yeast strain AH109 (as described in the BD Biosciences protocol).

2.4.2.2
Preparation of Competent Yeast Cells – LiAc Method

1. Inoculate fresh yeast strain AH109 (<4 weeks old, 2–3 mm diam.) into 3 ml of YPDA medium and incubate at 30 °C with shaking for 8 h.

2. Transfer 5 μl of the culture to a 250-ml flask containing 50 ml of YPDA. Incubate at 30 °C with shaking at 230–250 rpm for 16–20 h. The OD_{600} should reach 0.15–0.2.
3. Centrifuge the cells at 700 *g* for 5 min at room temperature. Discard the supernatant and resuspend the cell pellet in 100 ml of YPDA.
4. Incubate at 30 °C for 3–5 h (OD_{600} = 0.4–0.5). Centrifuge the cells at 700 *g* for 5 min at room temperature.
5. Discard the supernatant and resuspend the cell pellet in 60 ml of sterile, deionized H_2O. Centrifuge the cells at 700 *g* for 5 min at room temperature.
6. Discard the supernatant and resuspend the cells in 3 ml of 1.1× TE/LiAc solution.
7. Divide the resuspension between two 1.5-ml microcentrifuge tubes (1.5 ml per tube).
8. Centrifuge each tube at high speed for 15 s. Discard the supernatant and resuspend each pellet in 600 μl of 1.1× TE/LiAc solution.

Competent cells should be used for transformation immediately following preparation; however, if necessary they can be stored at room temperature for a few hours without significantly affecting the competency.

2.4.2.3
Transformation of Yeast Strain AH109 with dscDNA and pGADT7-Rec

1. In a sterile, prechilled, 15-ml tube combine the following: 20 μl dscDNA (from protocol Section IX.I, step 16), 6 μl pGADT7-Rec (0.5 μg/μl), 5 μg GBKT7/bait plasmid DNA, 20 μl herring testes carrier DNA, denatured*. (*Transfer ~50 μl of herring DNA to a microcentrifuge tube and heat at 100 °C for 5 min. Then, immediately chill the DNA by placing the tube in an ice bath. Repeat once more before adding the DNA to the 15-ml reaction tube.)
2. Add 600 μl of competent cells to the DNA. Gently mix by vortexing. Add 2.5 ml PEG/LiAc Solution. Gently mix by vortexing. Incubate at 30 °C for 45 min. Mix cells every 15 min.
3. Add 160 μl DMSO, mix, and then place the tube in a 42 °C water bath for 20 min. Mix cells every 10 min. Centrifuge at 700×*g* for 5 min.
4. Discard the supernatant and resuspend in 3 ml of YPD plus liquid medium.
5. Incubate at 30 °C with shaking for 90 min. Centrifuge at 700×*g* for 5 min. Discard the supernatant and resuspend in 6 ml of NaCl solution (0.9%).
6. Spread the co-transformation mixture on selection media. Transformants expressing interacting proteins were selected on triple dropout medium: SD/–His/–Leu/–Trp and quadruple dropout medium: SD/–Ade/–His/–Leu/–Trp.

Colonies become visible after 2–3 days, but plates should be incubated 5 days to allow slower growing colonies to appear.

To identify the gene responsible for a positive two-hybrid interaction, rescue the gene by plasmid isolation or by PCR colony-screening.

Further, AD/library cDNA insert can be sequenced using the AD LD-insert screening amplimer set, a T7 sequencing primer, or the 3' AD sequencing primer provided with the BD Matchmaker two-hybrid kit (Clontech, Calif., USA).

2.4.2.4
Northern Analysis

Total RNA from *L. bicolor*, subjected to interaction with *P. tremuloides* seedling roots for 6, 12, 24, 48, 72, 96 h, respectively, was electrophoresed on formaldehyde–agarose gels and transferred to Hybond-N membranes (Amersham Pharmacia Biotech, Piscataway, N.J., USA), as described by Kim et al. (1998). Total RNA from free-living *L. bicolor* was used as control. A 10-μg sample of RNA was loaded in each lane; and gels were stained with ethidium bromide (Sigma, USA) to determine equal loadings and intensity of RNA. The cDNA fragment coding for cyclophilin-like protein was labeled with ^{32}P-dCTP with the Rediprime DNA labeling kit (Amersham Pharmacia Biotech) and used as a probe in the hybridization analyses of the membrane-bound nucleic acids, as described previously (Sambrook et al. 1989; Kim et al. 1999).

2.4.3
Agrobacterium-Mediated Transformation in *Laccaria bicolor*

2.4.3.1
Preparation of Fungal Material

Laccaria bicolor mycelium was freshly cultured on a cellophane sheet overlaid on low glucose-MMN (2% glucose) agar medium at 22 °C for 1 week, as described by Balasubramanian et al. (2002).

2.4.3.2
Induction of *Agrobacterium*

Agrobacterium tumefaciens strain AGL1 containing plasmid pBGgHg or pBG-6.2 was grown overnight in 4 ml of minimal medium containing kanamycin at 50 μg/ml at 29 °C (until the cell density reached OD = 0.2).

Bacterial cells were collected by centrifugation (3000 *g* at 4 °C for 5 min) and resuspended in induction medium (200 µM AS plus kanamycin at 50 µg/ml) and grown for 6 h at 29 °C.

2.4.3.3
Transformation, Co-Cultivation, and Selection

After the fungal colonies reached 0.5 cm diameter, which took 7 days, the membranes with colonies were transferred to induction media plates with or without 200 µM AS.

Laccaria mycelium was then inoculated with 50 µl of induced *Agrobacterium*. The co-cultivation plates were incubated at 22 °C for 5 days in the dark.

The membrane with mycelia colonies were then transferred to MMN or Moser 6 selection plates (pH 7.5; containing antibiotics mix and hygromycin 300 µg/ml). The plates were incubated at 4 °C overnight and then shifted to 22 °C for 10 days in the dark.

After 2 weeks, the growing colonies were repeatedly subcultured on the selection plates containing antibiotics and tested to make sure no residual agrobacteria were present.

DNA isolation was performed from the putative transformant colonies growing on cellophane membrane using the methods described by Kim et al. (1998). The positive transformants were selected using primers which specifically amplify hygromycin (*hph*) or modified EGFP gene.

Further analyses of T-DNA integration were done by Southern analysis, as described by Kim et al. (1998).

1. Minimal medium: K_2HPO_4 10.5 g, KH_2PO_4 4.0 g, $(NH_4)_2SO_4$ 1.0 g, Na_3-citrate $2H_2O$ 0.5 g, $MgSO_4.7H_2O$ 0.2 g, thiamine-HCl 1.0 mg, glucose 2.0 g, plus 50 µg/ml kanamycin.
2. Induction medium: minimal medium, plus 40 mM MES and 0.5% glycerol, pH 5.3.
3. Induction agar: induction medium, plus 2% agar.
4. Antibiotics mix: cefotaxime (Sigma, USA) 100 µg/ml, ampicillin (Amresco, USA) 100 µg/ml, tetracyclin (Amresco, USA) 125 µg/ml, hygromycin (Roche, USA) 300 µg/ml.

2.4.3.4
Fungal DAPI Staining and Visualization of GFP Expression

Actively growing fungal hyphae from the edges of the colony were collected and transferred on the slides using a sterile needle or the forceps. Using the blunt end of forceps the cover slip was tapped gently to spread the mycelia uniformly.

Fungal mycelia were mounted in Vectashield plus 4,6-diamidino-2-phenylindole (DAPI) as per the manufacturer's instructions (Molecular Probes, USA). Hyphae were observed under a Nikon E600 microscope equipped with a Qi-Cam digital camera (Q Imaging, USA).

The GFP fluorescence was observed under the Nikon E600 microscope. The filters used were B2A (excitation filter wavelength: 450–490 nm) for green fluorescence and UV-2A (excitation filter wavelength: 330–380 nm) for DAPI stain at 100× magnification.

2.4.3.5
Fungal DNA PCR

DNA was extracted from the mycelia actively growing on selection medium containing hygromycin.

DNA was diluted to 10% and hot start PCR was performed using EGFP primers (Clontech, Calif., USA).

PCR consisted of 30 cycles of amplification on an Eppendorf Mastercycler gradient PCR machine. Each cycle consisted of 1 min of melting at 94 °C, 30 s of annealing at 55 °C, and 1 min of extension at 72 °C. Prior to the first cycle, the samples were heated to 94 °C for 3 min. The last cycle was followed by a final extension at 72 °C for 5 min.

Amplification products were detected by electrophoresis on 1.2 % agarose gels that were stained with ethidium bromide and were visualized with a UV trans-illuminator. The identity of PCR products was further confirmed by DNA sequence analysis.

Acknowledgements

Part of the work presented here is supported by NSF grant MCB to G.K.P.

References

Amey RC, Athey-Pollard A, Burns C, Mills PR, Bailey A, Foster GD (2002) PEG-mediated and *Agrobacterium*-mediated transformation in the mycopathogen *Verticillium fungicola*. Mycol Res 106:4–11

Balasubramanian S, Kim SJ, Podila GK (2002) Metabolism in early stages of mycorrhizal symbiosis – *Lb-MS*, a differentially regulated malate synthase from the ectomycorrhizal fungus *Laccaria bicolor*. New Phytol 154:517–527

Barker SJ, Tagu D, Delp G (1998) Regulation of root and fungal morphogenesis in mycorrhizal symbiosis. Plant Physiol 116:1201–1207

Bartel PL, Chein CT, Strenglanz R, Fields S (1993) Using the two-hybrid system to detect protein–protein interactions. In: Hartley DA (ed) Cellular interactions in development: a practical approach. Oxford University Press, Oxford, pp 153–179

Bendixen C, Gangloff S, Rothstein R (1994) A yeast mating-selection scheme for detection of protein–protein interactions. Nucleic Acids Res 22:1778–1779

Bundock P, Dulk-Ras A, Beijersbergen A, Hooykaas PJJ (1995) Trans-kingdom T-DNA transfer from *Agrobacterium tumefaciens* to *Saccharomyces cerevisiae*. EMBO J 14:3206–3214

Chein CT, Bartel PL, Strenglanz R, Fields S (1991) The two-hybrid system: a method to identify and clone genes for proteins that interact with a protein of interest. Proc Natl Acad Sci USA 88:9578–9582

Cole L, Orlovich DA, Ashford AE (1998) Structure, function, and motility of vacuoles in filamentous fungi. Fungal Genet Biol 24:86–100

Combier JP, Melayah D, Raffier C, Gay G, Marmeisse R (2003) *Agrobacterium tumefaciens*-mediated transformation as a tool for insertional mutagenesis in the symbiotic ectomycorrhizal fungus *Hebeloma cylindrosporum*. FEMS Microbiol Lett 220:141–148

De Camilli P, Emr SD, McPherson PS, Novick P (1996) Phosphoinositides as regulators in membrane traffic. Science 271:1533–1539

Dougherty MK, Muller J, Ritt DA, Zhou M, Zhou XZ, Copeland TD, Conrads TP, Veenstra TD, Lu KP, Morrison DK (2005) Regulation of Raf-1 by direct feedback phosphorylation. Mol Cell 17:215–224

Duplessis S, Courty PE, Tagu D, Martin F (2005) Transcript patterns associated with ectomycorrhiza development in *Eucalyptus globulus* and *Pisolithus microcarpus*. New Phytol 165:599–611

Feugey L, Strullu DG, Poupard O, Simoneau P (1999) Induced defense response limit Hartig net formation in ectomycorrhizal birch roots. New Phytol 144:541–547

Fields S (1993) The two-hybrid system to detect protein–protein interactions methods: a comparison. Methods Enzymol 5:116–124

Fields S, Song OK (1989) A novel genetic system to detect protein–protein interactions. Nature 340:245–246

Fields S, Strenglanz R (1994) The two-hybrid system: an assay for protein–protein interactions. Trends Genet 10: 286–292

Fitzgerald DJ, DeLuca C, Berger I, Gaillard H, Sigrist R, Schimmele K, Richmond TJ (2004) Reaction cycle of the yeast Isw2 chromatin remodeling complex. EMBO J 23:3836–3843

Gouka RJ, Gerk C, Hooykaas PJJ, Bundock P, Musters W, Verrips CT, Groot de MJA (1999) Transformation of *Aspergillus awamori* by *Agrobacterium tumefaciens*-mediated homologous recombination. Nat Biotechnol 17:598–601

Groot de MJA, Bundock P, Hooykaas PJJ, Beirjesbrgen AGM (1998) *Agrobacterium tumefaciens*-mediated transformation of filamentous fungi. Nat Biotechnol 16:839–842

Han L, Colicelli J (1995) A human protein selected for interference with Ras function interacts directly with Ras and competes with Raf1. Mol Cell Biol 15:1318–1323

Hanif M, Pardo AG, Gorfer M, Raudaskoski M (2002) T-DNA transfer and integration in the ectomycorrhizal fungus *Suillus bovinus* using hygromycin B as a selectable marker. Curr Genet 41:183–188

Hao W, Luo Z, Zheng L, Prasad K, Lafer EM (1999) AP180 and AP-2 interact directly in a complex that cooperatively assembles clathrin. J Biol Chem 274:22785–22794

Harrison MJ (1999) Molecular and cellular aspects of the arbuscular mycorrhizal symbiosis. Annu Rev Plant Phys 50:361–389

Kemppainen M, Circosta A, Tagu D, Martin F, Pardo AG (2005) *Agrobacterium*-mediated transformation of the ectomycorrhizal symbiont *Laccaria bicolor* S238N. Mycorrhiza 16:19–22

Khang CH, Park SY, Lee YH, Kang S (2005) A dual selection based, targeted gene replacement tool for *Magnaporthe grisea* and *Fusarium oxysporum*. Fungal Genet Biol 42:483–492

Kim SJ, Zheng J, Hiremath ST, Podila GK (1998) Cloning and characterization of a symbiosis-related gene from an ectomycorrhizal fungus *Laccaria bicolor*. Gene 222:203–212

Kim SJ, Bernreuther D, Thumm M, Podila GK (1999) LB-AUT7, a novel symbiosis-regulated gene from an ectomycorrhizal fungus, *Laccaria bicolor*, is functionally related to vesicular transport and autophagocytosis. J Bacteriol 181:1963–1967

Kulkarni RD, Thon MR, Pan H, Dean RA (2005) Novel G-protein-coupled receptor-like proteins in the plant pathogenic fungus *Magnaporthe grisea*. Genome Biol 6:24

Kunik T, Tzfira T, Kapulnik Y, Gafni Y, Dingwall C, Citovsky V (2001) Genetic transformation of HeLa cells by *Agrobacterium*. Proc Natl Acad Sci USA 98:1871–1876

Marmeisse R, Gay G, Debaud JC, Casselton LA (1992) Genetic transformation of the symbiotic basidiomycete fungus *Hebeloma cylindrosporum*. Curr Genet 22:41–45

Marsh L, Herskowitz I (1988) STE2 protein of *Saccharomyces kluyveri* is a member of the rhodopsin/beta-adrenergic receptor family and is responsible for recognition of the peptide ligand alpha factor. Proc Natl Acad Sci USA 85:3855–3859

Martin F, Tagu D (1999) Developmental biology of a plant–fungus symbiosis: the ectomycorrhiza. In: Varma AK, Hock B (eds) Mycorrhiza: structure, function, molecular biology and biotechnology. 2nd edn. Springer, Berlin Heidelberg New York, pp 51–73

Martin F, Duplessis S, Ditengou F, Lagrange H, Vioblet C, Lapeyrie F (2001) Development cross talking in the ectomycorrhizal symbiosis: signals and communication genes. New Phytol 151:145–154

Martin F, Tuskan GA, Difazio SP, Lammers P, Newcombe G, Podila GK (2004) Symbiotic sequencing for the *Populus* mesocosm. New Phytol 161:330–335

Paoluzi S, Castagnoli L, Lauro I, Salcini AE, Coda L, Fre S, Confalonieri S, Pelicci PG, Di Fiore PP, Cesareni G (1998) Recognition specificity of individual EH domains of mammals and yeast. EMBO J 17:6541–6550

Pardo AG, Hanif M, Raudaskoski M, Gorfer M (2002) Genetic transformation of ectomycorrhizal fungi mediated by *Agrobacterium tumefaciens*. Mycol Res 106:132–137

Pardo AG, Kemppainen M, Valdemoros D, Duplessis S, Martin F, Tagu D (2005) T-DNA transfer from *Agrobacterium tumefaciens* to the ectomycorrhizal fungus *Pisolithus microcarpus*. Rev Argent Microbiol 37:69–72

Pin JP, Kniazeff J, Liu J, Binet V, Goudet C, Rondard P, Prezeau L (2005) Allosteric functioning of dimeric class C G-protein-coupled Receptors. FEBS J 272: 2947–2955

Podila GK, Zheng J, Balasubramanian S, Sundaram S, Hiremath S, Brand J, Hymes M (2002) Molecular interactions in ectomycorrhizas: identification of fungal genes involved in early symbiotic interactions between *Laccaria bicolor* and red pine. Plant Soil 244:117–128

Riquelme M, Challen MP, Casselton LA, Brown AJ (2005) The origin of multiple B mating specificities in *Coprinus cinereus*. Genetics 170:1105–1119

Rolland S, Jobic C, Fevre M, Bruel C (2003) *Agrobacterium*-mediated transformation of *Botrytis cinerea*, simple purification of monokaryotic transformants and rapid conidia-based identification of the transfer-DNA host genomic DNA flanking sequences. Curr Genet 44:164–171

Sambrook J, Fritsch EF, Maniatis T (1989) Molecular cloning: a laboratory manual. Cold Spring Harbor Laboratory, Cold Spring Harbor

Smith SE, Read DJ (1997) Mycorrhizal symbiosis, 2nd edn. Academic, London

Sundaram S, Kim SJ, Suzuki H, Mcquattie CJ, Hiremath ST, Podila GK (2001) Isolation and characterization of a symbiosis-regulated ras from the ectomycorrhizal fungus *Laccaria bicolor*. Mol Plant Microbe Interact 14:618–628

Sundaram S, Brand JH, Hymes MJ, Hiremath ST, Podila GK (2004) Isolation and analysis of a symbiosis-regulated and Ras-interacting vesicular assembly protein gene from the ectomycorrhizal fungus *Laccaria bicolor*. New Phytol 161:529–538

Tagu D, Lapeyrie F, Ditengu F, Lag H, Larent P, Missoum N, Nehls K, Martin F (2000) Molecular aspects of ectomycorrhiza development. In: Podila GK, Douds DD (eds) Current advances in mycorrhiza research. American Phytopathological Society, St. Paul, pp 69–90

Tinland B (1996) The integration of T-DNA in plant genomes. Trends Plant Sci 1:178–184

Wright DP, Johansson T, Le Quere A, Soderstrom B, Tunlid A (2005) Spatial patterns of gene expression in the extramatrical mycelium and mycorrhizal root tips formed by the ectomycorrhizal fungus *Paxillus involutus* in association with birch (*Betula pendula*) seedlings in soil microcosms. New Phytol 167:579–596

3 Automated Fluoroscence Sequencing and Troubleshooting

S. Gochhait, D. Malhotra, E. Rai, and R.N.K. Bamezai

3.1 Introduction

A major step towards understanding the genetic basis of an organism is the complete sequence determination of all the genes in its genome (Sterky and Lundeberg 2000). On 26 June 2000, a landmark achievement was announced: the compilation of a working draft of the human genome by the Human Genome Project, with the first assembly (by Celera Genomics, Rockville, Md.) of a complete human genome sequence (Macilwain 2000). Three major milestones played a prominent role in achieving such a goal: the invention of sequencing reactions (Maxam and Gilbert 1997; Sanger et al. 1977), the polymerase chain reaction (PCR; Mullis et al. 1986; Mullis and Faloona 1987), and automated fluorescent DNA sequencers (Smith et al. 1985, 1986; Hood et al. 1987; Hunkapillar et al. 1991), which made it possible to streamline and automate most of the processes required for DNA analysis. The DNA sequencing technique can be helpful in routine molecular biology work to confirm recombinant plasmids, authenticate the orientation of the cloned fragment, and assess the success of the site-directed mutagenesis experiment. It has further enabled the highest resolution of the blueprint of an organism and facilitated the characterization of a given sequence in terms of polymorphisms (Malhotra et al. 2005), germline/somatic mutation detection (Mir et al. 2005); (Fig. 3.1a, b), and insertions–deletions. This has helped in establishing an association with simple or complex diseases and in understanding genetic diversity and evolution through comparative genomics and forensic sciences. This chapter discusses the evolution of the method of DNA sequencing from a manual process to the development

National Centre of Applied Human Genetics, School of Life Sciences,
Jawaharlal Nehru University (JNU), New Delhi-110067, India,
Fax: +91-011-26103211, email: bame0200@mail.jnu.ac.in; bamezai@hotmail.com

Soil Biology, Volume 11
Advanced Techniques in Soil Microbiology
A. Varma, R. Oelmüller (Eds.)

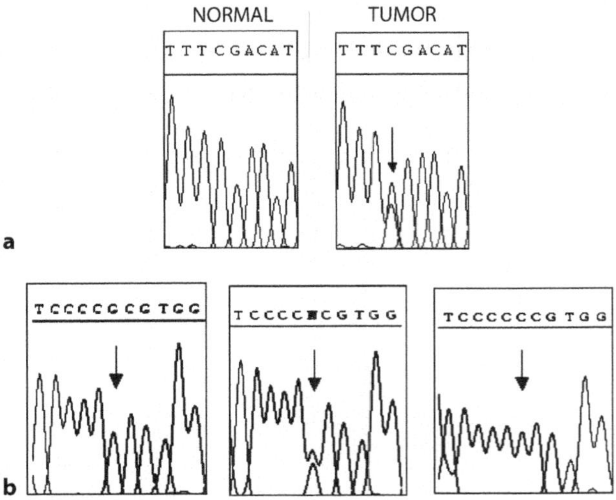

Fig. 3.1 Detection of somatic mutation in tumour samples (**a**) and polymorphism using automated fluorescence sequencing (**b**)

of automated fluorescent DNA sequencing strategy, and the troubleshooting, which encompasses common problems, associated with the technique.

3.2 Evolution of the Method

3.2.1 Manual Sequencing

One of the major methods of DNA sequencing is the chain termination dideoxynucleotide method by Sanger and coworkers.

Dideoxynucleotidetriphosphates (ddNTPs) lack a hydroxy (OH) group at the 3' position. This position is normally where one nucleotide attaches to another to form a chain. If there is no OH group in the 3' position, additional nucleotides cannot be added to the chain, thus interrupting chain elongation. It is an elegantly simple process and involves the following steps:

1. Reaction setup: the reaction takes place in four different tubes, each of which contains a different ddNTP. It also includes the following components:
   a. Template – whose DNA sequence needs to be determined. It may be a PCR product or a cloned fragment.
   b. Primer – which is a short fragment of DNA with an annealing site in the template which initiates DNA synthesis. The primer determines the start-

ing point of the sequence being read and the direction of the sequencing reaction.

c. Deoxynucleotidetriphosphates (dNTPs) – which extend the primer, forming a DNA chain. All four nucleotides (A, T, G, C in deoxynucleotide form) are added to the sequencing reaction.
d. DNA polymerase – incorporates the nucleotides and dideoxynucleotides into the growing DNA chain in a 5' to 3' direction.
e. Buffer – a solution that stabilizes the reagents and the products in the sequencing reaction.

2. The reaction is denatured at 95 °C for 10 s, primer is allowed to anneal the template at 50 °C for 5 s and finally extension is carried out by DNA polymerase at 60 °C for 4 min. This temperature cycle is repeated 25 times. Many copies of the template DNA are made by primer extension (adding nucleotides on to the primer). All copies have the same nucleotide sequence but vary in length because the ddNTPs incorporate randomly and stop extension. Thus the tube that contains ddATP has fragments that all terminate at an adenosine (A), the tube with ddGTP has fragments that end with a guanine (G), and so on.
3. The extension products from the four tubes are then purified and run in parallel lanes of a polyacrylamide gel.
4. The gel separates the sequencing products based on size; smaller fragments travel through the gel faster then longer fragments. The electrophoresed products of DNA sequencing can be visualized because the primer is tagged with a radioactive label (dATP, 35-s or 33-p). When exposed to X-ray film, the radioactivity shows up as a dark band. If the first lane contains the products from reaction containing ddCTP, every band that shows up in that lane represents a sequencing product that terminates at a C. The four lanes are read together in a horizontal hierarchy from bottom (smallest) to top (largest).

3.2.2 Automated Sequencing

The reaction in automated sequencing is essentially the same as in manual sequencing. There are two main differences: the labeling and reading. In automated sequencing, each ddNTP used is labeled with a different fluorescent dye; thus the products are fluorescently labeled rather than radioactively. Each fluorescent dye used, corresponding to a different ddNTP, emits a characteristic wavelength upon excitation: ddATP = green, ddTTP = red, ddCTP = blue, and ddGTP = yellow. Thus each fragment has a different color at its end depending on which is the terminating nucleotide (ddNTP). This allows the sequencing reaction to be put in a single tube and hence products of sequencing to be run on a single lane of a gel rather than in four parallel lanes. In addition, the sequencing of nucleotides is determined by the computer, rather than being read manually by a technician. As the samples pass through the gel, a laser excites the fluorescent labels and the emission wavelength of each fragment is detected by a charge-coupled device (CCD) camera. The data is compiled into a gel image, analyzed by specific computer software and the resulting sequence is written into a text file and a chromatogram file, in an automated fashion which is much faster and more efficient than manual sequencing.

Until recently, the most commonly used format was a horizontal or vertical slab gel. The ABI Prism 377 DNA sequencer from Applied Biosystems (Foster City, Calif.; http://www.appliedbiosystems.com) uses multicolor fluorescence labeling technology with four dyes and one-lane detection (Smith et al. 1985). A CCD camera is used for sequencing rates up to 200 bases per sample per hour. It can run 18, 36, 64, or 96 samples simultaneously per vertical gel. Gel plates come in four different lengths to optimize run times and sample resolution. Other gel-based sequencers which were developed as a result of a quest for better technology include ALF*express* (Amersham Pharmacia Biotech, Piscataway, N.J.), ASTRAL (O'Brien et al. 1998), ARAKIS (European Molecular Biology Laboratory, Heidelberg, Germany), 4200 (IR^2) (Li-COR, Lincoln, Neb.), BaseStation (GeneSys Technologies, Sauk City, Wis.), etc.

Years of research on capillary sequencers have yielded several recent commercial systems that significantly increase the throughput and decrease the time required to sequence. CEQ 2000 (Beckman Coulter, Fullerton, Calif.), the multicapillary electrophoresis (MCE) device (Max-Planck-Institute for Molecular Genetics, Berlin, Germany), and MegaBACE 1000 (Molecular Dynamics, Sunnyvale, Calif.) were some of the early capillary sequencers developed. Using experience gained with the ABI prism 310 genetic analyzer single capillary instrument, Applied Biosystems developed the ABI Prism 3700 DNA analyzer with 96 primary capillaries and eight reserve capillaries. It has a turnaround time of 2.5 h. for an average read length of 550 base pairs with single base resolution. Samples are automatically loaded from 96- or 384-well microplates by electrokinetic injection into 96 capillaries. At the detection end of the capillaries, the samples flow through a sheath flow cuvette detector (Swerdlow et al. 1991; Zhang et al. 1995; Service 1998) to eliminate capillary wall scattering and to provide continuous excitation and detection. Uncoated capillaries are used and are dynamically coated with a polymer such as POP 6 to control electroosmotic flow. The system uses two-dimensional CCD imaging and is capable of using five dyes. It has a signal-to-noise ratio that is ten times better than the ABI Prism 377 sequencer. The Applied Biosystems 3700 uses four dyes, one-lane simultaneous detection of 96 samples, and can run for 24 h unattended. The Applied Biosystems ABI Prism 3100 genetic analyzer, based on the ABI Prism 310 genetic analyzer and 3700 DNA analyzer, was recently introduced to support applications such as comparative sequencing and DNA fragment analysis. It has 16 capillaries, uses either POP 4 or POP 6 as the separation matrix, and can run unattended for 24 h too.

3.2.3 Pyrosequencing

Recently, pyrosequencing emerged as a new sequencing methodology. This technique is a widely applicable, alternative technology for the detailed characterization of nucleic acids. Pyrosequencing has the potential advantages of accuracy, flexibility, and parallel processing; and it can be easily automated. Furthermore, the technique dispenses with the need for labeled primers, labeled nucleotides, and gel electrophoresis (Ronaghi 2001). This method takes advantage of four enzymes cooperating in a single tube to determine the nucleotide composition of a DNA fragment in real time. Detection is based on the amount of visible light produced by coupling the pyrophosphate that is released during nucleotide incorporation with the enzymes sulfurylase and luciferase. Unincorporated nucleotides are degraded in the reaction mixture by the enzyme apyrase. A fully automated instrument called the PSQ96 System has been developed primarily for SNP analysis by Pyrosequencing AB (Uppsala, Sweden).

3.3 Methods

We would like to share our experience of handling the ABI Prism 3100 Avant genetic analyzer for sequencing, which involves the following basic steps:

3.3.1 Template Preparation

PCR products, cloned fragments in single-/double-stranded vectors, cosmids, BAC clones, and bacterial genomic DNA can be sequenced. PCR should be standardized so as to get a single band and high amplification with minimal dNTPs and primers. However, unused primers and dNTPs in the PCR reaction can be removed to some extent by Exo I (0.25 units/reaction) and Antarctic Phosphatase (0.50 units/reaction) treatment for 2 h at 37 °C followed by denaturation of enzymes at 72 °C for 15 min. High-quality plasmids are required for sequencing and thus should be isolated from fresh culture (grown for not more than 14 h) using commercially available kits (Sigma Aldrich, Qiagene have the tested kits). Table 3.1 below shows the quantity of different types of templates required for sequencing.

Table 3.1 Quantity of different types of templates required for sequencing

Template	Template quantity
PCR product:	
100–200 bp	1–3 ng
200–500 bp	3–10 ng
500–1000 bp	5–20 ng
1000–2000 bp	10–40 ng
>2000 bp	40–100 ng
Single-stranded	50–100 ng
Double-stranded	200–500 ng
Cosmid, BAC	0.5–1.0 μg
Bacterial genomic DNA	2–3 μg

3.3.2. Reaction Setup (BigDye Terminator Cycle Sequencing)

One of the main components of the sequencing reaction is the BigDye (Applied Biosystems), which consists of modified Taq DNA polymerase (AmpliTaq Gold), dNTPs, and fluorescence dye-labeled ddNTPs in buffer. Taq DNA polymerase enzyme functions at high temperatures, due to which it can be specially useful when dealing with GC-rich templates or templates that have extensive secondary structures. The thermostability of Taq polymerase also allows cycle sequencing to be used. This approach is similar to PCR (except only a single primer is used and is thus called cycle sequencing reaction or cyclization instead of PCR) and uses a high temperature to denature the duplex DNA strands. The primer is then annealed and extension performed. This allows starting with fewer templates of DNA. Modification of Taq DNA polymerase that enables it to incorporate the dye-labeled terminators more evenly, resulting in less pronounced peak height variability and consistently high accuracy, have made this the most commonly used automated sequencing method. Fresh aliquots of primer having annealing site in the template should be used. Table 3.2 shows separate cycle sequencing reactions setup for PCR products and plasmids. Also listed are recommended (ABI) and standardized (in our laboratory) protocols.

Mix well and spin briefly all reagents before proceeding to perform cycle sequencing reaction.

To prepare high-sensitivity BigDye-terminator reactions and further processing for BACs, PACs, YACs, cosmids, and bacterial genomic DNA, one can refer the Sequencing Chemistry Guide by Applied Biosystems.

Table 3.2 Sequencing reaction setup for PCR products and plasmids

Components	PCR products		Plasmids	
	Recommended	Standardized	Recommended	Standardized
Dye (2.5×)	8.00 μl (1×)	0.5 μl (0.125×)	8.00 μl (1×)	1.00 μl (0.25×)
Buffer (5×)	–	1.75 μl	–	1.50 μl
Primer	3.2 pmol	3.2 pmol	3.2 pmol	3.2 pmol
Template	Refer to Table 3.1	Refer to Table 3.1	Refer to Table 3.1	Refer to Table 3.1
Deionized water	q.s.	q.s.	q.s.	q.s.
Total volume	20 μl	10 μl	20 μl	10 μl

3.3.3 Performing Cycle Sequencing

The following are the universal cycle sequencing conditions used for PCR and single-/double-stranded vector templates. These are applicable to all primers.

First, place the plate in a thermal cycler (the GeneAmp PCR system 9700 has a rapid thermal ramp of 1 °C/s) and set the volume as required.

Next, repeat the following for 25 cycles: (a) rapid thermal ramp to 96 °C, (b) hold at 96 °C for 10 s (denaturation), (c) rapid thermal ramp to 50 °C, (d) hold at 50 °C for 5 s (primer annealing), (e) rapid thermal ramp to 60 °C, and finally (f) hold at 60 °C for 4 min (primer extension).

Then, rapid thermal ramp to 4 °C, spin down the contents of the plate, and store at 4 °C until ready to purify.

For short PCR products, a reduced numbers of cycles can be used (e.g., 20 cycles for a 300-bp or smaller fragment).

If the T_m of a primer is <50 °C, increase the annealing time to 30 s or decrease the annealing temperature to 48 °C.

For templates with high GC content (>70%), heat the tubes at 98 °C for 5 min before cycling, to help denature the samples.

For primers having annealing temperature higher than 60 °C, the annealing step can be eliminated.

Further processing should not be delayed, as the signal becomes weak with time.

3.3.4
Preparing Extension Products for Electrophoresis

Unincorporated dye-terminators must be completely removed before the samples can be analyzed by electrophoresis. Excess dye-terminators in sequencing reactions obscure data in the early part of the sequence and can interfere with base calling. Commercially available 96-well spin columns or ethanol precipitation methods can be used. The ethanol precipitation method given below is cheaper, but is more likely to leave unincorporated dye-labeled terminators that can obscure data at the beginning of the sequence.

1. Add 10 µl deionized water to the reaction, followed by 2 µl each of sodium acetate (3 M, pH 5.2) and EDTA (125 mM, pH 8.0).
2. Add 70 µl non-denatured 95% ethanol (the final ethanol concentration should be 60±3%).
3. Cover the plate with a rubber seal and invert the plate a few times to mix. Leave the plate at room temperature (23–25 °C) for 10–15 min to precipitate the extension products. Precipitation times <15 min result in the loss of very short extension products. Precipitation times >24 h increase the precipitation of unincorporated dye-terminators.
4. Place the plate in a table-top centrifuge with plate adaptor and spin for 25 min. at 3300 rpm at room temperature (23–25 °C). Proceed to the next step immediately. If not possible, then spin the plate for 2 min more immediately before performing the next step.
5. Without disturbing the precipitates, remove the rubber seal and discard the supernatant by inverting the plate onto a paper towel. Remove as much supernatant as possible.
6. Place the inverted plate with the paper towel in the table-top centrifuge and spin at 1000 rpm for 1 min.
7. Add 100 µl of 70% ethanol to each well. Cover the plate with a rubber seal and invert the plate a few times to mix.
8. Place the plate in a table-top centrifuge and spin for 10 min at 3300 rpm at room temperature.
9. Repeat the ethanol wash (steps 5–8). At this step, plates should not be inverted to mix after addition of 100 µl of 70% ethanol.
10. Remove the plate and discard the paper towel. Add 10 µl Hi-Di formamide (Applied Biosystems) followed by a brief spin to settle down the contents. Pellets may or may not be visible. Vacuum drying of the samples is not necessary.
11. Heat the contents at 95 °C for 4 min followed by snapchilling at 4 °C or it can be kept in ice and then loaded onto the autosampler. The sample file with appropriate dye set, mobility file, run and analysis module should be filled up. Also care should be taken to note the presence of the required amount of polymer in the reserve polymer syringe. Buffer and wash tanks should be filled with fresh 1× running buffer and deionized water, respectively. Before

placing the plates on the autosampler, one must centrifuge them to bring the samples down to the bottom of the tubes.

3.4 Trouble Shooting

3.4.1 Problem: Flat Line or "Dead On Analysis"

Reasons and remedies (Fig. 3.2):

1. Template – poor quality, poor concentration.
   Impurities like phenol, ethanol (>10%) or salt are undesirable and can inhibit the reaction completely. The processivity of Taq DNA polymerase is reduced by high salt concentration of sodium or potassium chloride, which will have a more severe effect than sodium acetate. However, a sodium acetate concentration >20 mM also severely inhibits the reaction. We have personal experience that classic methods of plasmid isolation like alkaline lysis, or boiling failed to give any sequencing results because of the poor quality of plasmids isolated by the above methods. Moreover strains like HB101 contain large amounts of carbohydrates and possess the endA locus, which produces an endonuclease, which might inhibit the reaction. Hence DH5α is the preferred host. Finally, it is essential that the template DNA be accurately quantified and the quality be checked by running on an agarose gel against a DNA of known concentration.
2. Poor primer annealing – T_m too high, primer concentration low, no primer-binding site.
   We highly recommend that a computer program be used during primer design in order to check for certain fatal design flaws. Numerous programs are capable of performing this analysis. We generally use "Oligo" (National Biosciences, Plymouth, Minn.), a program for the Macintosh that has produced excellent results in our hands. Two other programs one might consider are MacVector (Kodak/IBI) and the GCG suite of sequence analysis programs, but many others are available as well. The following are some guidelines for primer designing:
   a. The primers should be 20–30 bp long.
   b. T_m should be in the range 55–65 °C.
   c. GC content should be within the range of 50–55%.
   d. Long runs of single bases should be avoided, especially if that occurs at the 3' end.
   e. G or C at the 3' end is preferable so as to stabilize this end of the primer.

f. Discard candidate primers that show undesirable self-hybridization or forms secondary structures.
g. Verify the site-specificity of the primer by BLAST analysis. One should also check for duplicated regions, as this could lead to anomalous results from polymorphism study (Malhotra et al. 2005).
h. Primers should be dissolved in sterile-deionized water. The main stock should be stored at –80 °C while the working stock should be stored at –20 °C. Both main and working stock should be thawed on ice when necessary and care should be taken that they are not thawed repeatedly as this may degrade the primers.

3. Template base composition – GC-rich templates because of their tendency to form stable secondary structures. Try putting the reaction with 5–10% DMSO or formamide. Increase denaturing time.

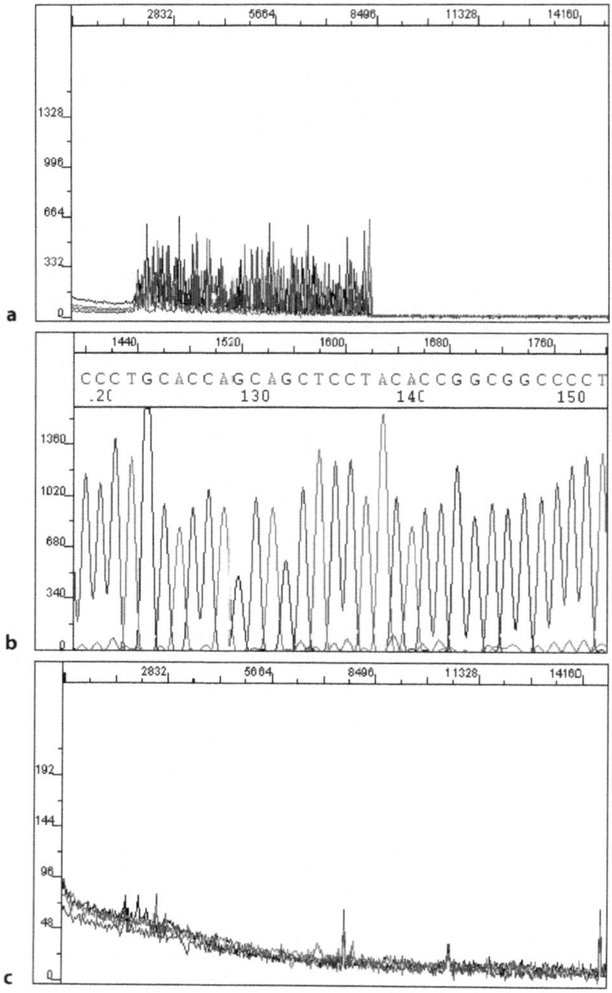

Fig. 3.2 Raw data (**a**) and corresponding electropherogram (**b**) of a good sequencing reaction. Peaks should be sharp, well defined, and scaled high in the first several panels. (**c**) represents raw data for a failed sequencing reaction

3.4.2
Problem: Noisy Data (Background)

Reasons and remedies: although base calling is easiest for the analysis software when the signal strength is high, a good signal strength does not always go hand-in-hand with high-quality data. Background noise can obscure the true sequence data.

1. Mixed templates – two or more different DNA are present in the reaction (a mixed template) and all possess a primer annealing site. Sequence data from each are superimposed in the electropherograms, giving a confused peak pattern (Fig. 3.3). It is a good idea to check each template preparation by agarose gel electrophoresis to determine its purity. When purifying recombinant plasmids in bacteria, plate out the transformants to obtain isolated colonies. Then select a single colony and restreak out on a plate to again select the colony for growth in cultures.
2. Enzyme slippage – this can occur in homopolymer regions, thus skipping a base or incorporating an additional one. The sequence beyond the homopolymer region may then be shifted by one or more bases, giving the appearance of multiple overlapping sequences in the electropherograms.

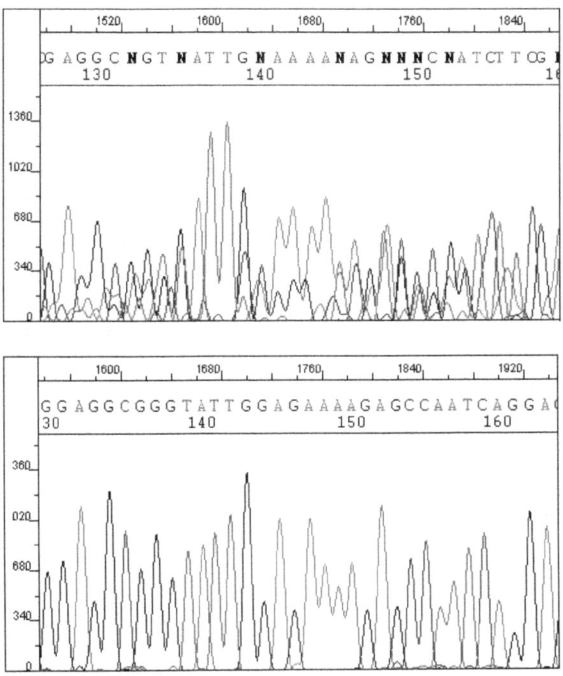

a

b

Fig. 3.3 Noisy data characterized by overlapping peaks due to non-specific amplification. **b** shows the sequencing reactions corrected for the problems

3. Multiple priming events – n–1 primers, contaminated primers, single primer annealing to more than one site. Truncated primers caused by poor synthesis or storage conditions have common 3' ends and different 5' ends. The products of sequencing reactions using such primers yield sequencing results containing a background "pre-read" where a small peak is seen before each correct peak. Repeated freeze–thaw of the primers should be avoided as it can lead to its degradation. It is advisable to use fresh aliquots of diluted primers for sequencing.
4. Quality of template – discussed above.
5. Primer concentration – the presence of primers used to amplify the PCR product results in excessive background due to sequence products generated by these primers. The PCR product should be treated with Exo I, which removes unused primers in the PCR reaction. Agarose gel extraction of the product or use of a size exclusion membrane are other alternatives.

3.4.3 Problem: Reading Near the Primer

Reasons and remedies: with any sequencing strategy there is a limit to how near to the primer it is possible to read. With standard dye-terminator reactions reading closer than 20–30 bases is not usually possible due to the presence of contaminating dye-terminator molecules that have not been incorporated into the extension products.

The terminators are present due to the method used to separate them from the extension products. The precipitation step is a differential precipitation and relies on precipitating the extension products while leaving the dye-terminators in the solution. For this reason, the precipitation and centrifugation steps are performed at room temperature, since this favors retention of the terminators in the supernatant (centrifugation at lower temperatures can cause excessive amounts of dye-terminators to be present in the sample; Fig. 3.4). To remove the dye-terminators more efficiently spin columns can be used but this increases the operational cost. In general, using a primer that is 50 bases or so away from the area where one wishes to start from is a safer and easier option.

3.4.4 Problem: Strong Terminator Peaks

Reasons and remedies: as mentioned above, the presence of terminator peaks can cause problems when the ability to read sequence near the primer is important (especially if the sequence signal is low). Additional problems that are associ-

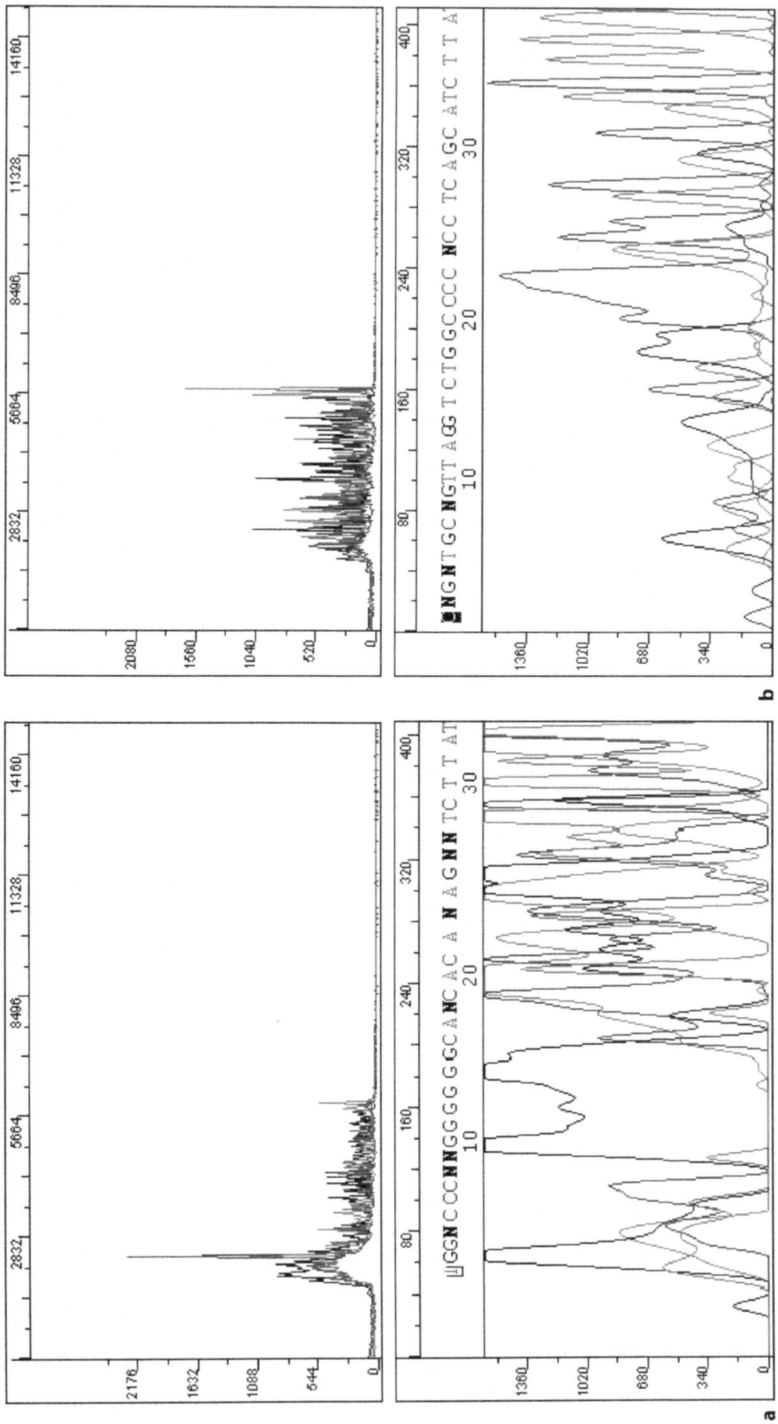

Fig. 3.4 Raw data (*upper panels*) and corresponding electropherograms (*lower panels*) showing the variation in sequencing output due to (**a**) inefficient and (**b**) efficient alcohol precipitation

ated with excessively strong terminator peaks (which are usually a result of inefficient washing after ethanol precipitation or precipitation under conditions that promote terminator precipitation) are generally low apparent signal strength and base "drop out".

To understand why these situations occur it is necessary to know a little about how the analysis program works. The sequence analysis program takes the output from the data collection program and attempts to extract sequence information. This process requires several distinct steps. Firstly, the program looks for where it thinks the sequence starts. This is the first significant fluorescent signal detected and is usually the dye-terminator peak that migrates fastest through the gel. Once the program has located this position, it then checks all the data points from there to the end of the run for fluorescence intensity and scales all the signals with respect to the strongest signal for each channel being detected. Usually the strongest fluorescence is associated with the dye-terminator peaks and so these are the signals to which all others are scaled. Excessively strong terminator signal causes the sequence peaks to be scaled down, thereby reducing their intensity and possibly resulting in inaccurate data. The obvious way around this is to ensure that excess dye-terminators are removed from the sequence sample before it is loaded on the gel. Proper ethanol precipitation conditions are the key factors here.

Once all the peaks are scaled proportionately, the base-calling program attempts to assign each peak a base value based on a number of criteria. Without going into specifics, this is principally achieved by assessing peak intensity (with respect to any minor peak at the same location) and peak shape. One problem that excess terminator peaks can cause is the loss of specific base signal, usually "C". This is caused by uneven recovery of terminators. As stated above, the signals are all scaled relative to the strongest signal for each channel. Hence, if an excess of "C" terminators is present, all the C signals are scaled down more than the other signals. This results in a loss of C signal strength. Fortunately this is easily rectified by reanalyzing the data from after the terminator peaks, thereby removing the artifacts.

3.4.5 Problem: Low Intensity of Shorter Products

Reasons and remedies:

1. Excess dNTPs from the PCR reaction promote longer extension products and lead to a reduction in intensity of shorter products. Treatment of PCR product by shrimp alkaline phosphatase (SAP) solves the problem. Agarose gel extraction of the product or use of size-exclusion membrane are other alternatives.
2. The precipitation step was not long enough, leading to the precipitation of only a small amount of these fragments.

3.4.6
Problem: Longer Fragments Missing

Reasons and remedies:

1. Primer dimer–this can be caused by a self-binding primer or a primer binding to other primers present in the template mixture. Short fragments are predominantly amplified. There may or may not be any sequencing following such an artifact.
2. High template DNA–the peaks at the start of the sample (first two panels) are all off-scale and then the data suddenly drops down in strength, finally dying off only 300–400 bases in to a run (ski-jump profile).
3. Presence of salt inhibits the reaction and gives raw data as if a high template was used. (Although this is usually not associated with the peaks that are off-scale at the start of the trace). The higher the salt concentration the fewer the data produced. DNA eluted in high salt buffer or not washed off properly either during the template preparation step or ethanol precipitation (during washing) may give rise to such a situation.
4. Homopolymer regions – Taq polymerase enzyme prematurely drops off or stops. Try using T7 DNA polymerase.
5. Long templates.

3.4.7
Problem: Presence of Spikes

Reasons and remedies:

Spikes can be present in between the data. This is characterized by sharp peaks of A and C followed by G and T. This can happen if the polymer has been in the syringes for a long time or if the polymer has passed its expiry date and crystallizes at 4 °C. Washing and refilling of fresh polymer should be done every 10 days.

3.4.8
Problem: Weaker Signals

Reasons and remedies:

1. Samples once resuspended in Hi-Di formamide should not be kept exposed to the air for long as Hi-Di formamide absorbs water from the atmosphere, which reacts slowly with formamide to form acid. The ionic products of their reaction can cause two problems – first, they compete significantly with the

larger DNA ions for injection into the capillary, resulting in weaker signals, and second, they react with the DNA, causing degradation of the sample. A pure form of Hi-Di formamide should be used and kept in aliquots to avoid repeated freeze–thawing. The samples resuspended in Hi-Di formamide should be kept sealed before loading onto the autosampler for sequencing.
2. Sample used is low in quantity. Increase the sample or increase the sample injection time from (15 s to 20 s in a 36-cm capillary). One should not change the sample injection voltage, as this affects the resolution across the array.
3. Spatial calibration – dislodge of the capillary should be followed by the spatial calibration. Peak intensity values of 15 000 and above with defined peaks and equal spacing between each capillary should be taken as a successful calibration run and selected for.

References

Hood LE, Hunkapillar MW, Smith LM (1987) Automayed DNA sequencing and analysis of the human genome. Genomics 1:201–212
Hunkapillar T, Kaiser RJ, Koop BK, Hood L (1991) Large-scale and automated DNA sequence determination. Science 254:59–67
Macilwain C (2000) World leaders heap praise on human genome landmark. Nature 405:983–985
Malhotra D, Relhan V, Reddy BS, Bamezai R (2005) TLR2 Arg677Trp polymorphism in leprosy: revisited. Hum Genet 116:413–415
Maxam AM, Gilbert W (1977) A new method for sequencing DNA. Proc Natl Acad Sci 74:560–564
Mir MM, Dar NA, Gochhait S, Zargar SA, Ahangar AG, Bamezai R (2005) *p53* mutation profile of squamous cell carcinomas of the esophagus in Kashmir (India) – a high incidence area. Int J Cancer 116:62–68
Mullis KB, Faloona FA (1987) Specific synthesis od DNA in vitro via a polymerase-catalyzed chain reaction. Methods Enzymol 155:335–350
Mullis KB, Faloona FA, Scharf SJ, Saiki RK, Horn GT, Erlich HA (1986) Specific enzymatic amplification of DNA in vitro: the polymerase chain reaction. Cold Spring Harb Symp Quant Biol 51:263–273
O'Brien KM, Wren J, Dave VK, Bai D, Anderson RD, Rayner S, Evans GA, Dabiri AE, Garner HR (1998) ASTRAL, a hyperspectral imaging DNA sequencing. Rev Sci Instr 69:2141–2146
Ronaghi M (2001) Pyrosequencing sheds light on DNA sequencing. Genome Res 11:3–11
Sanger F, Nicklen S, Coulson AR (1977) DNA sequencing with chain-termination inhibitors. Proc Natl Acad Sci 74:5463–5467
Service RF (1998) Picking up the pace of sequencing. Science 280:995
Smith LM, Fung S, Hunkapillar MW, Hunkapillar TJ, Hood L (1985) The synthesis of oligonucleotides containing an aliphatic amino group at the 5' terminus: synthesis of fluorescent DNA primers for use in DNA sequence analysis. Nucleic Acids Res 13:2399–2412
Smith LM, Sanders JZ, Kaiser RJ, Hughes P, Dodd C, Conell CR, Heiner C, Kent SBH, Hood LE (1986) Fluorescence detection in automated DNA sequence analysis. Nature 321:674–679

Sterky F, Lundeberg J (2000) Sequencing genes and genomes. J Bacteriol 76:1–31

Swerdlow H, Zhang JZ, Chen DY, Harke HR, Grey R, Wu S, Fuller C, Dovichi NJ (1991) Three DNA sequencing methods using capillary gel electrophoresis and laser-induced fluorescence. Anal Chem 63:2835–2841

Zhang JZ, Fang Y, Hou JY, Ren HJ, Jiang R, Roos P, Dovichi NJ (1995) Use of non-cross linked polyacrylamide for four-colour DNA sequencing by capillary electrophoresis separation of fragments up to 640 bases in length in two hours. Anal Chem 67:4589–4593

4 mRNA Quantitation Using Real Time PCR

S. Gochhait, S.I. Bukhari, and R.N.K. Bamezai

4.1 Introduction

Many cellular decisions concerning survival, growth, and differentiation are reflected in altered patterns of gene expression; and the ability to quantitate transcription levels of specific genes has always been central to any research to understand gene function (Zamorano et al. 1996). In the recent past, the use of real time PCR has become an important tool in determining the expression of drug resistance markers in tumor cells (Ramachandran and Melnick 1999), monitoring responses to chemotherapy (Desjardin et al. 1999), providing a molecular assessment of tumor stage (Bustin and Dorudi 1998), and detecting bacterial (Hill 1996) and viral pathogens (Holodniy 1994; Jothikumar et al. 2005), suggesting its wide range of applications in clinical diagnostics.

There are four methods commonly used for quantification of transcription: Northern blotting and in situ hybridization (Parker and Barnes 1999), RNAse protection assays (Hod 1992; Saccomanno et al. 1992), and the reverse transcription–polymerase chain reaction (RT-PCR; Weis et al. 1992). A fifth method, cDNA arrays, is of limited use because of cost considerations. Of all these, RT-PCR is the most sensitive. It involves an in vitro method for enzymatically amplifying defined sequences of RNA (Rappolee et al. 1988), which after gel electrophoresis allows target quantitation by comparison of the intensities of ethidium bromide stained control and target bands (Raeymaekers 1999). Conventional RT-PCR provides an endpoint detection method, but it is fraught with limitations. Poor precision, low resolution, non-automated, labor-intensive, involves post-PCR processing are some of the limitations which makes it

National Centre of Applied Human Genetics, School of Life Sciences,
Jawaharlal Nehru University (JNU), New Delhi-110067, India, Fax: +91-011-26103211,
email: bame0200@mail.jnu.ac.in, bamezai@hotmail.com

Soil Biology, Volume 11
Advanced Techniques in Soil Microbiology
A. Varma, R. Oelmüller (Eds.)

unsuitable for routine use, especially when several targets from many samples require quantification.

Constant efforts to enhance the accuracy of quantitation of transcription level led to the development of real time quantitative PCR, which allows cycle-by-cycle detection of accumulated PCR products by combining thermal cycling, fluorescence detection, and application-specific software. The seminal papers describing real time PCR were published in the year 1996 (Gibson et al. 1996; Heid et al. 1996). Today a search in Pubmed (NCBI data base) with the keyword "real time PCR" alone shows thousands of publications, suggesting the widespread acceptability that this method enjoys. However, obtaining accurate and reproducible information about the use of real time PCR for gene expression studies requires a thorough understanding of the nuances of the technique and the probable pitfalls associated with it. In the present chapter we discuss the basic steps involved in real time PCR assay development along with some important considerations for experimental design and data analysis.

4.2 Methods

4.2.1 Chemistry and Primer/Probe Design

Based on the fluorescent reporter molecule used for detection, real time PCR can be categorized into two types: (i) non-specific detection, using DNA binding dyes, e.g. Syber green, (ii) specific detection, using target specific probes, e.g. molecular beacons, FRET hybridization probes, Taqman probes and Scorpion probes. We describe here the most commonly used real time PCR chemistry, the "Taqman assay" (Fig. 4.1). This utilizes the 5' nuclease activity of the DNA polymerase to hydrolyze a hybridization probe bound to its target amplicon. After the reverse transcription step, successful quantification requires the annealing of three oligonucleotides to the DNA. Two template-specific primers define the endpoints of the amplicon and provide the first level of specificity. The additional specificity of this assay is provided by the use of a third oligonucleotide probe that hybridizes to the amplicon during the annealing/extension phase of the PCR. The probe is labeled at its 5' end with a fluorescent reporter dye and at the 3' end with a quencher molecule that can be either fluorescent or non-fluorescent. Although the labels do not necessarily have to be on the 5' and 3' nucleotides, this is the most common and effective probe design. To prevent the probe from acting as a primer in the PCR, the 3' end of the oligonucleotide is blocked with a phosphate group instead of a 3' hydroxyl group. When the probe is in solution and single-stranded, it is able to adopt conformations whereby the

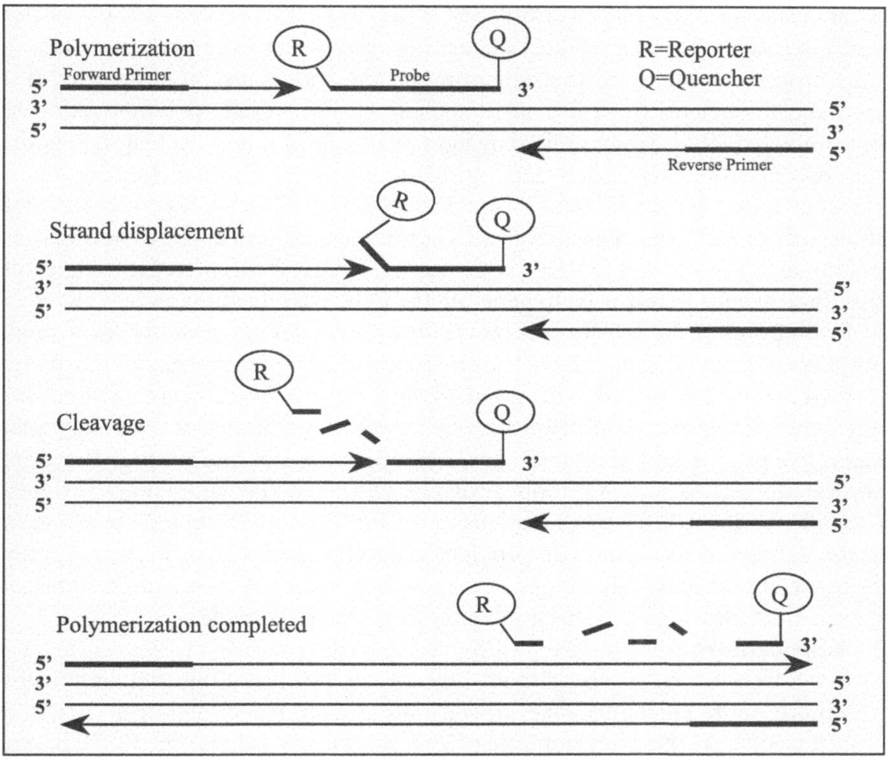

Fig. 4.1 Stepwise representation of the forklike structure-dependent, polymerization-associated, 5' to 3' nuclease activity of Taq DNA polymerase activity on a fluorogenic probe during one extension phase of PCR

5' and 3' dyes are very close together. When the probe is in these conditions, any light absorbed by the reporter dye is transferred by fluorescence resonance energy transfer (FRET) to the quencher dye. If the quencher is a fluorescent molecule, then the transferred energy is released as light of a longer wavelength, whereas non-fluorescent quenchers dissipate the energy as heat. In either case, the shorter wavelength fluorescence that would typically be emitted by the reporter dye is quenched when the probe is intact and single-stranded. At the beginning of a PCR reaction, there is very little specific probe target present, and therefore almost all the probe molecules are single-stranded and quenched. As the PCR proceeds, target sequences are amplified and accumulate exponentially. During the annealing/extension phase of each cycle, probe molecules bind to the homologous strand of the target sequences and form short stretches of double-stranded (ds)DNA. The polymerase extends the primers until it reaches the hybridized probe, where it displaces its 5' end to hold it in a forked structure. The enzyme continues to move from the free end to the bifurcation of the duplex, where cleavage takes place (Lyamichev et al. 1993). This separates the reporter and quencher dyes and releases quenching of reporter fluorescence emission. This process repeats until the remaining probe is too short to remain hybridized and dissociates from the target sequence, allowing the polymerization to be completed. Thus, when excited by light of an appropriate wavelength, it emits fluorescence that can be detected and distinguished from quencher dyes fluorescence based on its different emission wavelength. In this way, reporter dye fluorescence increases at each cycle in a cumulative fashion and is therefore proportional to the amount of PCR product generated. It is this increase in reporter dye fluorescence that is used to generate the PCR amplification plots.

Primer and probe selection is based on estimated T_m, the desire for a small amplicon size and the location of the primer/probe. The Taqman system provides its own primer/probe design program, Primer Express. The current version usually generates appropriate primer/probe sets, but it contains several bugs that can make primer design difficult, and it requires manual fine-tuning. There are several alternative software tools available: Oligo (Molecular Biology Insights, Cascade, Colo., USA; http://oligo.net) and the Beacon Designer are some of them. Applied Biosystems also provides a standardized 20× primer/probe mix as Assay On Demand (AOD) gene expression systems for a wide range of target genes. The following are some guidelines for primer/probe design.

1. Amplicon size

   Maximum amplicon size should not exceed 150 bases. Shorter amplicons amplify more efficiently than longer ones and are more tolerant to reaction conditions. As the extension rate of Taq polymerase is between 30–70 bases/s, it also means that a polymerization time as short as 15 s is sufficient to replicate the amplicon, making the amplification of genomic DNA contaminants less likely and reducing the time it takes to complete the assay.

2. Primer design

   Primer length should be 15–30 bp with 30–80% GC content and a melting temperature (T_m) of 58–60 °C. The T_m of both the primers should be equal or the difference in T_m should be <2 °C.

   Runs of four or more consecutive identical bases, e.g. four Gs should be avoided.

   The five 3' bases of the primers should consist of no more than two or three G or C residues. This helps to reduce non-specific priming.

3. Probe design

   Length of the probe should be of 9–40 bases with 30–80% GC content and a T_m 10 °C above those of the primers (68–70 °C). As amplification primers are extended as soon as they bind to their targets, the hybridization target sequence is rapidly masked with newly synthesized DNA. Therefore, the T_m of the probes must be significantly greater than that of the primers to ensure that they hybridize before the primers. The fluorescence of exonuclease probes is correlated with probe hydrolysis, and they must be available for cleavage during the polymerization step, to allow fluorescence measurement.

It is better to position the probe closer, i.e. within 1–4 bases, if possible, to the forward primer.

Runs of four or more consecutive identical bases, e.g. four Gs should be avoided.

The presence of more Cs than Gs is desirable and a G at the 5' end should be avoided, because such an arrangement quenches reporter fluorescence, even after cleavage.

4. False-positive results are obtained due to amplification of contaminating genomic DNA, if primer/probe are designed in exons only. Thus, it is preferable to have primer/probe spanning exon–exon junctions (Godfrey and Kelly 2005).
5. For each primer and probe sequence, run a separate Blast search as well as a search with the whole PCR amplicon, to rule out any homology with a nonspecific sequence.
6. In the relative gene expression study, variations in the quality and quantity of RNA used in the reaction as well as those introduced by the reverse transcription step can lead to a considerable amount of error in the final gene expression profiles of the target genes. To minimize such an error, the expression of the target genes need to be normalized with the expression of an endogenous control, which ideally should not vary from sample to sample, or in experimental conditions (see note 7, Section 4.3). This is analogous to the use of Actin or GAPDH as a control for loading and transfer for Western and Northern analysis. An appropriate gene should be selected to serve as control for the experiments and primer/probe for the same should also be designed.

4.2.2 RNA Isolation from the Sample

Several total RNA isolation methods in the form of kits/reagents are available from suppliers, and the appropriate selection of method depends on the users' laboratory set up. TRIzol reagent (Invitrogen, Sigma) or kits from Qiagen, Sigma work well in our experience. RNA isolation is a critical step in the whole process and it is essential that the quantity and quality of the RNA preparation be checked by UV spectroscopy and 1% agarose gel electrophoresis (Fig. 4.2). A peak at 270 nm indicates phenol contamination while that of protein is reflected in low 260/280 ratios (<1.8). The pH of the RNA solutions may significantly affect the 260/280 ratio, therefore measurements should be taken in 10 mM TE buffer, pH 7.5. Mix 0.5–2.0 μg total RNA in denaturing sample loading buffer. Denature at 65 °C for 2 min. Then, briefly spin and place on ice before loading onto the native agarose gel. An RNA ladder is loaded as a marker. The appearance of sharp 28s and 18s rRNA which migrate at approximately 5 kb and 1.9 kb markers, respectively, relative to a single stranded RNA ladder indicates

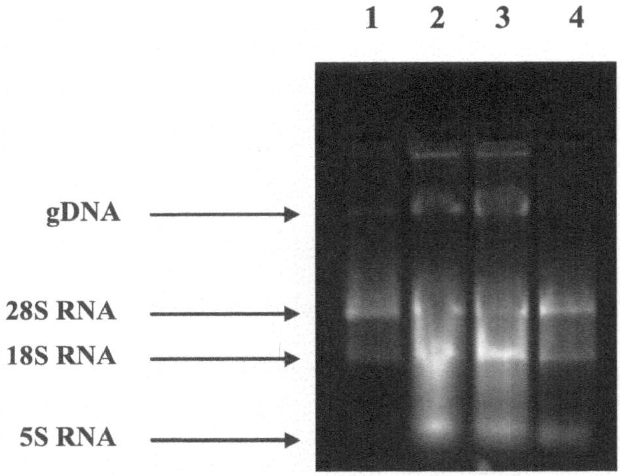

Fig. 4.2 Gel analysis of total RNA prepared from a HeLa cell line. The smears in wells *2* and *3* indicate degraded RNA, while the distinct bands of 28s and 18s in wells *1* and *4* show good-quality RNA. Low-mobility bands found in wells *1*, *2* and *3* show gDNA contamination in these samples

the integrity of the isolated RNA; whereas diffused and smeared bands are an indication of serious degradation. Low molecular weight 5s rRNA and tRNA may sometimes appear as faster migrating diffuse bands. The appearance of sharp and distinct high molecular weight bands running much slower than the 28s rRNA band is indicative of possible DNA contamination. The following are some points to be considered during RNA isolation to avoid commonly faced problems:

1. All materials and reagents must be RNase free.
2. RNA pellet should be completely resuspended by heating it at 55–60 °C for 10 min.
3. A low UV 260/280 ratio indicates phenol or protein contamination. It requires an additional phenol/chloroform/ethanol step.
4. Follow the manufacturer's recommendations strictly.
5. gDNA contamination can be avoided by careful separation of the aqueous phase and by not disturbing the interphase containing DNA. If DNA contamination still persists then the RNA should be subjected to DNase I digestion (Sigma, USA) followed by re-precipitation of RNA.

4.2.3 Reverse Transcription

1. Using the positive control RNA, set up on ice an RT and a no-RT control as shown in Table 4.1.
2. For the no-RT reaction, use water in place of Omniscript reverse trancriptase (Qiagen).
3. In the thermal cycler, incubate the reaction at 37 °C for 60 min and hold at 4 °C.

Table 4.1 Reaction set up for reverse transcription

Component	Volume/reaction	Final concentration
Master mix		
10× buffer RT	2.00 μl	1×
dNTP mix (5 mM each dNTP)	2.00 μl	0.5 mM each dNTP
Random hexamer (100 μM)	2.00 μl	10 μM
RNase inhibitor (40 units/μl)	0.25 μl	10 units per 20 μl reaction
Omniscript reverse transcriptase (4 units/μl)	1.00 μl	4 units per 20 μl reaction
RNase-free water	Variable	–
Template RNA	Variable	Up to 2 μg per 20 μl reaction
Total volume	20.00 μl	–

4. Store cDNA at 4 °C or –20 °C if long-term storage is required with infrequent access.
5. If using oligo-dT, then final concentration should not be less than 1 μM and optimization is required when specific primer are used (see note 6).

4.2.4 Real Time PCR Set Up

Before proceeding to this step, one should have an idea about the array of experimental designs that the process can be subjected to. The following are some important considerations for experimental designs:

1. If the assays on demand (AOD) primer/probe mix (Applied Biosystems, USA) is not used then one has to optimize the PCR, check the efficiency of the primer/probe (Godfrey and Kelly 2005).
2. One can follow the standard curve method or comparative C_T method (see Sections 4.2.5, 4.2.6; PE Applied Biosystems 1997; Livak and Schmittgen 2001), each of which again can be put in a single tube or multiplexed.
3. Running the target and the endogenous control amplification in separate tubes and using the standard curve method of analysis requires the least amount of optimization and validation.
4. To use the comparative C_T method, a validation experiment must be run to show that the efficiencies of the target and endogenous control amplification are approximately equal. A sensitive method for assessing this is to look at

how ΔC_T (target gene C_T – endogenous control gene C_T) varies with template dilution (see Sections 4.2.5, 4.2.6). The absolute value of the slope of log input amount of RNA vs ΔC_T should be <0.1 (Benoy et al. 2004). The advantage of the comparative C_T method lies in eliminating the need for a standard curve. This increases throughput because wells no longer need to be used for standard curve samples. Moreover, while using the single-tube comparative C_T method, the same amount of cDNA should be used for amplifying the target gene and endogenous control.

5. To amplify the target and endogenous control in the same tube, limiting primer concentrations (see note 8) must be identified and shown not to affect the C_T value. By running the two reactions in the same tube, the throughput is increased and the effects of pipetting errors are reduced. A drawback of using the multiplex PCR is in introducing some errors into the final results due to multicomponenting.

Table 4.2 shows the reaction set up for single-tube PCR amplification using ABI reagents.

Spin down the reaction to avoid air bubbles. Seal the tubes with caps and perform PCR in the ABI 7000 sequence detection system (Applied Biosystems, USA) following conditions provided in Table 4.3. Ensure that sample details (e.g. name, position, detector, total volume of reaction) are appropriately provided to the machine before starting the run.

Table 4.2 Reaction set up for single-tube PCR amplification using ABI reagents

Component	Volume/reaction	Final concentration
Universal Taqman PCR master mix (2×)	12.50 μl	1×
AOD primer/probe mix (20×)	1.25 μl	1×
or		
Primer (forward)	1.25 μl	900 nM
Primer (reverse)	1.25 μl	900 nM
Probe	0.5 μl	250 nM
cDNA target	5.00 μl	
Water	q.s.	–
Total volume	25.00 μl	–

Table 4.3 Cyclization conditions used in real time PCR (ABI 7000)

Initial steps		Each of 40 cycles	
		Melt	Anneal/extension
Hold		Cycle	
50 °C for 2 min[a]	95 °C for 10 min[b]	95 °C for 15 s	60 °C for 1 min

[a] The 2 min hold at 50 °C is required for optimal AmpErase UNG activity (see note 9, Section 4.3)
[b] The 10 min hold at 95 °C is required for Ampli Taq Gold DNA polymerase activation

4.2.5 Instrumentation

In recent years, many real time PCR machines have been introduced which are cost effective. The ABI prism 7700/7000 (PE Applied Biosystems, Foster City, Calif., USA), which is comparatively costlier, has become a part of the core facility in all leading research centers around the world. It contains a built-in thermal cycler with 96 well positions, and is able to detect fluorescence between 500 nm and 660 nm. It can be used for assays based on DNA binding dyes, molecular beacons, and hydrolysis probes. RT-PCR reactions typically take 2 h to complete.

During PCR, light from a tungsten–halogen lamp gets focused simultaneously to each well on the microplate. The light passes through the opening of the sample well and the light excites the fluorescent dyes present in each well of the consumable. The resulting fluorescence emission is collected from each well during the extension phase of the PCR reaction. A system of lenses, filters, and a dichroic mirror focuses the fluorescence emission into a charge-coupled device (CCD) camera. The filters separate the light (based on wavelength) into a predictably spaced pattern across the CCD camera. The software collects the fluorescent signals from the CCD camera and applies data analysis algorithms by which the composite spectra is separated from the raw spectrum, the contribution of each dye is determined, the background contribution is eliminated, the reporter signal is normalized with respect to passive reference dye, and finally the software displays the cycle-by-cycle changes in the normalized reporter signal (ΔR_n) as an amplification plot. The passive reference dye is a component of the PCR master mix and hence it is present at the same concentration in all the wells of the plate. By normalizing the data using the passive reference, the software can account for minor variations in signal strength caused by pipetting inaccuracies and make better well-to-well comparisons of reporter dye signal. The graph of normalized reporter (R_n) versus cycle number during PCR appears to have three stages. Initially, R_n appears as a flat line because the fluorescent signal is below the detection limit of the sequence detector. In the second stage, the signal can be detected as it continues to increase in direct proportion to the increase in the products of PCR. As PCR product continues to increase, the ratio of Taq polymerase to PCR product decreases. When the template concentration reaches 10^{-8} M, the PCR product ceases to grow exponentially. This signals the third stage of R_n change, which is roughly linear and finally reaches a plateau at about 10^{-7} M (Figs. 4.3, 4.4).

After the run is finished, one needs to set the threshold and baseline (see note 10). This is done to deduct the background noise. The software then calculates the C_T value for each amplification curve, which for a given amplification curve occurs at the point that the fluorescent signal grows beyond the value of the threshold setting. The C_T represents a detection threshold for the machine and is dependent on two factors:

1. Starting template copy number.
2. Efficiency of DNA amplification on the PCR system.

Thus the system creates quantifiable relationships between test samples on the number of cycles elapsed before achieving detectable levels of fluorescence. Test samples containing a greater initial template number cross the detection threshold at a lower cycle and have a lower C_T value than samples containing a lower initial template.

► **Fig. 4.3** Typical displays of the contribution of each component spectrum when (**a**) FAM- and (**b**) VIC-labeled probes are used separately in a real time PCR reaction in the presence of FAM, VIC, and ROX dye detector **b** *see next page*

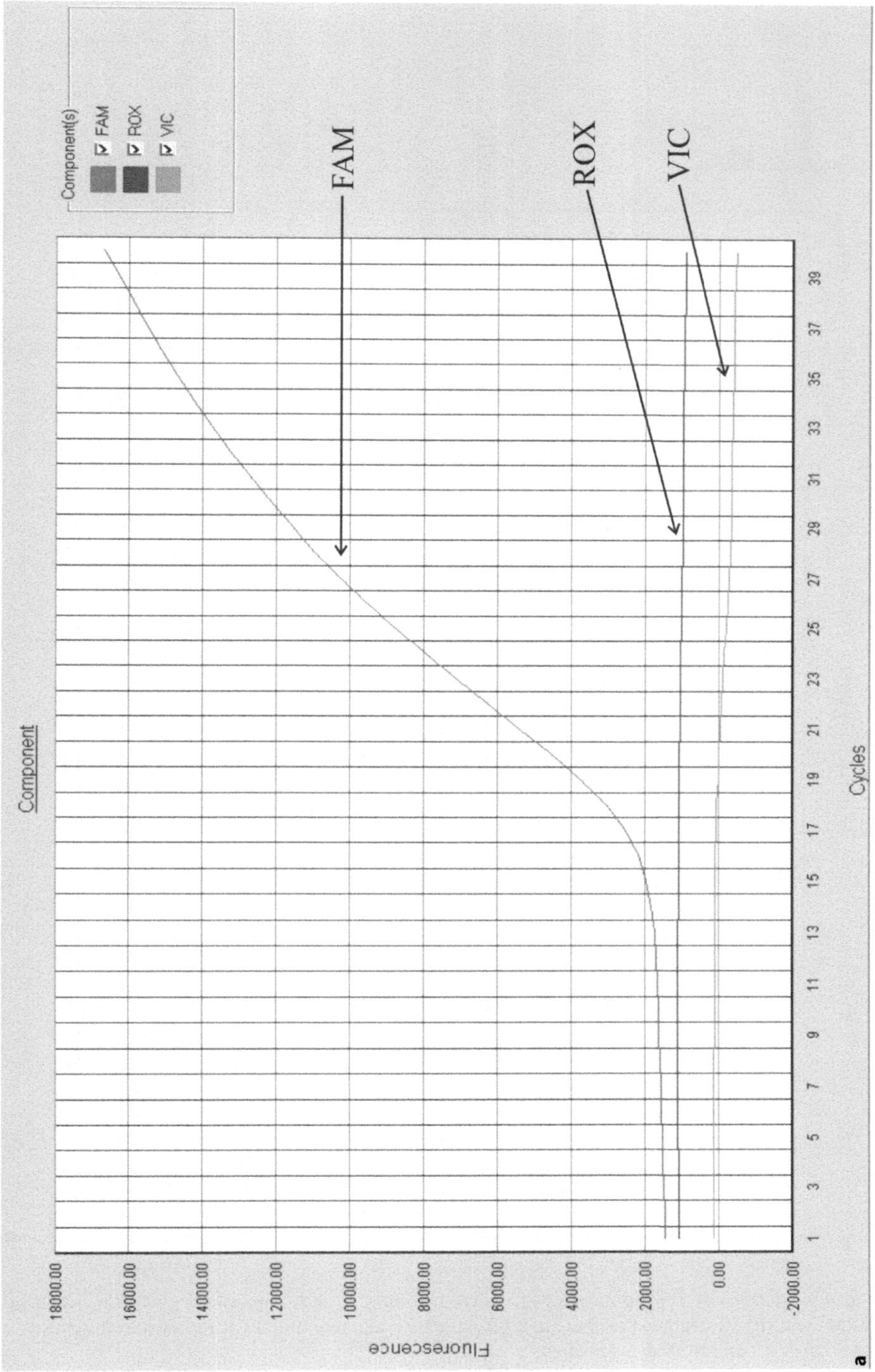
Component
Component(s)
FAM
ROX
VIC
FAM
ROX
VIC
Fluorescence
18000.00
16000.00
14000.00
12000.00
10000.00
8000.00
6000.00
4000.00
2000.00
0.00
-2000.00
Cycles
1
3
5
7
9
11
13
15
17
19
21
23
25
27
29
31
33
35
37
39
a

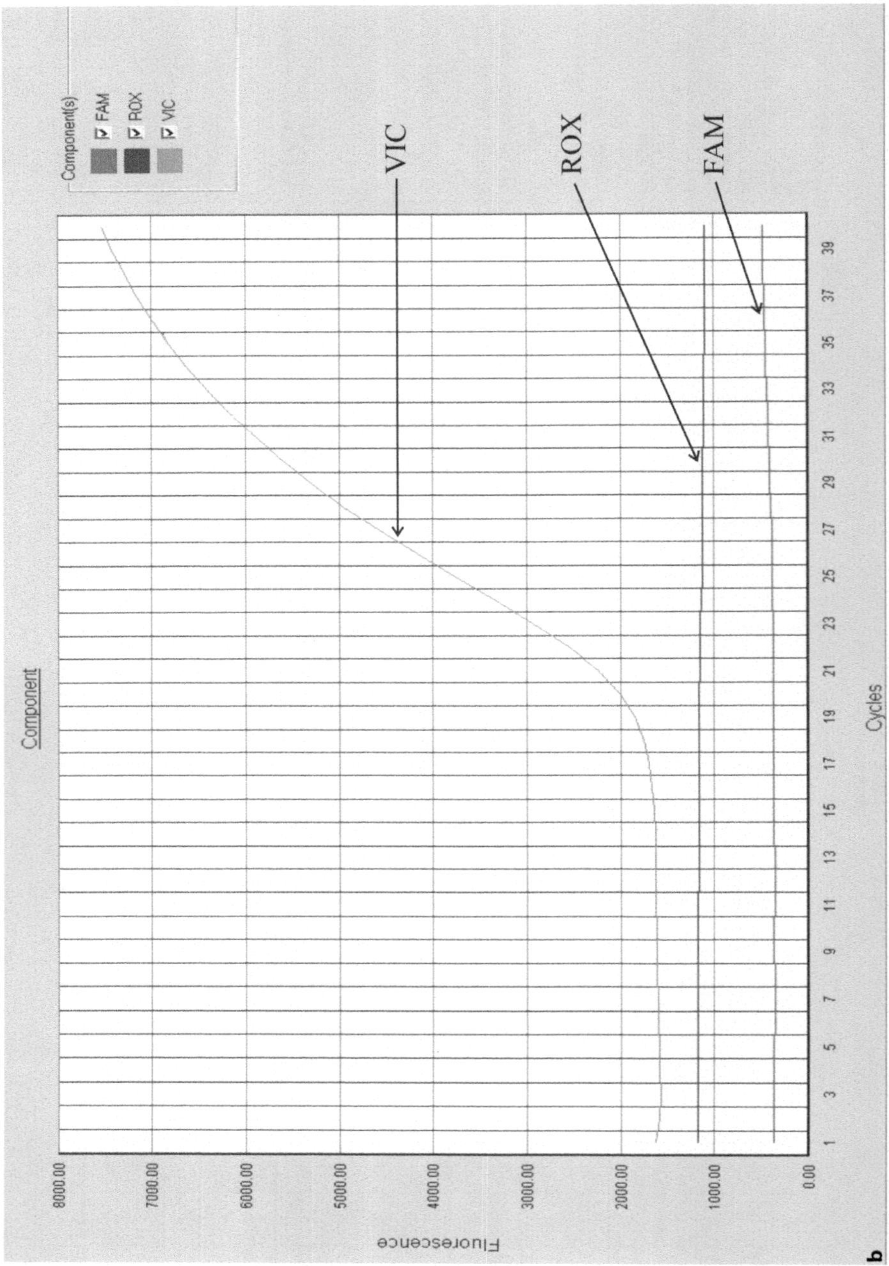

Fig. 4.3 (*continued*) Typical displays of the contribution of each component spectrum when (**a**) FAM- and (**b**) VIC-labeled probes are used separately in a real time PCR reaction in the presence of FAM, VIC, and ROX dye detector

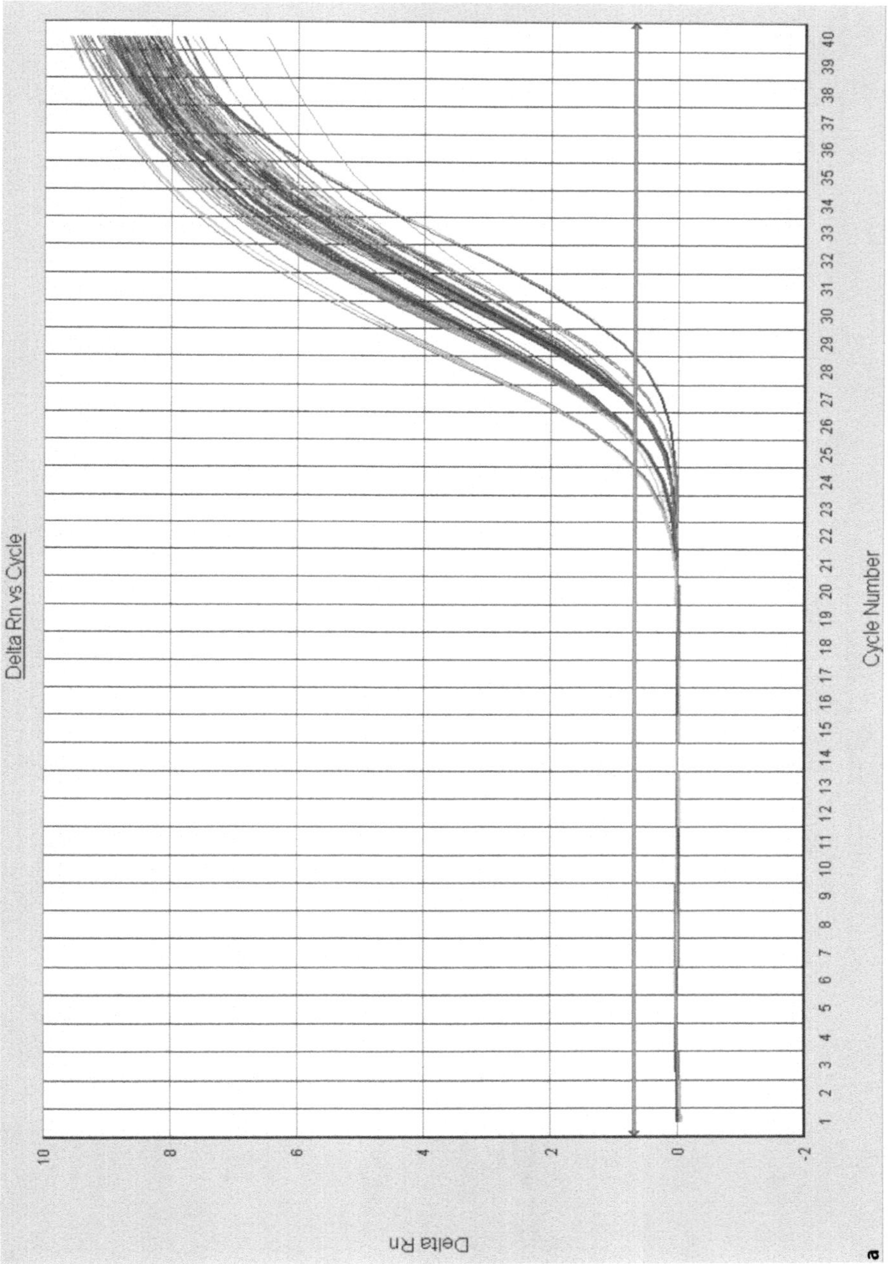

Fig. 4.4 Real time amplification depicted as (**a**) linear plot and (**b**) log plot, showing baseline, threshold, and C_T **b** *see next page*

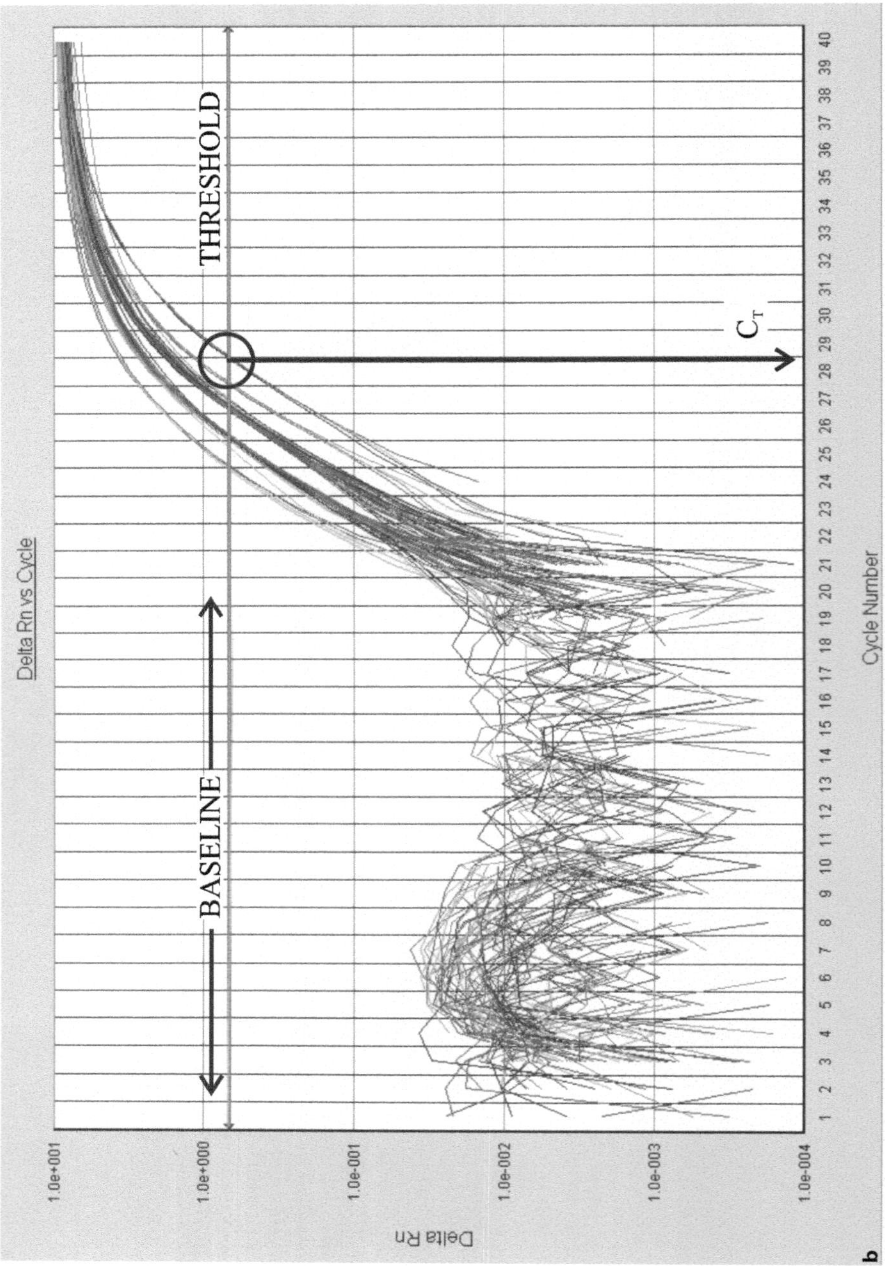

Fig. 4.4 (*continued*) Real time amplification depicted as (**a**) linear plot and (**b**) log plot, showing baseline, threshold, and C_T

4.2.6
Data Analysis

Table 4.4 shows the calculation for a set of experiments, which consists of chemically treated or untreated tissue cell line culture. The target gene and the endogenous control gene have been amplified separately and the relative gene expression is calculated by the comparative C_T method. The final result is illustrated in Fig. 4.5.

Table 4.4 Calculation for a set of experiment consisting of chemically treated or untreated tissue cell line culture. *SD* Standard deviation

Chemical treatment (μm)	Target gene C_T	Endogenous control gene C_T	ΔC_T (target gene C_T - control gene C_T)	$\Delta\Delta C_T$	Relative expression
Untreated (0 μm)	19.97	13.74	6.19±0.08	0	1 (0.95–1.06)
	19.88	13.81			
	19.99	13.72			
Average±SD	19.95 ±0.06	13.76±0.05			
Treated (1 μm)	18.91	15.09	3.86±0.07	−2.33	5.03 (4.79–5.28)
	18.95	14.99			
	18.85	15.05			
Average±SD	18.9±0.05	15.04±0.05			

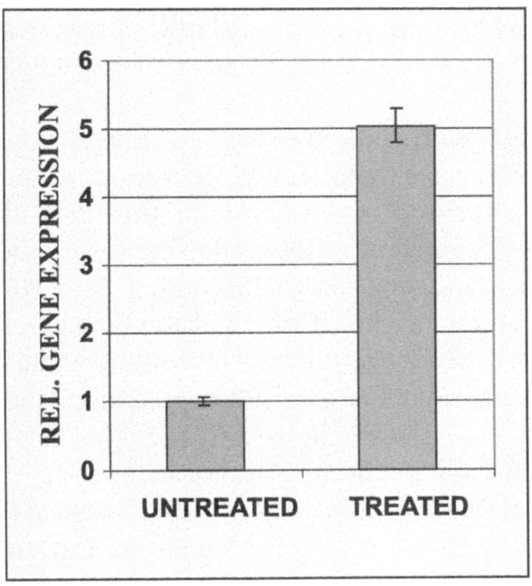

Fig. 4.5 Graphical representation of relative gene expression data. The normalized target gene expression in the treated sample is increased 5-fold as compared with the untreated sample (calibrator)

4.3 Notes

1. Work area should be clean and dust-free.
2. To avoid contamination of the stock primers and probes, it is important to use clean, fresh TE buffer for the reconstitution as well as diluting the primers/probes. The stock primers and probes must be stored at –20 °C while the working solutions can be kept at 4 °C for 1–2 months to avoid repeated thawing and freezing, to avoid damage.
3. Data variability and pipetting errors can be greatly reduced by making master mixes of all common reagents whenever possible.
4. Run appropriate controls along with each batch of samples. Each RT-PCR run should contain duplicate or triplicate reactions for NTC (no-template controls) and positive PCR controls. In most cases, we also run no-RT (without RT enzyme) controls, even if we know that the assay design is cDNA-specific, as this control also detects RT reagent contamination if it is present. The positive control should be a cDNA from an RNA sample known to express the target gene(s). The purpose of this control is to show that the PCR step was set up properly and functioned as expected. This is particularly useful when the samples show low or negative expression of the target gene. The NTC is used to detect PCR reagent contamination.
5. Be consistent with every step of the assay. Always try to perform each part of the RT-PCR assay the same way, i.e. use the same RNA isolation procedure, quantification, same amount of RNA for RT reaction (by spectroscopy measurement), same amount of target (cDNA) for the real time PCR; do not change reagent brands without ensuring that the data does not change; set the same threshold for at least the same gene. This allows one to identify the problem when something goes wrong at any step.
6. RTs extend a double-stranded oligonucleotide in a 5' to 3' direction. As such, the synthesis of a cDNA strand from a single-stranded mRNA requires a primer (with DNA better than RNA). Traditionally, three primer systems have been used for this purpose:
   a. The first is hexamer oligonucleotides with random sequences. These primers bind to the RNA randomly and reverse transcribe all RNA proportionately. The advantages of using this method are several. First, this method reverse transcribes ribosomal RNA as well as mRNA, allowing for use of 18S rRNA as the endogenous invariant gene control. Second, it does not require prior planning for the targets of choice for PCR amplification. Last, this approach is insensitive to sequence polymorphism and mutations and can be somewhat beneficial when there is RNA degradation. The main disadvantage to this system is that one is limited to 1–3 μg of total RNA input before the reaction saturates and becomes non-quantitative.
   b. The second primer system for cDNA synthesis is the use of oligo dT (15–20 base poly-T primer). This binds to the poly A sequence of mRNA

and reverse transcribes mRNA from its 3' poly A tail in a 5' direction. The potential advantage of this system is that it can support a larger input of RNA, because it reverse-transcribes only mRNA, which accounts for 2–5% of the total RNA. Similar to the hexamer method, this approach also allows one to examine any mRNA transcript, as well as evaluate the expression of new genes as they become important at future dates. However, the greatest disadvantage of this approach is that the results can vary based on the quality of the RNA. As more of the RNA is degraded, there are fewer full-length transcripts. Therefore, the evaluation of gene expression using sequences at the 5' end of the message can be unreliable and even artifactual, especially when the possibility exists that the different samples may have slightly different RNA integrity (as is the case with most clinical samples).

c. The third and the last method utilizes sequence-specific RT primers. Here RT can be performed using gene-specific primers that are reverse-complemented to a region 3' of the region of interest for PCR. Also, because mRNA is single-stranded, only one primer is needed. Owing to the relatively small number of mRNA molecules undergoing RT by this approach, RNA input in excess of 10 μg (dependent on the level of gene expression) can often be used in a single RT reaction using optimized protocols. Additionally, multiplex RT with several gene-specific primers can also be used without a loss in quantitation. The practice of using the reverse PCR primer in the RT should be discouraged, because the high T_m of the PCR primer (~60 °C) will result in non-specific priming during the relatively low-temperature RT and regions 3' of the RT primer will not be reverse transcribed and thus cannot be used in future.

7. Actin, GAPDH, and 18sRNA are the most commonly used endogenous control genes but histone 3, phosphoglycerate kinase1, β-2 microglobulin, cyclophilin A and β-glucoronidase have been also used in a few instances. Out of these, particularly, GAPDH has been under a scanner for use as a normalizer (Bustin et al. 2000). Reports show that its expression changes under a variety of circumstances, suggesting a complex regulatory mechanism for the gene expression. Its expression level changes during the cell cycle (Mansur et al. 1993), therefore it should not be used as an endogenous control when doing experiments on animal tissue culture cell line; alternatively harvesting of the cells has to be done at the same confluence level. The same is correct when comparing gene expression between normal and tumor tissue, as the latter contains more actively dividing cells. Still, it is being used as an endogenous control owing to the fact that almost all the genes in the eukaryotic system invariably involve a complicated gene expression modulation. There is also a trend of using total RNA input (by accurate spectroscopy reading) as normalizing factor but it has its own limitations. Contaminating protein, free nucleotides, genomic DNA contamination, and other impurities influence the spectroscopic reading of RNA at 260 nm. However, it is safe to consider more than one control for experimental purposes until and

unless it is proved that a particular control shows most stable expression in one's system. Thus there is an urgent need for a search of relevant and universal normalizer for accurate quantification of gene expression.

8. Multiplex PCR is the use of more than one primer pair in the same tube. This can be used in relative quantitation when one primer pair amplifies the target and the other primer pair amplifies the endogenous reference in the same tube, provided that the reporter dyes used in the probes have different emission wavelength maxima. The ABI prism 7700 sequence detection system has multicomponenting software, which uses a mathematical algorithm to determine the contribution of each dye using a pure dye spectra reference. But this can introduce some errors and the best way to get rid of this is to have reporter dyes having their largest differences in the emission wavelength maxima, e.g. 6-FAM, λ_{maxima} = 518 nm, JOE, λ_{maxima} = 554 nm. Reactions to amplify two different segments in the same tube share common reagents and, if the initial copy number is different than the more abundant species, will use up common reagents and ultimately hamper the amplification of rare species. Thus for accurate quantification, the primers of the abundant species can be limited. Fig. 4.6. shows amplification with the same amount of target but with different primer concentrations. It demonstrates that a lower primer concentration forces the reaction to enter the plateau phase at a lower level of product but C_T remains the same (except for 5 nM). Thus the strategy for performing two independent reactions in the same tube is to adjust the primer concentrations such that C_T values are obtained after which the exhaustion of primers leads to the end of reaction for the majority species so that the common reactants are available for amplification of the minority species. When both have the same or unknown mRNA abundance, then the limiting primer concentration needs to be defined for both amplicons. It can be defined by running a matrix of forward and reverse primer concentration. The desired concentrations are those that show a reduction in ΔR_n but have little effect on C_T.

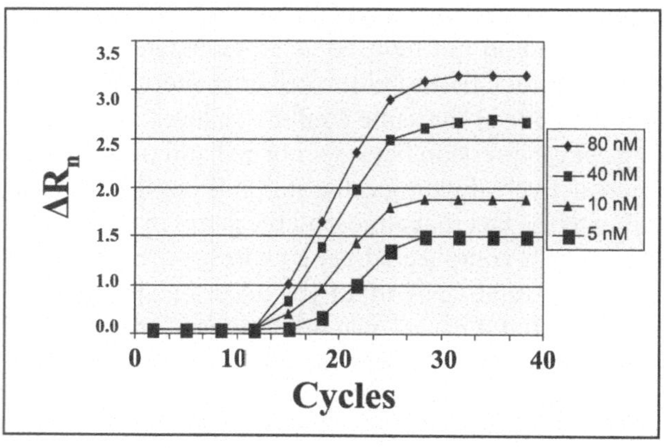

Fig. 4.6 PCR amplification with decreasing primer concentration

9. Uracil-N-glycosylase (UNG) is the active ingredient in AmpErase. UNG recognizes and catalyzes the destruction of DNA strands containing deoxyuridine but not the DNA containing thymidine. Incorporation of AmpErase into the master mix allows for the selective destruction of carryover products (containing deoxyuridine) from previous amplification reactions. UNG catalyzes the cleavage of deoxyuridine-containing DNA at the deoxyuridine residues by opening the deoxyribose chain at the C1 position. When heated in the first thermal cycle step at the alkaline pH of the master mix, the amplicon DNA chain breaks at the position of the deoxyuridine, thereby rendering the DNA non-amplifiable.
10. Baseline start and end cycles are chosen so that they correspond to the initial cycles which show no amplification. Usually the narrowest linear part of the baseline in the linear amplification plot is selected and the end cycle selected is approximately five cycles before the first amplification starts. Baseline correction sets each curve at the origin. The threshold should be placed in linear part of the amplification plot above the noise. The run is analyzed once the baseline and the threshold are selected. The new SDS 1.1 version for data analysis by Applied Biosystems has auto baseline and threshold settings, which allows uniform correction.

References

Benoy IH, Elst H, Van der Auwera I, Van Laere S, Van Dam P, Van Marck E, Scharpe S, Vermeulen PB, Dirix LY (2004) Real-time RT-PCR correlates with immunocytochemistry for the detection of disseminated epithelial cells in bone marrow aspirates of patients with breast cancer. Br J Cancer 91:1813–1820

Bustin SA (2000) Absolute quantitation of mRNA using real-time reverse transcription polymerase chain reaction assays. J Mol Endocrinol 25:169–193

Bustin SA, Dorudi S (1998) Molecular assessment of tumour stage and disease recurrence using PCR-based assays. Mol Med Today 4:389–396

Desjardin LE, Perkins MD, Wolski K, Horn S, Teixeira L, Chen Y, Johnson JL, Ellner JJ, Dietze R, Bates J, Cave MD, Eisenach KD (1999) Measurement of sputum mycobacterium tuberculosis messenger RNA as a surrogate for response to chemotherapy. Am J Resp Crit Care 160:203–210

Gibson UE, Heid CA, Williams PM (1996) A novel method for real time quantitative RT-PCR. Genome Res 6:995–1001

Godfrey TE, Kelly LA (2005) Development of quantitative reverse transcriptase PCR assays for measuring gene expression. In: Keohavong P, Grant SG (eds) Molecular toxicology protocols. (Methods in molecular biology, vol 291) Humana, Totowa, pp 423–445

Heid CA, Stevens J, Livak KJ, Williams PM (1996) Real time quantitative PCR. Genome Res 6:986–994

Hill WE (1996) The polymerase reaction: applications for the detection of foodborne pathogens. Crit Rev Food Sci Nut 36:23–173

Hod Y (1992) A simplified ribonuclease protection assay. Biotechniques 13:852–854

Holodniy M (1994) Clinical application of reverse transcription–polymerase chain reaction for HIV infection. Clin Lab Methods 40:335–349

Jothikumar N, Cromeans TL, Robertson BH, Meng XJ, Hill VR (2005) A broadly reactive one step real-time RT-PCR assay for rapid and sensitive detection of hepatitis E virus. J Virol Methods 131:65–71

Livak KJ, Schmittgen TD (2001) Analysis of relative gene expression data using real-time quantitative PCR and the 2(–Delta Delta C(T)) method. Methods 25:402–408

Lyamichev V, Brow MA, Dahlberg JE (1993) Structure specific endonucleolytic cleavage of nucleic acids by eubacterial DNA polymerases. Science 260:778–783

Mansur NR, Meyer-Siegler K, Wurzer JC, Sirover MA (1993) Cell cycle regulation of the glyceraldehydes-3-phosphate dehydrogenase/uracil DNA glycosylase gene in normal human cells. Nucleic Acids Res 21:993–998

Parker RM, Barnes NM (1999) mRNA detection by in situ and northern hybridization. Methods Mol Biol 106:247–283

PE Applied Biosystems (1997) Relative quantitation of gene expression. (User bulletin 2) Applied Biosystems, Foster City

Raeymaekers L (1999) General principles of quantitative PCR. In: Kochanowski B, Reischl U (eds) Quantitative PCR protocols, Humana, Totowa, pp 31–41

Ramachandran C, Melnick SJ (1999) Multidrug resistance in human tumours – molecular diagnosis and clinical significance. Mol Diagn 4:81–94

Rappolee DA, Mark D, Banda MJ, Werb Z (1988) Wound macrophage express TGF alpha and other growth factors in-vivo; analysis by mRNA phenotyping. Science 241:708–712

Rychlik W, Spencer WJ, Rhoads RE (1990) Optimization of the annealing temperature for DNA amplification in vitro. Nucleic Acids Res 18:6409–6412

Saccomanno CF, Bordonaro M, Chen JS, Nordstrom JL (1992) A faster ribonuclease protecrtion assay. Biotechniques 13:846–850

Schutz E, Ahsen von N (1999) Spreadsheet software for thermodynamic melting point prediction of oligonuleotide hybridization with and without mismatches. Biotechniques 27:1218–1224

Weis JH, Tan SS, Martin BK, Wittwer CT (1992) Detection of rare mRNAs via quantitative RT-PCR. Trends Genet 8:263–264

Zamorano PL, Mahesh VB, Brann DW (1996) Quantitative RT-PCR for neuroendocrine studies. A minireview. Neuroendocrinology 63:397–407

5 Laboratory Practice for the Production of Polyclonal and Monoclonal Antibodies

S. Khurana, S. Bhaskar, and A. Mukhopadhyay

5.1 Introduction

Antibody is defined as an immunoglobulin capable of specifically binding with the antigen that caused its production in a susceptible animal. Antibodies are produced by the plasma cell in response to the invasion of foreign molecules in a host. Immunoglobulins are glycoproteins and are present in five distinct classes in higher mammals: IgG, IgA, IgM, IgD, and IgE. They differ in size, charge, amino acid composition, carbohydrate content, and functions. IgG and IgA have different subclasses.

Immunoglobulins have a basic unit of two light chains and two heavy chains held together by disulfide bonds. The light chain has molecular weight of 25 kDa, whereas the heavy chain molecular weight varies from 50 kDa to 77 kDa, depending upon the class of immunoglobulin. The chains are folded into discrete regions called domains; light chains have two domains and four to five domains are present in heavy chains. Half of the N-terminals of both light and heavy chains show much sequence variability and are known respectively as V_L and V_H domains (Fig. 5.1). Papain digestion generates two Fab fragments (antigen-binding sites) and one Fc fragment (antibody receptor-binding site) from each IgG molecule (Roitt et al. 1996).

Antibodies are of two types, monoclonal and polyclonal. As the name suggests, a monoclonal antibody is a homogeneous population of antibodies produced by a single clone, termed a hybridoma. A hybridoma produces many copies of a gene exactly for the same antibody that recognize one epitope on the antigen. Since monoclonal antibodies are highly specific, they are vulnerable to loss of antigen recognition if the epitope is modified through chemical treatments. In contrast, polyclonal antibodies are a heterogeneous mixture,

National Institute of Immunology, Aruna Asaf Ali Marg, New Delhi-110067, India,
email: ashok@nii.res.in

Soil Biology, Volume 11
Advanced Techniques in Soil Microbiology
A. Varma, R. Oelmüller (Eds.)

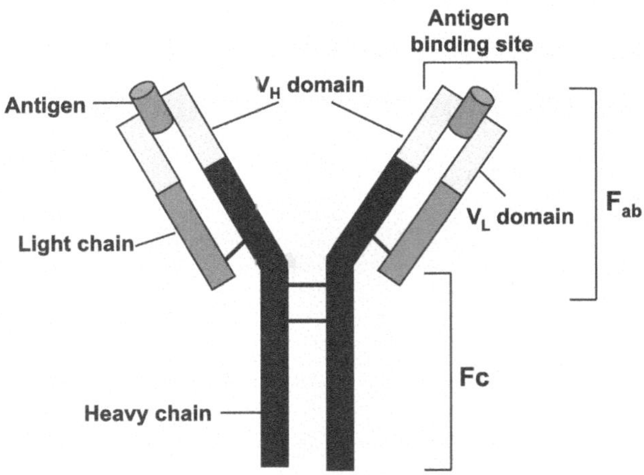

Fig. 5.1 Basic structure of IgG_1. It is composed of two heavy and two light chains, joined together by disulfide bonds. V_H and V_L domains are responsible for antigen detection and binding. Fc domain is involved for binding antibody with its cell surface receptor

which recognizes a variety of epitopes on the same antigen. Because these polyclonal mixtures of antibodies react with multiple epitopes of the antigen, they are relatively less specific and more tolerant of minor changes in the epitope.

5.2 Production of Polyclonal Antibodies

Raising polyclonal antibodies is much simpler than making monoclonal antibodies. The production of polyclonal antibodies consists of a few basic steps, like immunization of a suitable host, withdrawal of blood to check the antibody levels, final collection of blood, and purification of antibodies. It is analogous to the immunization of humans against certain diseases, where the humoral immune response (antibody response) is adequate to neutralize infection. The quantity and quality of the polyclonal antibodies produced from a given immunogen depend on a few important factors, as detailed in the next sections.

5.2.1 Choice of Animal and Method of Immunization

The choice of animal is governed by the amount of antibody required, the amount of immunogen/antigen available, and the phylogenetic relationship between the animals from which the antigen is obtained. For obvious reasons, phylogenetically unrelated species are considered to raise antibodies against an antigen. A good antibody response is generally obtained if slowly released antigen is processed by antigen-presenting cells (APCs) and subsequently presented to the T-cells. The antigen is mixed with adjuvant, and injected by either intradermal (ID), or intramuscular (IM), or subcutaneous (SC) routes in multiple sites (6–10) of the animal. Further, booster (secondary) immunization become necessary for a high antibody titer.

5.2.2 Preparation for Immunization

Irrespective of the type of antigen injected, preparation for the production of polyclonal antibody remains unaltered. The general details are given here.

5.2.2.1 Animal

Rabbit, sheep, goat, and chicken are commonly used as hosts to produce polyclonal antibodies.

5.2.2.2 Adjuvant

Freund's complete adjuvant (FCA), Freund's incomplete adjuvant (FIA), alum, muramyl dipeptide, and monophosphoryl lipid A can be used as an adjuvant. However, the most potent antibody response is obtained with FCA. Except Freund's adjuvant, other antigen-adjuvant preparations can be stored for several months at 4 °C. Antigen preparation with Freund's adjuvant should be made at the time of immunization (within 1–2 h). If FCA is used, it is recommended to make a uniform suspension of the adjuvant before mixing with the antigen. As a general rule, adjuvant and antigen solution are mixed in equal volumes. To prepare a uniform suspension, first Adjuvant (say 1 ml) is taken in a siliconized glass-vial, and to it 1 ml of detergent-free antigen solution in normal saline is added drop-wise with vigorous mixing. The final concentration of antigen is maintained at 0.2–2.0 mg/ml, depending upon the application.

5.2.2.3 Antigen and Immunization Schedule

The quantity of antigen to be injected, the duration, and the number of booster injections are determined by the nature of antigen and the host animal. The exact quantity of antigen to be injected and the ideal host for the specific antigen are decided by trial and error. In the case of rabbits, primary immunization with 50–200 μg antigen (non-rabbit source) in FCA is injected at several sites on the back of the animals (100–150 μl in each site). For each booster, the same dose of antigen in FIA is similarly injected 3–4 weeks after primary immunization. Primary immunization with 0.5–10.0 mg antigen (non-sheep/non-goat sources) in FCA is recommended for sheep and goats. The booster immunization is exactly similar to rabbits, except with a higher amount (0.5–10.0 mg) of antigen (Johnstone and Thorpe 1996). Chickens are immunized similar to rabbits, except that the booster is given 2–3 weeks after primary immunization.

5.2.3 Production of Polyclonal Antibodies

The procedure for manufacturing polyclonal antibodies in rabbits are given below:

1. Take healthy rabbits (age: 4–6 months, weight: 0.5–0.6 kg) and be sure that the animals were not used in an immunization program in the past 6 months. Bleed the animals from an ear vein (2–3 ml blood) to check the total IgG level (pre-immunized) in the blood, and also to determine the presence of specific antibodies for which immunization has been planned.
2. Prepare fresh FCA-antigen suspension, clean the back of the animals with 70% ethanol, and inject 100 μl of suspension in 6–10 sites through the ID route. Maintain the rabbits as usual.
3. At 3–4 weeks after primary immunization, a booster dose of antigen is injected, same as above.
4. Bleed the animals through an ear vein 2 weeks after booster, and determine the antibody titer. When the sera contain a high level of antibody, it is advisable to sacrifice the animal and collect the entire blood through a heart puncture. About 60–70 ml blood can be collected from a rabbit. Alternately, weekly bleeding for 1–2 months is performed through jugular vein. In each bleed, about 15 ml of blood is collected.
5. Antibodies are present in the serum fraction of the blood, so blood cells are separated soon after collecting blood. Otherwise, lysed cells contribute protein contamination in the antibodies; also, proteolytic enzymes may degrade the antibodies.
6. Allow blood to clot at room temperature for 1 h, detach clot from the walls of the container, and separate clot-free serum.
7. Centrifuge clot for 30 min at 2000 *g* at 4 °C to remove trapped serum and mix with the clot-free serum.
8. Further, centrifuge the pooled serum at 1500 *g* for 20 min at 4 °C. Store the serum at −20 °C or −70 °C, if not immediately subjected to purification steps.

5.2.4 Materials and Equipment

Polyclonal antibodies can be easily raised, as the production steps do not require any sophisticated laboratory facility. Besides the animal, it requires the adjuvant and the antigen, a bench-top centrifuge, glass/plastic containers for storing sera, and a freezer at −20 °C or −70 °C.

5.3 Production of Monoclonal Antibodies

Mice and rats have the ability to make antibodies which are able to recognize virtually all antigenic determinants and even discriminate between similar epitopes. These make monoclonal antibodies a most attractive tool to target many molecules found in wide systems, such as receptors or other molecules found on the surface of normal cells, molecules specifically expressed on the surface of cancer cells, etc. In 1975, Köhler and Milstein described the process of cell fusion, "hybridoma technology", where B cells confer antibody production capability, while myeloma cells enable hybridomas to divide indefinitely and grow well in culture. A single clone producing the desired antibody at high titer can be selected for large-scale culture for monoclonal antibody production (Köhler and Milstein 1975). There are many critical steps for generating hybridomas and producing monoclonal antibodies.

5.3.1 Immunization of Mice or Rats

Monoclonal antibodies specific for human antigens are generally raised in mice, and those of mice are raised in rats. The most important steps for the immunization of mice are described below:

1. Immunize 6–8 week-old Balb/c mice with 5–20 µg antigen (98–99% pure) isolated from humans or any non-mice species in the presence of appropriate adjuvant.
2. Booster is injected 3–4 weeks after primary immunization. Before immunization, pre-bleed the mice for ELISA negative control. The handling and processing techniques followed to separate sera are the same as for rabbits, except that a maximum of 200 µl of blood can be collected from each mouse from either tail vein or retro-orbital plexus.
3. Bleeding is performed every week after injecting the booster dose.

4. Once the antibody level in sera is significantly high, the mouse is aseptically sacrificed and the spleen removed.
5. The connective tissue and fat is removed, and the spleen is taken in a 35-mm Petri plate containing 2 ml of RPMI-1640 containing 10% FCS (A1).
6. The spleen is teased with the help of a blunt-headed sterile forceps to obtain a single cell suspension. Cell clumps are separated, and the centrifuged cell pellet is suspended in 5 ml of ice-cold hemolytic reagent.
7. After 5 min of treatment, 5 ml of ice-cold RPMI-1640 is added and the cell suspension is immediately centrifuged at 1500 rpm for 5 min. The cell pellet is washed in the same medium two times. About 50–60×10^6 cells can be recovered from a spleen.
8. The spleen cells are suspended in A1 medium at a density of 10×10^6 cells/ml. The cell suspension is kept at room temperature until used for fusion.

5.3.2 Myeloma Cell Culture

The myeloma (tumor) cells used for making hybridomas should not secrete a paraprotein. Ideally, myeloma cells should not produce immunoglobulin light chains; otherwise this will combine with the heavy chains of the monoclonal antibody to produce undesirable hybrid molecules (Johnstone and Thorpe 1996). Commonly used hypoxanthine, aminopterin, and thymidine (HAT)-sensitive myeloma cell lines are shown in Table 5.1. Suitable myeloma cells are cultured one week prior to fusion in roller bottles containing RPMI-1640 with 10% FCS (A1). Cells are cultured up to mid-log phase (2–3×10^6 cells/ml), harvested, washed two times in A0 medium, and finally resuspended in A1 medium at a density of 5×10^6 cells/ml.

Table 5.1 Rodent B myeloma cell lines commonly used for making hybridoma (taken from Johnstone and Thorpe 1996)

Cell line	Animal	Synthesis
NS1	Mice (Balb/c)	Light chain
NSO	Mice (Balb/c)	Nothing
Y3	Rat (Lou)	Light chain
Y2B	Rat (Lou)	Nothing

5.3.3
Setup for Fusion of Myeloma with Spleen Cells

In general, a 1:5 proportion of myeloma and spleen cells is used for fusion. Take about 20×10^6 and 100×10^6 myeloma and spleen cells, respectively, in a 50-ml conical tube and centrifuge at 500 *g* for 10 min. Discard the supernatant, resuspend the cell pellet in 30 ml A0 medium at 37 °C, equilibrate cell suspension at the same temperature and further centrifuge. The cell pellet is used in the following steps to carry out fusion.

1. Set the timer for 6 min, and disrupt the cell pellet by tapping the tube. Start the timer, add 1 ml of 50% PEG (M.W. 4000) solution to the pellet over a 1 min period with constant shaking. Mix for another 1 min, stop fusion by slowly adding 1 ml of A0 medium over a period of 1 min with constant stirring, and add 3 ml of A0 medium over 1 min and then 10 ml of A0 medium over a period of 2 min.
2. Incubate for 5 min in a 37 °C water bath, then slowly add 30 ml of medium A0.
3. Centrifuge cell suspension at 500 g for 10 min, resuspend the pellet in 50 ml of HAT medium by gently swirling the tube, and culture the cells in 2×T-75 flasks for 2 days.
4. Transfer the cells from the flask into 2×50-ml tubes and centrifuge at 500 *g* for 10 min. Resuspend the cell pellet in 2×20 ml HAT medium, mix with 1×106 peritoneal macrophages per tube, and distribute 100 μl suspension in each well of 4×96-well plates. Alternatively, resuspend the cell pellet in 2×20 ml HAT medium supplemented with 50 μg/ml of LPS and 20 μg/ml of dextran sulfate. Hybridomas grow faster and produce more clones if the spleen cells respond well to these mitogens.
5. Maintain the culture for 10–14 days, with 50% replacement of the medium every 3 days. Examine the plates for the presence of colonies and, for each well having colonies, test the culture supernatant for the presence of antibodies.

5.3.4
Selection and Cloning of Hybridoma

When some of the wells show positive in antibody tests, it is important to reclone the hybridoma as soon as possible. This is done to avoid potential loss of the positive clone due to overgrowth of non-secreting cells. To ensure that the antibody is monoclonal, cloning should be done 2–3 times before selecting the final hybridoma clone. Cloning can be accomplished by either growing the hybridoma in soft agar or by a limiting dilution method.

5.3.4.1 Cloning by Limiting Dilution Method

1. Collect cells from each positive well, and dilute cells at 50 cells/ml, 30 cells/ml, 10 cells/ml, and 5 cells/ml in A1 medium.
2. Plate 100 μl cell suspension of each dilution in a 96-well plate (24–36 wells for each dilution). Maintain the cells in a CO_2 incubator at 37 °C for 2–3 weeks with 50% medium replacement twice a week (care should be taken to avoid cell loss during replacement of medium).

If there is cell growth in ≤5 wells in each dilution, the odds are greater than 95% that the clones will produce monoclonal antibody. If <80% of the clones tested are positive, the hybridoma should be recloned for the second time. If the cell dilutions mentioned above give too much or too little growth, an appropriate adjustment in cell dilution is needed to obtain either a lower or higher cell count. The cloning efficiency of hybridomas can be greatly enhanced by the addition of 2000–3000 peritoneal macrophages per well or by including LPS and dextran sulfate in the cloning medium. The fastest growing clones producing a high antibody titer are selected for clonal expansion. At this point it is necessary to store a few vials of clone in liquid nitrogen.

5.3.5 Production of Monoclonal Antibodies

There are two techniques by which monoclonal antibodies are produced. Highly concentrated (5–10 mg/ml) antibodies can traditionally be produced in mouse peritoneal cavity; however, this is now banned in most countries. The second option for the production of monoclonal antibodies is cell culture based.

5.3.5.1 Production in Ascitic Fluid

Hybridomas are grown in the peritoneal cavity of the same strain of mice used as a donor of myeloma or spleen cells. This is to avoid rejection of hybridomas by host animals. If the NS-1 myeloma cell line is used for fusion with spleen cells of another mouse strain, it is recommended that F1 crossbreeds between Balb/c and the donor mouse strain of spleen cells are used to prepare ascitic fluid. Ascitic fluid is produced as below:

1. Each mouse is initially primed by injecting 0.5 ml of pristane via intraperitoneal (IP) injection. Pristane is a C_{14} branched oily hydrocarbon which induces

an oil-granuloma in the peritoneal cavity of the mouse. This environment is optimal for the acceptance and growth of hybridomas for the production of a high antibody titer in ascitic fluid.
2. At 5–10 days after priming, freshly grown hybridoma cells (10×10^6 per 0.5 ml) are washed, suspended in normal saline, and injected IP into each mouse.
3. At 7–21 days later, the ascitic fluid is collected by inserting a 20-gauge needle into the swollen peritoneum in the inguinal area. About 3–6 ml of ascitic fluid can be withdrawn from each mouse. Ascitic fluid is tapped every week for a few successive weeks.
4 The ascitic fluid is allowed to clot, and cells and fibrin are removed by centrifugation at 1000 *g* for 10 min. The clear fluid is stored in –70 °C freezers, if antibody is not purified immediately.

5.3.5.2
Production in Cell Culture

Monoclonal antibodies can be produced in culture; in fact large-scale antibodies are produced in this way. However, by this process the yield of antibodies (5–50 μg/ml) is much lower, as compared with ascitic fluid. Cell culture based monoclonal antibody production can be scaled-up in much larger volumes in spinner flasks, as well as in bioreactors. A high concentration of monoclonal antibodies can be achieved in culture using a hollow fiber bioreactor. Two of the most useful strategies for improving yield of antibodies are: optimization of culture medium with low serum concentration, and high cell density culture.

5.3.6
Materials and Equipment

Besides an animal house, an established tissue culture laboratory is essential for making hybridomas and monoclonal antibodies; the details of the materials and facility requirements are as follows:

5.3.6.1
Media and Materials

- A0 medium is RPMI-1640; A1 medium is RPMI-1640 supplemented with 10% heat inactivated FCS.
- Stock hypoxanthine/thymidine (HT) solution (100×): dissolve 135 mg hypoxanthine to approximately 60 ml of distilled water containing 1.2 ml of

1 M NaOH by stirring. If hypoxanthine is not completely dissolved, add additional 0.1 ml NaOH. Dissolve 38.6 mg thymidine into it, bring the volume to 100 ml, filter sterilize, and store at 4 °C. The solution is stable for several months.

- Stock hypoxanthine/aminopterin/thymidine (HAT) solution (100×): dissolve 1.91 mg aminopterin in 100 ml of stock HT.
- Dextran sulfate solution (200×; not needed if macrophages are used): dissolve 40 mg of dextran sulfate (MW=500 000) in 10 ml of distilled water, filter sterilize, and store at 4 °C. The solution is stable for several weeks.
- Fusion reagent (50% polyethylene glycol-4000 solution): dissolve 10 g melted PEG and 1 ml DMSO in 9 ml of 0.15 M HEPES buffer (pH 7.5), filter sterilize, and store at room temperature. The stock is stable for several months.
- Hemolytic agent: dissolve 0.2 g Tris base and 0.83 g NH_4Cl in 60 ml MilliQ water, adjust to pH 7.2 with HCl, make up the volume up to 100 ml, filter sterilize, and store at 4 °C. The stock is stable for several months.
- LPS concentrate (100×; not needed if macrophages are used): aseptically add 20 ml of distilled water to 100 mg vial of *E. coli* lipopolysaccharide, aliquot 1 ml per tube, and store frozen at –20 °C. The stock solution is stable for several months.
- Other materials: phosphate buffered saline (sterilize by autoclaving), pristane (2,6,10,14-tetramethyl pentadecane), B-tumor cells, immunized animal.

5.3.6.2
Laboratory Equipment

CO_2 incubator, table-top centrifuge, 37 °C water bath, inverted microscope, sterile equipment for dissecting animal, plastic ware, etc.

5.4
Purification of Antibody

Antibody is a glycosylated protein, which is produced in relatively low amounts as compared with the other proteinacious contaminants present in either ascitic fluid or in cell culture supernatant. Because of this, purifying a small quantity of antibody from a large volume of dilute solution is carried out following a combination of different techniques, normally used in protein chemistry.

5.4.1 Purification of IgG by Precipitation with Ammonium Sulfate

The addition of ammonium sulfate in ascitic fluid or hybridoma culture supernatant causes precipitation (salting-out) of IgG, which can be directly used in many applications or further purified by chromatography techniques. The precipitated IgG is usually very stable and can be stored long-term. About 40% pure antibody can be obtained by the ammonium sulfate precipitation technique. Ammonium sulfate precipitation steps are as follows:

1. Centrifuge ascitic fluid/culture supernatant at 2000 *g* for 15 min at 4 °C, and collect the clear supernatant.
2. Add saturated ammonium sulpfate solution drop-wise to produce 35–45% final saturation (alternatively, directly add 2.7 g of ammonium sulfate per 10 ml of fluid to obtain 45% saturation). Stir the mixture at 4 °C for 4–10 h.
3. Centrifuge at 2000 *g* for 15–20 min at 4 °C, collect the pellet, and dissolve in one-tenth volume of PBS. The crude antibody solution is dialyzed against PBS for overnight with 3–4 buffer changes to obtain an ammonium sulfate-free preparation. Alternatively, dialyze in a buffer recommended in subsequent purification steps.

5.4.1.1 Materials and Equipment

Saturated ammonium sulfate solution: dissolve excess $(NH_4)_2SO_4$ into double distilled water (900 g in 1 l final volume), filter through 0.45 μm filter and store at 4 °C PBS.

Table-top centrifuge, dialysis membrane (20 kDa MW cut-off), magnetic stirrer, plastic/glass beaker.

5.4.2 Purification of IgG by DEAE-Sepharose Chromatography

Crude antibody from the previous step or directly from the ascitic fluid/culture supernatant can be purified by DEAE-Sepharose chromatography (Page and Thorpe 1996). The antibody is purified based on the principle that IgG has a higher or more basic isoelectric point than most serum proteins. The solution pH is kept below the isoelectric point of antibodies and, since IgG do not bind to the DEAE column (anion exchanger), they are thus separated out from the majority of the protein contaminants bound to the anion exchanger. Highly purified (>90%) antibodies can be obtained by this technique. DEAE-Sepharose-based purification steps are as follows:

1. Extensively dialyze the crude antibody/ascitic fluid/serum/culture supernatant against 50 mM sodium phosphate buffer with 3–4 changes over a period of 24 h.
2. Apply the dialyzed sample to the DEAE-Sepharose column, previously equilibrated in the same buffer, and collect the flow-through. Wash the column with 2 column volumes of the same buffer until the absorbance of the eluate at 280 nm (A_{280}) falls to base line. IgG is present in the wash, which is mixed with the flow-through.
3. Elute the adsorbed protein contaminants and regenerate the column by passing 2–3 column volumes of the phosphate buffer containing 1 M NaCl. Wash the column in 2–3 column volumes of the 50 mM phosphate buffer, and store in the same buffer containing 0.1% NaN_3.

5.4.2.1 Materials and Equipment

DEAE Sepharose Cl-6B (Pharmacia, Uppsala, Sweden), 50 mM sodium phosphate buffer (pH 5.3), 1 M NaCl, sodium azide.

Chromatography column, standard chromatography unit (feeding pump, UV-monitor, fraction collector, chart recorder), dialysis membrane (20 kDa MW cut-off), magnetic stirrer, plastic/glass beaker.

5.4.3 Purification of IgG Using Immobilized Protein A

Protein A is a group-specific ligand that binds with the Fc region of IgG-type antibodies. In the immobilized form, protein A is extremely useful in the purification of antibodies, because it is easy to use and a high-capacity protein adsorbent. The recombinant Protein A-Sepharose (BioChain Institute, Calif.) is an affinity chromatographic matrix with recombinant protein A, immobilized by the epoxy method to a Sepharose 6B fast flow base matrix (BioChain Institute 2006). The capacity of IgG binding to Protein A could be up to 25 mg of human IgG in 1 ml of wet gel. One-step purification of IgG of about 98% purity can be achieved using a Protein A-Sepharose column. Despite these enormous advantages, protein A cannot be used for all classes of IgG, as for example human IgG_3, mouse IgG_3, sheep IgG_1, etc., as they do not bind with the matrix. The purification steps are as follows:

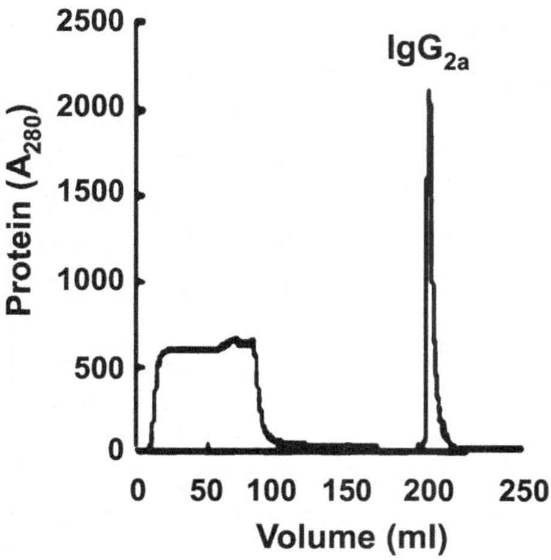

Fig. 5.2 Separation of mouse IgG_{2a} by protein A-Sepharose column chromatography. Chromatogram shows distinct peak of IgG_{2a} (Biochain Institute, Hayward, Calif.)

1. Equilibrate the column with 5–10 column volumes of binding buffer (20 mM sodium phosphate, pH 7).
2. Apply the sample (equilibrated with binding buffer) to the column using a syringe or pump. Wash the matrix with 5–10 column volumes of binding buffer until no protein appears in the effluent.
3. Elute IgG with 2–5 column volumes of elution buffer (0.1 M sodium citrate, pH 4). The eluted fraction is immediately neutralized by adding 50-100 µl of 1 M Tris-HCl, pH 9.0, in 1 ml of eluted fraction.
4. Regenerate the column with 5 column volumes of regeneration buffer containing 0.1 M sodium citrate (pH 3).
5. Wash the column with 5–10 column volumes of distilled water and finally with 20% ethanol. The column is stored at 4–8 °C.

The chromatogram of mouse IgG2a sub-type, purified in protein A-Sepharose chromatography, is shown in Fig. 5.2.

5.4.3.1 Materials and Equipment

rProtein A-Sepharose (BioChain Institute, Calif.), chromatography column, buffers (for binding, washing, elution, as mentioned in the protocol), standard chromatography unit (feeding pump, UV-monitor, fraction collector, chart recorder).

5.5 Analysis of Purity of IgG by Electrophoresis

After purification of IgG using the above techniques, it is necessary to obtain some index of purity of the product. Sodium dodecyl sulphate–polyacrylamide gel electrophoresis (SDS-PAGE) is used to separate proteins and hence to determine their purity. The molecular weight of native IgG is 150 kDa, but in reducing SDS-PAGE it degrades into molecules of two different sizes, comprising heavy (50 kDa) and light (25 kDa) chains. These degraded heavy and light chains can be easily detected by staining with a suitable protein dye. If contaminating proteins are present in the preparation, they are also detected in the same gel, depending upon their quantity and the sensitivity of staining technique. The IgG sample, to be electrophoresed, is heated at 90–95 °C for 5 min in the presence of reducing sample buffer (Laemmli 1970). The sample buffer contains SDS and 2-mercaptoethanol (2-ME) or dithiothreitol (DTT). The anionic detergent SDS denatures IgG, binds to the uncoiled molecule, and confers a uniform negative charge. Whereas, being a reducing agent, 2-ME/DTT reduces the disulfide bonds of IgG to free sulfhydryl groups, and forms lower molecular weight proteins. The sequential steps for reducing SDS-PAGE analysis of protein are given below:

1. Cast 10% and 5% (w/v) resolving and stacking gel, respectively, in a mini gel apparatus (10 cm height). Fill the anode and cathode reservoirs with running buffer.
2. Prepare samples by boiling antibody solution (~1 mg/ml) with sample buffer (3:1). Load 15–20 μl sample(s) in each well, also load appropriate molecular weight standard.
3. Electrophorese the gel at 30 mA constant current (60 mA for two gels) until the dye-front reaches the bottom-most portion of the gel. The average gel running time is 1.5–2.0 h.
4. Remove the gel carefully from glass plates and stain with Coomassie brilliant blue R-250 for 5–10 min by gently rocking the content.
5. Pour off the stain, wash the gel with tap water, and destain it with a methanol–acetic acid mixture with a couple of changes until the gel background is clear.

Coomassie brilliant blue staining of purified IgG subunits are shown in Fig. 5.3. The sensitivity of the Coomassie blue staining is about 1 μg to 2 μg protein/band; below that level a protein band cannot be detected. The silver staining method is more sensitive, and protein as low as 100 ng/band can be detected. Therefore, silver staining is a more preferred technique to identify protein than Coomassie blue staining in order to detect minor protein contaminants. Silver staining steps can be followed from step 4 in the above. It is recommended to use gloves when handling gel, otherwise finger impressions may appear on the gel.

1. Remove gel (resolving gel only) carefully from the glass plates and wash with an excess of gel fix solution (gel can be stored in this solution for several days with no effect on the quality of the staining).

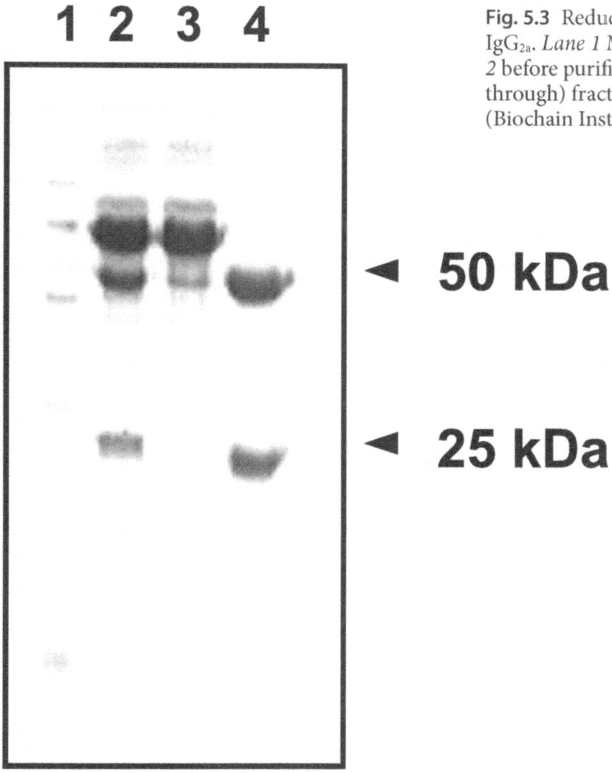

Fig. 5.3 Reducing SDS-PAGE analysis of IgG_{2a}. *Lane 1* Molecular weight marker, *lane 2* before purification, *lane 3* unbound (flow-through) fraction, *lane 4* purified IgG_{2a} (Biochain Institute, Hayward, Calif.)

2. Wash gel with 50% ethanol three times for 20 min each. Submerge the gel in sodium thiosulfate solution for exactly 1 min (long exposure in sodium thiosulfate gives more background staining).
3. Rinse the gel with double-distilled water three times for 20 s each. Again, submerge the gel for 20 min in the silver nitrate solution (gel may appear yellowish). Rinse the gel with distilled water three times for 20 s each to remove excess silver nitrate.
4. Visualize the bands by incubating the gel (20–25 °C) for 10 min in developing solution. When the desired bands appear, terminate the developing process by washing the gel with water.
5. Soak the gel for 10 min in gel fix solution, wash in 50% methanol for 20 min, and finally stored in 50% methanol at 4 °C.

Note: While developing the gel, a faint brown precipitate may appear that can be dissolved by agitating the content. The intensity of the band increases with the time of development, so care must be taken to avoid over- or under-staining the gel. A band of 500 ng protein typically arrives within 30 s of development, a band of 50 ng protein takes 2 min to be visualized (Rosenberg 1996).

5.5.1 Materials and Equipment

1. Electrophoresis: acrylamide solution (30%; 29.2 g acrylamide, 0.8 g bisacrylamide per 100 ml; stock solution is filtered and stored at 4 °C in a dark-colored bottle), 1.5 M Tris-HCl (pH 8.8), 1 M Tris-HCl (pH 6.8), SDS solution (10%), ammonium persulfate solution (10%), *N,N,N',N'*-tetramethylene-diamine (TEMED), double-distilled water, 10% resolving gel (10 ml; prepared from 4.0 ml water, 3.3 ml of 30% acrylamide, 2.5 ml of 1.5 M Tris, 0.1 ml of 10% SDS, 0.1 ml of 10% APS, 4 µl TEMED), 5% stacking gel (3 ml; prepared from 2.1 ml water, 0.5 ml of 30% acrylamide, 0.38 ml of 1.0 M Tris, 0.03 ml of 10% SDS, 0.03 ml of 10% APS, 3 µl TEMED), running buffer (10×: dissolve 30.3 g Tris, 144.2 g glycine, 10 g SDS in 1000 ml distilled water), sample buffer [4×: 2.0 ml 2-ME, 0.8 g SDS, 2.5 ml Tris-HCl (1 M, pH 6.8), 4.0 ml glycerol, 0.1 mg bromophenol blue, distilled water to make 10 ml; warm the mixture at 37 °C for complete solubilization of SDS, aliquot in 1.5 ml tube, store at –20 °C], molecular weight marker (prepare by dissolving powdered marker proteins standard in sample buffer, or use prestained markers in 20–100 kDa range).
2. Coomassie blue staining: Coomassie brilliant blue R-250 (0.25% in destaining solution), destaining solution (400 ml methanol, 100 ml glacial acetic acid in 1000 ml water).
3. Silver staining: gel fixing solution (50% methanol in 12% acetic acid solution), sodium thiosulfate solution (dissolve 0.2 g in 1000 ml of double-distilled water), silver nitrate solution (dissolve 2 g in 1000 ml of double-distilled water containing 0.5 ml formaline), developing solution (dissolve 60 g sodium carbonate, 4 mg sodium thiosulfate, 0.5 ml formalin in 1000 ml water), ethanol solutions (50%, 30%), methanol solution (50%).
4. Equipment: power pack, electrophoresis tank and gel apparatus, micro-syringe, hot plate, microfuge, rocking platform, glass/plastic container.

5.6 Enzyme-Linked Immunosorbent Assay

A number of alternative enzyme-linked immunosorbent assay (ELISA) protocols are available, allowing the qualitative detection or quantitative measurement of either antigen or antibody in sera or culture supernatant. ELISA is broadly classified in three groups: (a) indirect ELISA, (b) sandwich ELISA, and (c) competitive ELISA. These assays can be used for quality control of the culture supernatants containing antibody/antigen, and the results are expressed in relative values (qualitative) or in exact concentrations using a standard curve based on known concentrations of the antibody or antigen in preparation (quantitative).

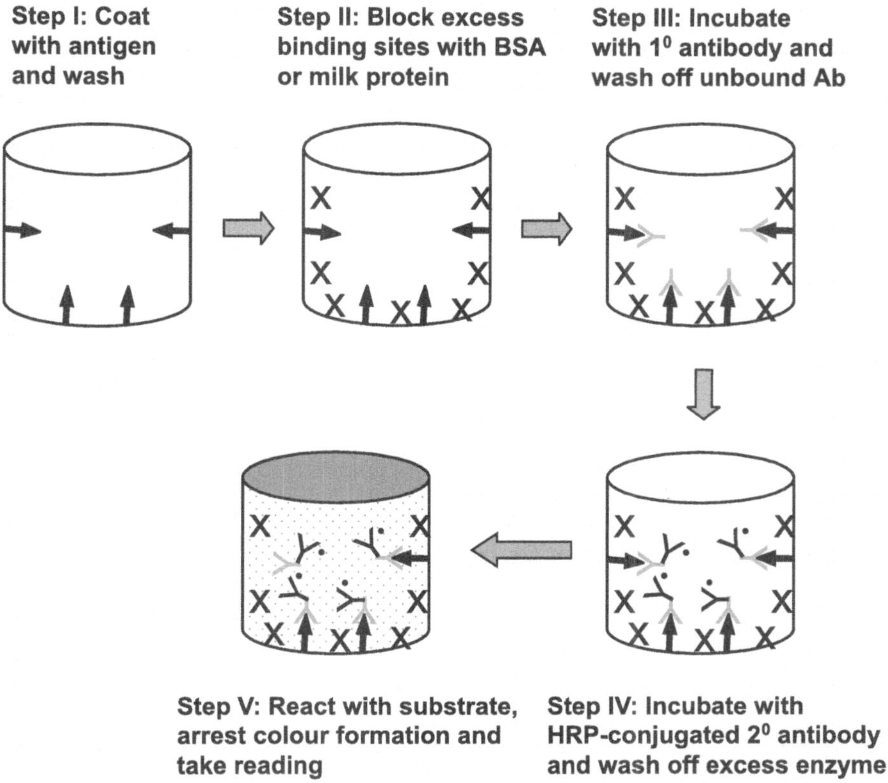

Fig. 5.4 Schematics of indirect antibody ELISA. Antigen (*arrow*), blocking (X), primary antibody (Y), HRP-conjugated secondary antibody (Y·)

Antibody can be detected or quantitatively determined with an indirect ELISA. Here, microtiter plate wells are coated with antigen of one particular concentration and different dilutions of antibody/antibodies (polyclonal) are trapped on it. Bound antibody is then detected by reaction with the labeled second antibodies (Fig. 5.4). The amount of label bound is proportional to the concentration of the antibody in the test solution. The different steps involved in indirect antibody ELISA are described below:

1. Coat 96-well plate with 100 μl of antigen (0.2 μg/ml in PBS) and incubate overnight at 4 °C.
2. Wash the plate with 3×200 μl of PBS-T per well with constant shaking. Block the wells with 200 μl of blocking solution by incubating at 37 °C for 1 h.
3. Discard the blocking solution, add 100 μl of the primary antibody (1^0) of different dilutions in triplicate wells and incubate at 37 °C for 1 h.
4. Wash the plate with 3×200 μl of PBS-T per well with constant shaking. Add 100 μl of horseradish peroxidase (HRP)-coupled anti-mouse IgG secondary antibody (2^0) from 1:5000 to 1:10 000 dilutions in blocking solution, and in-

cubate for 30–45 min at 37 °C (if primary antibody is rabbit polyclonal, use anti-rabbit IgG/HRP as secondary antibody). In place of HRP, alkaline phosphatase or β-galactosidase conjugated 2^0 antibody can be used (with respective substrate).
5. Wash the plate with 3×200 μl of PBS-T per well with constant shaking. Incubate the plate with 100 μl of developing solution until a color change is evident. Stop the reaction by adding 50 μl of 2.5 M H_2SO_4 per well, and measure the absorbance at 492 nm.

5.6.1 Materials and Equipment

Standard antigen solution (1–10 μg/ml), diluted monoclonal antibody solution/ culture supernatant, HRP-conjugated secondary antibody (anti-mouse), PBS (20 mM, pH 7.5, containing 150 mM NaCl), PBS-T (0.05% Tween 20 in PBS), blocking solution (2% BSA or 5% non-fat milk in PBS-T), developing solution [10 mg o-phenylenediamine (OPD) in 25 ml of citrate buffer at pH 5.5 (20 ml of H_2O, 5.16 ml of 0.5 M Na_2HPO_4, 645 μl of 1.5 M citric acid), plus 12 μl of 30% H_2O_2], 2.5 M H_2SO_4. The developing solution is unstable and should be freshly prepared. Sodium azide is an inhibitor of HRP, hence it should be avoided in reagents.

Incubator (37 °C), micropipettes, ELISA plates, ELISA plate washer, ELISA plate reader, 4 °C refrigerator, plastic ware.

5.7 Conclusion

The antibody is the workhorse in immunology and cell biology research to detect specific cell types in the body. The major use of antibodies has been limited to the research applications. However, in the recent past, therapeutic monoclonal antibodies have become available to treat certain diseases in humans. Antibodies are characterized in terms of affinity, avidity, and bioneutralization capacity, which are not discussed in this Chapter.

References

BioChain Institute (2006) rProtein A-Sepharose product literature. BioChain Institute, Hayward

Johnstone A, Thorpe R (1996) Immunochemistry in practice, 3rd edn. Blackwell Science, Oxford

Köhler G, Milstein C (1975) Continuous cultures of fused cells secreting antibody of predefined specificity. Nature 256:495–497

Laemmli UK (1970) Cleavage of structural proteins during the assembly of the head of bacteriophage T4. Nature 227:680–685

Page M, Thorpe R (1996) Purification of IgG using DEAE-sepharose chromatography. In: Walker JM (ed) The protein protocols handbook. Humana, Totowa, pp 725–726

Roitt I, Brostoff J, Male D (1996) Immunology, 4th edn. Mosby, London

Rosenberg IM (1996) Protein analysis and purification – benchtop techniques. Birkhauser, Boston

6 Modern Techniques for Analyzing Immunological Responses

Satish Khurana, Sangeeta Bhaskar, and Asok Mukhopdhyay

6.1 Introduction

The immune system is a versatile defense method that has evolved to protect animals from invading pathogenic microorganisms. The immune response is the way body recognizes and defends itself against bacteria, viruses, and substances recognized as foreign and potentially harmful to the body. Functionally, an immune response can be divided into two related activities – recognition and response. The immune system is able to discriminate one foreign pathogen from another and also between foreign molecules and the body's own cells and proteins. Once a foreign organism/antigen has been recognized, the immune system recruits a variety of cells and molecules to mount an appropriate response, called an effector immune response, to eliminate or neutralize the organism/antigen. Later exposure to the same foreign organism/antigen induces a memory immune response, characterized by a more rapid and heightened immune reaction that serves to eliminate the pathogen and prevent diseases (Goldsby et al. 2000).

6.2 Type of Immune Responses

The immune system is a complex set of cellular elements comprising different forms of lymphocytes and antigen-presenting cells to protect against infections.

National Institute of Immunology, Aruna Asaf Ali Marg, New Delhi-110067, India, email: ashok@nii.res.in

Soil Biology, Volume 11
Advanced Techniques in Soil Microbiology
A. Varma, R. Oelmüller (Eds.)

The immune response is often divided into two types: the innate and the adaptive.

6.2.1 Innate Immune Response

This provides the first line of defense against infection, which includes cellular and molecular components that recognize a wide spectrum of conserved pathogenic components. It has broad reactivity and is uniform in all members of a species.

6.2.2 Adaptive Immune Response

This provides long-lasting and specific protection against known pathogens. It has a high degree of antigen specificity and memory. The major agents of adaptive immunity are lymphocytes, antibodies, and other molecules they produce. The adaptive immune system, also called the acquired immune system, ensures that most mammals that survive an initial infection by a pathogen are generally immune to further illness caused by that same pathogen. This chapter describes specifically various techniques to study adaptive immune responses developed in mammals. The immune system operates throughout the body. However, there are certain sites where the cells of the immune system are organized into specific structures. These are classified as central lymphoid tissue (bone marrow, thymus) and peripheral lymphoid tissue (lymph nodes, spleen, mucosa-associated lymphoid tissue). The location of various lymphoid tissues in mice is shown in Fig. 6.1. Immune cells are formed in the bone marrow and are grouped into two major classes: lymphocytes and antigen-presenting cells (APC). Once the lymphocytes are initially formed, some continue to mature in the bone marrow and become B cells. Other lymphocytes finish their maturation in the thymus and become T cells. Once mature, some lymphocytes stay in the lymphoid organs, while others travel continuously around the body through the lymphatic vessels and bloodstream.

6.3 Adaptive Immune System

In mammals, the adaptive immune system is divided into two classes: humoral and cellular.

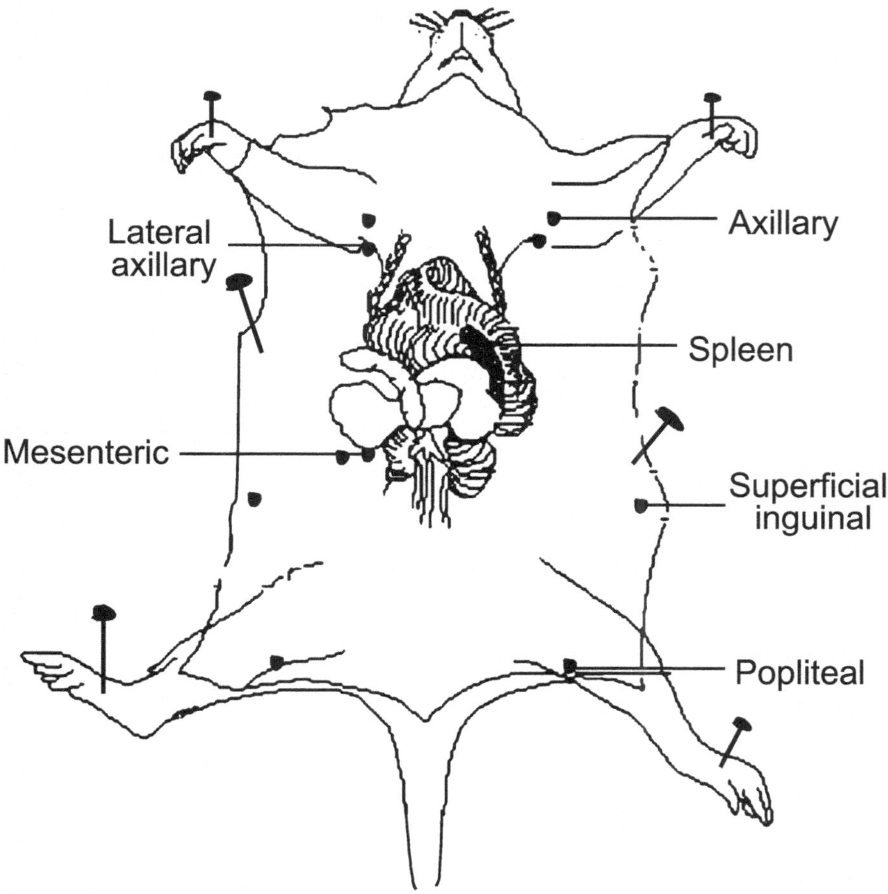

Fig. 6.1 Lymphoid tissues in mice: central lymphoid tissue (bone marrow and thymus, not shown in the figure), and peripheral lymphoid tissues (lymph nodes: mesenteric, popliteal, superficial inguinal, axillary, lateral axillary; spleen)

6.3.1 Humoral Immune System

This acts against bacteria and viruses in the body (blood) by means of eliciting an antibody (immunoglobulins) response. Antibodies are the antigen-binding proteins produced by differentiated B cells, known as plasma cells. Antigen receptors on B cells consist of membrane-bound immunoglobulin heavy and light chains. Following the interaction of cell-surface Ig with its specific antigen and in the presence of T cells, B cells differentiate into plasma cells, which secrete antibodies. Secreted antibodies circulate in the blood, where they serve as effectors of humoral immunity by searching out and neutralizing antigens or marking them for elimination. Most antigens are complex and contain many different antigenic determinants, and the immune system usually responds by producing antibodies to several epitopes on the antigen. Hence, there is a heterogeneous serum (polyclonal) antibody response to an immunizing antigen.

6.3.2 Cellular Immune System

This is involved in destroying bacteria and virus-infected cells with the help of T cells. T cells express a unique binding molecule on its membrane, known as the T cell receptor. Unlike antibodies, which can recognize antigen alone, T cell receptors can recognize only antigen that is bound to cell membrane proteins called MHC molecules. Both T and B cell compartments display enormous heterogeneity in function and antigen specificity. There are two well defined subpopulations of T cells: T cytotoxic cells (Tc) and T helper cells (Th). T helper and T cytotoxic cells can be distinguished from one another by the presence of either CD4 or CD8 membrane glycoproteins on their surfaces, respectively. After a Th cell recognizes and interacts with an antigen–MHC class II molecular complex, the cell is activated and secretes cytokines. The secreted cytokines play an important role in activating B cells, cytotoxic T cells, macrophages, and various other cells that participate in the immune response.

6.4 Different Assay Systems to Study the Adaptive Immune Response

The humoral immune response is studied by determining specific antibodies produced against antigens. The concentration of antibodies is determined by radio immuno assay (RIA) or ELISA techniques. The general procedures of direct ELISA for antigens/antibodies are described in Chapter 5. In this chapter we discuss assay systems used for measuring cell-mediated immune (CMI) responses.

6.4.1 Mixed Lymphocyte Proliferation Assays

The lymphocyte proliferation assay measures the memory response of T cells. Lymphocytes placed in short-term tissue culture undergo clonal proliferation, when stimulated *in vitro* by a foreign molecule or antigen with which they are primed before. $CD4^+$ lymphocytes proliferate in response to recognition of antigenic peptides in association with class II major histocompatibility complex (MHC) molecules on APCs. This proliferative response of lymphocytes to antigen in vitro occurs only if the mouse is immunized with the same antigen. Antigen-specific T-cell proliferation is a major technique for assessing the functional capacity of $CD4^+$ lymphocytes to respond to various stimuli. The degree of proliferation is assessed by adding ^{3}[H]-thymidine to the culture medium and monitoring uptake of that into DNA in the course of repeated cell divisions. Spleen cells are mixed population of different subsets of T and B cells. Spleen also contains major APCs. Therefore, mixed lymphocyte proliferation assay can be conducted from spleenocytes isolated from immunized mice. The details of the proliferation assay are as follows:

1. Aseptically remove the spleen from the immunized mice, and place on a sterile petri plate containing 2 ml of RPMI 1640. Crush the spleen with the help of blunt forceps to release the cells. Collect cells by centrifugation and lyse the erythrocytes by treatment with Gey's solution for 90 s. Stop the reaction of Gey's solution by diluting 10 times volume of sterile PBS, centrifuge the cells, resuspend the pellet in 3 ml of complete medium and enumerate the cell number on a hemocytometer. About $50–60\times10^6$ mononuclear cells can be recovered from a spleen.
2. Take 0.5×10^6 cells in each well of a 96-well plate, add 2-fold serially diluted antigen (adjuvant-free) in triplicate wells, by which the mouse was immunized. Antigen concentrations may vary from 0.2 μg/well to 2.0 μg/well. As a negative control, take the same number of cells in triplicate wells, but without antigen. For a positive control, add PHA (1 μg/ml) in triplicate wells. The volume of the suspension is maintained at 200 μl. Culture the cells for about 54 h in a CO_2 incubator at 37 °C.
3. Add 0.5 μCi ^{3}[H]-thymidine in each well and further incubate for 18 h. Observe the cells under a microscope to check proliferating cells. Cells exposed to antigen or PHA will show signs of proliferation (clumps of growing cells).
4. Harvest the cells from the 96-well plate onto a glass fiber filter mat. Remove the mat from the cell harvester, insert in a plastic bag, add 10 ml of scintillation fluid and seal the bag. Measure the incorporation of ^{3}[H]-thymidine in the DNA of multiplying cells on a β-counter. The higher the proliferation, the higher is the cell count per minute (cpm).

6.4.1.1 Materials and Equipment

Immunized mice, antigen, Gey's solution [dilute 20 ml of sterile solution A (3.5 g NH_4Cl, 0.185 g KCl, 0.15 g Na_2HPO_4, 0.012 g KH_2PO_4, 0.5 g glucose, 2.5 g gelatin in 100 ml water) with 5 ml of sterile solution B (0.42 g $MgCl_2$ $6H_2O$, 0.14 g $MgSO_4$ $7H_2O$, 0.34 g $CaCl_2$ in 100 ml water) and 5 ml of sterile solution C (2.25 g $NaHCO_3$ in 100 ml water) in 70 ml of sterile water]; complete medium

(RPMI 1640 containing 10% FCS), PHA (1 mg/ml), ^{3}[H]-thydimine (specific activity ~ 6.5 Ci/mmol).

Sterile surgical apparatus, CO_2 incubator, inverted microscope, hemocytometer, cell harvester (including glass fiber filter), β-counter, plastic ware.

6.4.2 Detection of Type of T Helper Responses (Th1/Th2)

Two distinct classes of helper T cells, namely Th1 and Th2, are formed in response to immunogens. Th1 cells participate in cell-mediated immunity by secreting IL-2, TNF-α, and IFN-γ for activation of macrophages and Tc. In addition to that, they inhibit both Th2 subset cell activity and the humoral immune responses. Th2 cells participate in humoral immunity by providing help to B cells to produce antibodies, which are needed to control extracellular pathogens. Moreover, they are also involved in the inhibition of cell-mediated responses. They secrete IL-4, IL-5, and IL-10, induce a class switch to IgE and IgG1, and support eosinophils and mast cells. Traditionally, the Th1/Th2 response is determined by measuring individual cytokines secreted by ELISA. In the recent past the cytometric bead array (CBA) technique has been introduced by BD Pharmingen, in which a series of particles with discrete fluorescence intensities are employed for the simultaneous detection of multiple cytokines. The CBA is combined with flow cytometry to create a powerful multiplexed assay. Each bead in a CBA provides a capture surface for a specific protein and is analogous to an individually coated well in an ELISA plate. This capture bead mixture is in suspension, hence it allows for the detection of multiple analyses in a small sample volume (BD Biosciences 2006). Five bead populations with distinct fluorescence intensities have been coated with capture antibodies specific for IL-2, IL-4, IL-5, IFN-γ, and TNF-α proteins. The cytokine capture beads are mixed with the PE-conjugated detection antibodies and then incubated with recombinant standards or test samples to form sandwich complexes. The samples are analyzed on a flow cytometer. CBA has several advantages compared with conventional ELISA: (a) it requires one-fifth sample volume as compared with ELISA technique, and (b) it takes less time than ELISA. The distinct steps for the CBA assay are as follows:

1. Preparation of cytokine standards: reconstitute the lyophilized standards in assay diluent (buffer). The standards are diluted in the ranges from 1:2 to 1:256 in the same assay buffer. The assay buffer is taken as a negative control.
2. Preparation of cytokine capture beads: vigorously vortex each capture bead suspension for a few seconds, take 10 µl of each capture bead for each assay tube, and transfer 50 µl of mixed beads to each assay tube.
3. Assay procedure: add diluted standards and test samples to the appropriate sample tubes (50 µl/tube). Then add PE detection reagent (50 µl/tube), incubate for 2 h at room temperature in the dark. Wash samples in 1 ml of wash

buffer and centrifuge. Add 300 µl of wash buffer to each assay tube and analyze by flow cytometer.

The ranges of cytokines which can be estimated by the CBA technique varies from 20 pg/ml to 5000 pg/ml. For details, read the BD-CBA technical literature (BD Biosciences 2006).

6.4.2.1 Materials and Equipment

Mouse Th1/Th2 cytokine CBA kit (Becton Dickinson, Calif., USA), samples for examination.

Flow cytometer equipped with a 488 nm laser capable for detecting and distinguishing fluorescence emissions at 576 nm and 670 nm, BD Cell Quest software, sample acquisition tubes, microfuge, vortex mixture.

6.4.3 Cytotoxic T Lymphocyte Activity

Under the influence of Th cell-derived cytokines, and on recognition of the antigen–MHC class I molecule complex, the Tc cell proliferates and differentiates into an effector cell, called a cytotoxic T lymphocyte (CTL). In contrast to the Tc cell, the CTL generally does not secrete many cytokines and instead exhibits cell-killing or cytotoxic activity. The CTL has a vital function in monitoring the cells of the body and eliminating virus-infected cells, tumor cells, and cells of a foreign tissue graft. They release granzymes to trigger apoptotic death of the target (infected) cells. In general, CTLs are $CD8^+$ and are therefore class I MHC restricted, although in rare instances $CD4^+$ class II restricted T cells have been shown to function as CTLs. Three experimental systems are followed for measuring cell-mediated cytotoxic responses.

6.4.3.1 Cell-Mediated Lympholysis

In the cell-mediated lympholysis (CML) assay, suitable target cells (infected cells or altered self-cells, e.g. tumor cells) are labeled intracellularly with chromium-51(^{51}Cr) by incubating the target cells with $Na_2{}^{51}CrO_4$. Chromium diffuses into the cells and binds to cytoplasmic proteins, thus reducing passive diffusion of the same out of the cell. When specifically activated CTLs (effector cells) are incubated with such labeled target cells, the latter undergo lysis and intracellular ^{51}Cr is released. The amount of ^{51}Cr released correlates directly with the number of target cells lysed by the CTLs (Brunner et al. 1968). By comparison with the ^{51}Cr release of the control cells, the corrected percent lysis is calculated for each concentration of effector cells. A non-radioactive flow cytometry-based technique is also followed for the CML assay (Derby et al. 2001). The details of the ^{51}Cr release assay are as follows:

1. Preparation of target and effector cells:
   a. Wash the target cells in complete medium and resuspend the cells in 5 ml of the same medium. Allow cell clumps to settle down under gravity or pass the cell suspension through a single layer of 100-μm nylon mesh, collect unsettled or filtered cells, and enumerate the viable cell number by trypan blue exclusion.
   b. Centrifuge the cells for 5 min and gently resuspend the cell pellet in 2–3 ml of complete medium. Add 0.2 ml of ^{51}Cr solution (1 mCi/ml) and 20 μl of FBS. Mix gently and incubate in a loosely capped 15-ml conical tube for 45 min at 37 °C in a CO_2 incubator.
   c. Wash ^{51}Cr-labeled target cells 2–3 times with complete medium and collect the supernatant in a radioactive waste container. Resuspend labeled target cells in complete medium to a density of 10^6 cells/ml.
   d. Prepare a single-cell suspension of effector cells (spleen cells of immunized mice) in complete medium. Activate the effector cells with Concanavalin A (2 μg/ml) for 2–3 days to sensitize the cells; unactivated cells are used as control.
   e. Add 100 μl of the effector cell suspension to triplicate wells of a 96-well plate for each effector cell density (effector:target cell ratio may vary from 1:1 to 10:1).
2. Co-culture of target cells with CTL:
   a. Add 100 μl of ^{51}Cr-labeled target cells to wells containing effector cells or control lymphocytes or medium, for a final volume of 200 μl/well.
   b. Centrifuge the 96-well plate for 30 s at 200 *g* to enhance the contact between the effector and the target cells. Incubate the plate for 3–6 h at 37 °C in a CO_2 incubator.
   c. Centrifuge the cells for 5 min at 200 *g*, lyse the cells by adding 100 μl of 2% Triton X-100 to the control target cells alone (without effector cells) to measure maximum releasable ^{51}Cr. Harvest 100 μl of each supernatant into ^{51}Cr counting tubes.
   d. Count ^{51}Cr in a γ-scintillation counter (1–2 min/sample). Calculate the corrected percent lysis for each concentration of effector cells, using the mean counts for each set of replicate wells.

Test refers to effector cells with CTL activity and *control* refers to non-lytic cells or cell-free medium; and CTL activity is calculated as:

CTL (%) = [(Test ^{51}Cr released –Control ^{51}Cr released) / (Maximum ^{51}Cr released –Control ^{51}Cr released)] ×100.

6.4.3.1.1
Materials and Equipment

Single-cell suspension of target cells, control target cells, effector cells, control effector cells, complete medium (RPMI 1640 supplemented with 10% ECS), sensitization medium (RPMI medium containing 1 mM sodium pyruvate and 1× non-essential amino acids), concanavalin A (2 μg/ml), $Na_2{}^{51}CrO_4$ (1 mCi/ml), FBS, 2% Triton X-100.

CO_2 incubator, inverted microscope, ^{51}Cr counting tubes (Skatron), γ-counter.

6.4.3.2
Mixed-Lymphocyte Reaction

The mixed-lymphocyte reaction (MLR) is an in vitro method for assaying the proliferation of T cells in a cell-mediated response. Functional CTLs can be generated by co-culturing allogeneic spleen cells (e.g. rat lymphocytes co-cultured with mouse lymphocytes) in a MLR. The T cells in a MLR undergo extensive blast transformation and proliferation. Both populations of allogeneic T lymphocytes proliferate in a MLR unless one population is rendered unresponsive by treatment with mitomycin C or lethal irradiation by X-rays (Lightbody and King 1974). In the latter system (unresponsive), called one-way MLR, the unresponsive population provides stimulator cells that express alloantigens foreign to the responder T cells. Within 24–48 h, the responder T cells start dividing in response to the alloantigens of the stimulator cells, and by 72–96 h, an expanding population of functional CTLs is generated, after which their activity is assessed with various effector assays (Ashwell et al. 1984). The assay method is as follows:

1. Take a single-cell suspension of 1×10^7 responder cells/ml of sensitization medium in a 15-ml conical tube.
2. Prepare single-cell suspension of 1×10^7 stimulator cells/ml in complete medium in a 15-ml conical tube. Red blood cells may be removed, but this is not strictly necessary.
3. Add mitomycin C to the stimulator cell suspension up to 25 μg/ml final concentration. Incubate cells for 20 min at 37 °C in a CO_2 incubator.
   Note: mitomycin C treatment blocks cell division in the stimulator cells. This is particularly relevant for a MLR because the stimulator cells can also recognize alloantigens on the responder cells. Although syngeneic stimulator cells (such as those used for anti-viral and anti-TNP responses) do not recognize responder cells as foreign, blocking the division of stimulator cells is recommended for providing a clear distinction between responder and stimulator cells. Recovery of cells after mitomycin C treatment may be as low as 50%. Alternatively, treat the cells by γ-irradiation at 2000 rad (tumor cells may require very high doses of about 10 000 rad). Irradiation at higher than 2000 rad inhibits the antigen-presenting activity of B cells, but not that of macrophages or dendritic cells.

4. Add 1 ml each of responder and stimulator cells to the wells of a 24-well microtiter plate. Final cell concentration (i.e. sum of responder and stimulator cells) should not exceed 12×10^6 cells/well in a volume of 2 ml. Cell recovery after 5 days is generally 50–100% of the responder cells initially plated.
   Note: addition of IL-2 (~10 units/ml) may enhance the generation of CTL activity.
5. Culture cells for 5 days at 37 °C in a CO_2 incubator.
6. Transfer non-adhered effector cells in a sterile 15-ml conical tube, centrifuge for 5 min at 200 *g*, and resuspend cell pellet in complete medium. Maintain cells at room temperature until CTL activity is assayed by ^{51}Cr release assay, as mentioned in Section 6.4.3.1.

6.4.3.2.1 Materials and Equipment

Responder cells, stimulator cells, sensitization medium, complete medium (RPMI 1640 supplemented with 10% ECS), mitomycin C (0.5 mg/ml in HBSS).

6.4.3.3 Graft Versus Host Reaction

The graft versus host (GVH) reaction in experimental animals provides an in vivo system for studying cell-mediated cytotoxicity. The GVH reaction develops when immuno-competent lymphocytes are injected into an allogeneic recipient whose immune system is compromised. Because the donor and recipient are not genetically identical, the grafted lymphocytes begin to attack the host and the host's compromised state prevents an immune response against the graft. The grafted lymphocytes generally are carried to a number of organs, including the spleen, where they begin to proliferate in response to the allogeneic MHC antigens of the host. This induces an influx of host cells and results in visible spleen enlargement, or splenomegaly. The intensity of a GVH reaction is assessed by determining the increase in the weight of the spleen as compared with the control spleen.

6.5 Flow Cytometric Analysis of Immune Cells

This technique is predominantly used to measure fluorescence intensity produced by the fluorescent-labeled antibodies or ligands that bind to specific

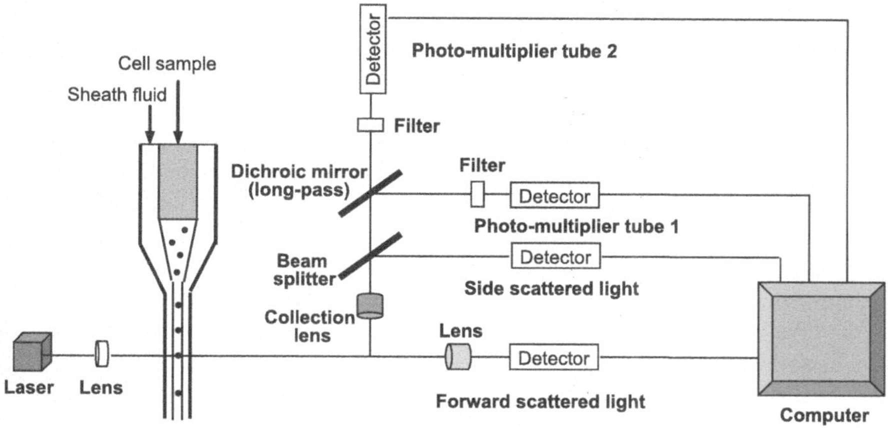

Fig. 6.2 Optical system of a simplified four-color-parameter flow cytometer. The diagram shows, in schematic form, the arrangement of mirrors, optical filters, and detector photomultiplier tube. A single-cell suspension is forced through the nozzle of the machine under pressure. The cells, confined to the axis of the resultant fluid stream by a concentric sheath of cell-free fluid, pass through a laser beam, which is focused onto the stream

cell-associated molecules. Thus, the flow cytometric technique is used to characterize different cell types present in heterogeneous populations. Morphologically all lymphocytes are almost uniform, being small round cells with a dense nucleus and little cytoplasm. The different subpopulations of lymphocytes can be identified on the basis of expression of specific cell-surface proteins, using monoclonal antibodies targeted to the protein molecules. Flow cytometry is the most widely used technique in immunology and cell biology.

Flow cytometer detects individual cells passing in a stream of fluid through a laser beam. Every time a cell passes through the laser beam, it deflects the light from the detector, and this interruption of the laser signal is recorded. If some of these cells are labeled with specific monoclonal antibodies tagged with fluorescent dyes, the labeled cells experience excitation by the laser and emit light that is recorded by a second detecting system (photomultiplier tubes) located at a right angle to the laser beam (Fig. 6.2). Depending upon the fluorescent dye used, the photomultiplier tubes are changed, as each dye has a different intensity of emitted light. The simplest form of the instrument counts the cell numbers and records the level of fluorescence by individual cells. By using a standard program, a computer can generate a plot (histogram) of cell number versus fluorescence intensity. The results can be generated in the form of a dot-plot, where each dot represents a single cell. By using flow cytometer, a mixture of more than five different cell types labeled with cell-specific antibodies tagged with five different fluorescent dyes can be detected and analyzed. The advanced version of flow cytometer, called a fluorescence-activated cell sorter (FACS), is used both for analysis and for sorting one type of cells from a heterogeneous population, again depending upon the stained fluorescence intensity. The preparation of a lymphocyte subpopulation by flow cytometric techniques is described by Manson et al. (1987). The general staining technique and analysis of two different subtypes of T cells isolated from lymph nodes of immunized mice is described below:

1. Prepare a single-cell suspension from lymph nodes of immunized mice and enumerate the viable cell number using the trypan blue exclusion method.
2. Take 50 μl of cell suspension (about 0.5×10^6 to 1.0×10^6 cells) into four wells of a 96-well round-bottom microtiter plate, spin down to obtain cell pellet.
3. Dislodge the cell pellet by hand-tapping and leave the plate on a ice bath. Add appropriately diluted labeled antibodies in each well as follows:
   a. Well 1 (sample for control; auto-fluorescing cells), add 50 μl of PBS-BSA-sodium azide (AZ).
   b. Well 2 (sample for $CD4^+$ cells), add 50 μl of anti-CD4 antibody conjugated with PE.
   c. Well 3 (sample for $CD8^+$ cells), add 50 μl of anti-CD8 antibody conjugated with FITC.
   d. Well 4 (sample for $CD4^+CD8^+$ cells), add 50 μl each of both anti-CD4 and anti-CD8 antibodies.

   Incubate cells in ice bath for 45–60 min in dark.

4. Spin down the cells, discard supernatants, and wash the cells three times each with PBS-BSA-AZ and PBS-AZ. Finally, resuspend the cells in 400 μl of PBS-AZ, and analyze by flow cytometer (the intensity of emission of fluorescent dye is quenched on exposure to light, so always store the sample tubes in the dark at 4–8 °C). If the samples are not analyzed on the same day, they must be fixed with 0.2–0.5% paraformaldehyde (final concentrations).

To exclude dead cells from the assay, propidium iodide (PI) solution is sometimes added to unfixed samples (fixing with paraformaldehyde leads to the

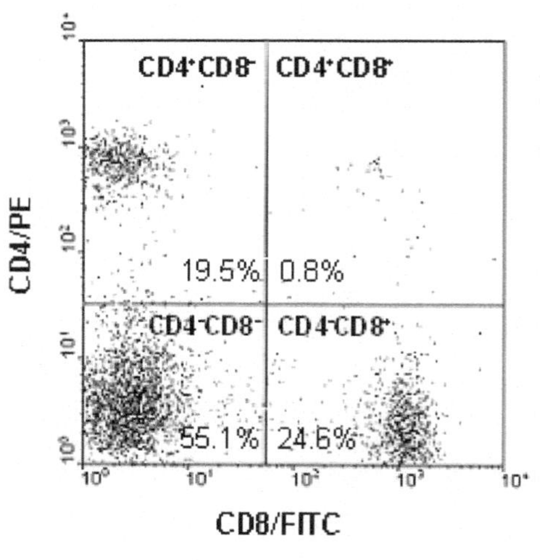

Fig. 6.3 Dot-plot of flow cytometry analysis of lymphocytes. Cells are labeled with CD4 and CD8 antibodies conjugated with PE and FITC, respectively. Data shown in the figure are not actual

death of cells). Dead cells are stained with PI, thus while analyzing the samples in a flow cytometer, the PI^+ cells (dead cells) are excluded.

Figure 6.3 shows a typical dot-plot of CD4 and CD8 stained cells. The lower left quadrant represents autofluorescence, which means these cells (55.1%) express neither CD4 nor CD8 molecules on their surface. The autofluorescence is due to intracellular constituents. Cells in the upper left quadrant (19.5%) represent $CD4^+$ cells. The average fluorescence intensity of these cells is increased more than the normal cells (autofluorescence) due to selective binding of PE-conjugated anti-CD4 antibody. Similarly, cells in the lower right quadrant (24.6%) represent $CD8^+$ cells. It may be seen that about 0.8% cells are labeled with both anti-CD4/PE and anti-CD8/FITC antibodies (upper right quadrant), as these cells are $CD4^+CD8^+$ (all these percentage values are not actual, they are just used to interpret the data of flow cytometry experiments).

6.5.1 Materials and Equipment

Cell samples (single-cell suspension), anti-CD4/PE, anti-CD8/FITC, PBS-BSA-AZ mixture (0.5% BSA and 0.1% sodium azide in PBS), PBS-AZ mixture (0.1% sodium azide in PBS), paraformaldehyde solution (5%, w/v, solution in PBS; paraformaldehyde is dissolved by warming the suspension at 50 °C for 1 h; the clear solution is filtered-sterilized and stored at 4 °C; the solution should be used within 1 month of preparation), Propidium iodide (PI) solution (2 mg/ml; optional).

Refrigerated centrifuge with swing-out bucket, micropipettes, round-bottom 96-well plate, flow cytometer.

6.6 Magnetic Activated Cell Sorting

Other than FACS, a sub-set of immune cells can also be sorted (purified) by using magnetic activated cell sorting (MACS) technology. The principle of sorting is based on super-magnetic microbeads coupled with specific monoclonal antibodies. Depending upon the requirements, microbeads are added to the cell suspension and get attached to the corresponding cell population on the basis of their antibody-tags with them. When this mixture is passed through a MACS column in the presence of a high magnetic field, the cells previously attached with the magnetic particles are retained on the column, allowing other cells to pass through the column (Funderud et al. 1987; Miltenyl Biotec 2006). Later, the column-retained cells are eluted by detaching from the magnetic field, followed

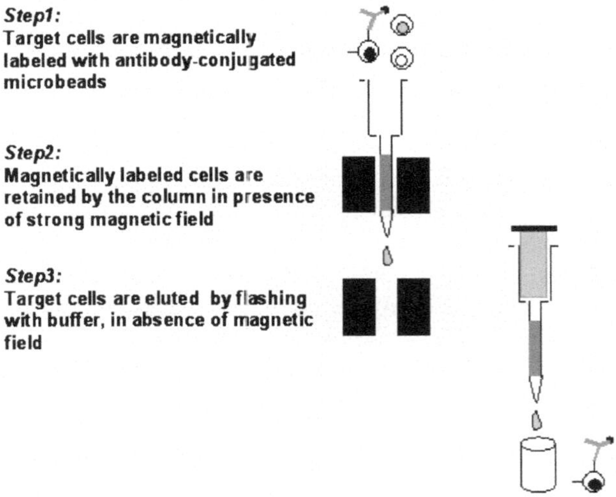

Fig. 6.4 Magnetic activated cell sorting. Positively selected cells are collected on the column, which are later eluted (adapted from Miltenyl Biotec, Bremen, Germany)

by washing the column with buffer (Fig. 6.4). Two different strategies are followed to sort the cells:

1. Positive selection. This means that the target cells are magnetically labeled and isolated directly as the positive cell fraction attached on the column. The positive selection is used for enrichment of rare cells at the highest purity level. The recovery of cells is of a high order and the selection process is much faster than alternative methods.
2. Negative selection. This means that unwanted cells are magnetically labeled and eliminated from the mixture of cells; the unattached cells are the target cells. Negative selection is used for the removal of unwanted cells, and when specific antibody to the target cells is not available.

The procedure for the positive selection of $CD8^+$ cells from the spleen of an immunized mouse is given below:

1. Prepare a single-cell suspension from spleen of immunized mice, as mentioned in the earlier section and enumerate the cell number. Take a fixed number of cells (~20×10^6) in a tube and wash with washing buffer. Resuspend the pellet in 100 µl of same buffer and place the cells in an ice bath. Add the required amount of CD8 microbeads to the cell suspension, mix the content, and incubate for 20–30 min.
2. In the meantime, place the MACS column in the magnetic field of a suitable MACS separator, prepare the column by washing with the buffer. After the incubation of cells in step 1, wash the cells once in the same buffer, resuspend in 1 ml of buffer, and apply onto the column.
3. Allow unbound cells pass through the column, and discard those cells. Wash the column with the same buffer to remove unbound cells.

4. Detach the column from the magnet and place on a test tube stand on the top of a collection tube. Flush the column with the buffer to elute positively selected cells.
5. To determine the purity of $CD8^+$ cells, take an aliquot of cells and label with secondary antibody conjugated with FITC/PE, wash the cells, and analyze in a flow cytometer.

6.6.1 Materials and Equipment

Cell sample (single-cell suspension), filter-sterilized MACS buffer (PBS containing 0.5% BSA and 2 mM EDTA), CD8 microbeads, secondary antibody conjugate (FITC/PE).

MACS separation column, magnet and stand, sample collection tubes.

6.7 Isolation of Mononuclear Cells from Peripheral Blood

Often mononuclear cells are isolated from the peripheral blood for analyzing subsets of immune cells (T-, B-, NK cells, macrophages, etc.). Experimental animals are generally sacrificed to recover these cells from spleen or lymph nodes. However, in humans peripheral blood is the only source of these cells for immunological studies. Mononuclear cells (lymphocytes) can be separated from peripheral blood by Ficoll–Hypaque density gradient centrifugation ($\rho = 1.077$). This method is based on the fact that the different cell types present in the peripheral blood have different densities. When applied to density gradient centrifugation, mononuclear cells accumulate in the buff-colored layer, near the Ficoll–Hypaque and aqueous medium interphase, whereas eythrocytes and granulocytes, the denser cells fractions, are collected in the bottom of the centrifuge and in the Ficoll-Hypaque, respectively. Careful collection of the buff-colored layer yields highly purified mononuclear cells. In order to remove platelets, the mononuclear cells are centrifuged through a FBS cushion gradient, which allows the penetration of mononuclear cells but not platelets. The lymphocytes can be further purified from monocytes by plastic adherence of the mononuclear cells (monocytes adhering on the plastic plate). The various steps involved in isolation of mononuclear cells are described below:

1. Collect blood (~2 ml, in the case of humans) in a heparinized tube, add an equal volume of sterile PBS and mix well.
2. In a separate tube (15 ml), take 3–4 ml of Ficoll–Hypaque solution, and slowly layer the cell suspension over it with the help of a pipette.

3. Centrifuge the tube in a swing-out rotor at 600 *g* for 30 min at room temperature.
4. With the help of a 1-ml pipette tip, suck-out the buff layer and transfer it to a similar tube containing HBSS. Wash the cell pellet two times in the same buffer to remove traces of Ficoll–Hypaque.
5. Finally, resuspend the cell pellet in complete medium and enumerate the viable cell number.

6.7.1 Materials and Equipment

Heparinized blood, PBS, Ficoll–Hypaque solution, filter-sterilized HBSS (pH 6.4; 5.4 mM KCl, 0.3 mM Na_2HPO_4 $7H_2O$, 0.4 mM KH_2PO_4, 4.2 mM $NaHCO_3$, 1.3 mM $CaCl_2$, 0.5 mM $MgCl_2$ $6H_2O$, 0.6 mM $MgSO_4$ $7H_2O$, 137 mM NaCl, 5.6 mM D-glucose, 0.2% phenol red in water), FBS, complete medium.

Conical tube (15 ml), table-top centrifuge with swing-out rotor, plastic ware.

6.8 Conclusions

The immune system has been developed to protect against the threat of pathogens. When invaded by foreign molecule(s), each mammal elicits a unique immune response. The primary function of such immune response is two-fold: first to counteract the pathogen, second to educate the immune system to take care of any similar threat in the future. The immune response means either the formation of neutralizing antibodies, or the development of cytotoxic T cells to destroy infected cells to prevent further spread of the disease. The immune response is specific for each pathogen/foreign molecule. With the development of understanding on various immune cells and the technological advancements of assay systems, it has become possible to analyze immune response qualitatively and quantitatively, and a few of them have been discussed in this chapter.

References

Ashwell JD, DeFranco AL, Paul WE, Schwartz RH (1984) Antigen presentation by resting B cells: radiosensitivity of the antigen-presentation function and two distinct pathways of T cell activation. J Exp Med 159:861–869

BD Biosciences (2006) Technical literature of BD cytometric bead array: mouse Th1/Th2 cytokine CBA. BD Biosciences, San Diego

Brunner KT, Mauel J, Cerottini J-C, Chapuis B (1968) Quantitative assay of the lytic action of immune lymphoid cells on 51Cr labeled allogenic target cells in vitro: inhibition by isoantibody and by drugs. Immunology 14:181–196

Derby E, Reddy V, Baseler M, Malyguine A (2001) Flow cytometric assay for the simultaneous analysis of cell-mediated cytotoxicity and effector cell phenotype. BioTechniques 31:660–665

Funderud S, Nustad K, Lea T, Vartdal F, Gaudernack G, Stenstad P, Ugelstad J (1987) Fractionation of lymphocytes by immunomagnetic beads. In: Klaus GGB (ed) Lymphocytes. IRL Press, Oxford, pp 55–65

Goldsby RA, Kindt TJ, Osborne BA (2000) Kuby immunology, 4th edn. Freeman, New York

Lightbody J, King YC (1974) Comparison of 137Cs irradiation and mitomycin C treatment of stimulator cells in the mixed lymphocyte culture reaction. Cell Immunol 13:326–330

Manson DW, Penhale WJ, Sedgwick JD (1987) Preparation of lymphocyte subpopulations. In: Klaus GGB (ed) Lymphocytes. IRL Press, Oxford, pp 35–54

Miltenyl Biotec (2006) Technical literature of MACS magnetic cell sorting. Miltenyl Biotec, Bremen

Reeves JP, Reeves PA (1992) Selection of surgical procedures. In: Coligan JE, Kruisbeek ADF, Margulier DH, Shevach EM, Strober W (eds) Current protocols in immunology, vol 1. Wiley, New York

7 Transcriptome Analysis

S.K. Yadav, S.L. Singla-Pareek, and A. Pareek

7.1 Introduction

The phenotype of a living organism depends on the expression of its genetic material. Most living organisms, except some viruses, have DNA as genetic material. All cells of an organism have a similar genome structure and organization but there is a highly specialized regulatory network, which brings about the specificity in their expression. At a given point in time, in a given cell, only a specified part of its genome is active transcriptionally. In this modern era of "omes" and "omics" a new name – *transcriptome* – has been suggested to represent the set of all mRNA molecules (or transcripts) in one or a population of biological cells for a given set of environmental circumstances. Therefore, unlike the genome, which is fixed for a given organism (apart from genetic polymorphisms), the transcriptome varies depending upon the context of the experiment. As not all the genes in the genome are transcribing at a given time, the transcriptome is less complex than the genome of an organism. The transcriptome, partially or fully, eventually gets translated to give rise to a proteome. Again, as some of the transcripts within the transcriptome never get translated into proteins, the proteome seem to be even less complex than the transcriptome. However, the complexity of the proteome increases due to its post-translational modifications. These differently modified proteins function in the synthesis of various

Sudesh Kumar Yadav: Biotechnology Division, Institute of Himalayan Bioresource Technology, CSIR, Palampur-176061 (HP), India

Sneh L. Singla-Pareek: Plant Molecular Biology, International Centre for Genetic Engineering and Biotechnology, Aruna Asaf Ali Marg, New Delhi-110067, India

Ashwani Pareek: Stress Physiology and Molecular Biology Laboratory, School of Life Science, Jawaharlal Nehru University, Aruna Asaf Ali Marg, New Delhi-110067, India, email: ashwanip@mail.jnu.ac.in

Soil Biology, Volume 11
Advanced Techniques in Soil Microbiology
A. Varma, R. Oelmüller (Eds.)

primary and secondary metabolites, the total make-up of which is known as a metabolome. The combination of these intricate and interlinked information of the genome, RNA, proteins, and metabolites, are vital for the biological activity of an individual.

The era of genomics started as early as in 1995 when the entire genome sequence of the self-replicating organism, *Haemophilus influenzae*, was first described. This success opened up a new direction and within a decade, we had more than 100 genomes being sequenced or already sequenced, including higher plants such as *Arabidopsis thaliana*. The economically important crop plant rice has also been reported to be completely sequenced by the commercial sector (Arabidopsis Genome Initiative 2000; Butler and Pockley 2000; Davenport 2001). By the end of 2004, there were as many as 163 genomes which had been completely sequenced. Many more such efforts are currently in progress – including representatives from Bacteria, Archaea, and Eukarya. A list of the genomes which have been completely sequenced or are in progress is available at: http://www.tigr.org/tdb/mdb/mdb.html. This illustrative information on genome sequences can be exploited in several ways. One of the most important uses of the information available from genome sequencing is for the understanding of the regulation of a functional genome, i.e. the transcriptome. This has been made possible with the advent of novel tools and techniques of molecular biology which aid in isolation, and thus, the analysis of biological material from very small amount of tissues. With the advancement of our understanding of these techniques, transcript analysis (transcriptomics), and protein profiling (proteomics), the most advanced strategies have been developed to provide an answer to the question: how does the expression of a gene respond variably to a particular external or developmental stimulus?

A gene for a specific trait can be identified through analysis of complementary DNA (cDNA) or copies of messenger RNA (mRNA). This requires the isolation of pure RNA, and in some cases, mRNA. Messenger RNAs represent only a small percentage of the total RNA (about 1–3% in eukaryotes). However, mRNA contains valuable information and is directly responsible for protein translation. To assess the expression of a gene, it is quite important to analyze the amount of mRNA corresponding to that particular gene. Therefore, transcriptome analysis is one of the important tools to assess the expression of a gene. There are several ways to quantify RNA. Among these, the commonly used gene expression and quantitation assay methods include Northern blot analysis, dot/slot blot hybridization, in situ hybridization, RT-PCR, and microarray analysis. In this chapter, we have attempted to provide an insight into these contemporary techniques of transcriptome analysis. For brevity sake, we have not provided details related to the development of these techniques over time, but have presented the information in a manner in which the reader can gather basic information about these techniques and learn the principle along with the essential steps in the experiment. Readers are encouraged to explore the detailed references for individual techniques, as provided in the reference list.

7.2 RNA Preparation

For each of the analysis techniques described in this chapter, one of the prerequisites is to have a good quality RNA preparation. Thus before describing each of these techniques, it is worthwhile to suggest some measures to obtain good preparation of RNA. To begin with, one needs to have a suitable extraction method for the RNA isolation from the source. If the sample is a plant, animal or microbial source, it can be fresh or frozen tissue. However, it has been seen that fresh tissue is a better source of good quality RNA. As little as 50–100 mg of tissue is suitable for total RNA extraction. If the source is a microorganism, grow cells and harvest at the stage when the maximum number of cells are at their exponential phase of growth. The basic principle of RNA extraction has been well summarized by Sambrook et al. (1989). This crude RNA preparation is well suited for its analysis by most techniques. However, if the RNA is to be used in enzymatic reactions, better and more reproducible results can be obtained when the RNA preparation is further cleaned. A fast and easy method relies on commercially available columns (e.g. Qiagen columns). In these methods, the RNA preparation is conditioned with a buffer provided by the manufacturer. The solution is applied to a column on which the RNA is specifically retained. The column is then washed with several buffers before the RNA is eluted and obtained in a highly pure form. The RNA quantity is measured with a spectrophotometer and its integrity verified on a 0.8–1.0% agarose gel. This highly purified RNA can be commonly used for transcriptome analysis by employing techniques such as microarray and reverse transcriptase (RT)-PCR.

7.3 Northern Analysis

7.3.1 Principle

Northern blot analysis is the simplest and most commonly used technique for detection and quantification of specific RNA species from a particular cell or tissue type. In this method, total RNA is isolated and separated by electrophoresis through an agarose/formaldehyde gel, which separates the RNA by size. The distance of migration of the RNA molecule is inversely proportional to the size of RNA molecule. After its separation on agarose, RNA is stained with ethidium bromide and visualized using UV light. In those gels having total RNA, the 28S and 18S ribosomal subunits are conveniently visible due to their high abundance and act as convenient size markers (approx. 4.8 kb and 1.9 kb, respectively). To detect the mRNA of interest, we perform Northern blotting. For this purpose, it is necessary to transfer the RNA from the agarose/formaldehyde gel to a nitrocellulose or nylon membrane. On the membrane, RNA is detected by hybridization using a labeled probe. The probe may be a DNA or RNA molecule, which is labeled either chemically or radioactively. During Northern blotting, one should keep the following precautions in mind:

1. As formaldehyde is toxic and a potential carcinogen, use it inside the fume hood and do not inhale.
2. Formamide and ethidium bromide are also toxic, and hence should be handled with gloves.
3. UV light can damage eyes if not protected, therefore, wearing a facemask or goggles is recommended.
4. Agarose can also cause nasty burns, so handle with care (Sambrook et al. 1989; Trayhurn et al. 1994).

7.3.2 Procedure

After isolating the RNA, prepare the sample for electrophoresis. Take 10–20 μg of RNA in a sterile Eppendorf tube, while maintaining the sample on ice. The volume of RNA should be increased to 15 μl by the addition of DEPC-treated water. To this mixture, add de-ionized formamide, formaldehyde and running buffer (10× stock), and 2.5 μl loading buffer. Next, denature samples by heating at 60 °C for 20 min, then snap-cool on ice. Prepare a 0.8–1.0% agarose in 10× running buffer by boiling (for example 1.2 g in 15 ml of buffer and make its volume up to 150 ml with DEPC-treated water at the end). Add an appropriate concentration of iodoacetamide, formaldehyde, and ethidium bromide. Mix all the contents, make-up to its final volume by adding DEPC water, and pour into a sealed gel tray (do not forget to place the comb to form the wells just after pouring the gel). After 30 min, load the samples into the wells and run at 100 V for 2 h.

After electrophoresis, analyze the gel under UV light. Wearing gloves, remove the gel from the tank, carefully place the gel on a transilluminator and either take a photograph or see it in a gel-documentation system, which is fitted with a camera and monitor. This picture of gel aids later in matching the position of the ribosomal markers on the photograph with the final X-ray film of the hybridized mRNA. The use of a Saran wrap is recommended while transporting gel from tank to UV transilluminator. Now, transfer the RNA from the gel to the nylon membrane (nitrocellulose or Hybond N nylon). The nylon membrane should be the same size as the gel. Take a glass tray of an appropriate size and pour transfer buffer into it. Place a glass plate on this tray and a Whatman paper over the glass plate, ensuring that two ends of the Whatman paper dip into the transfer buffer. Put the gel over the Whatman paper and the nylon membrane over the gel. Fi-

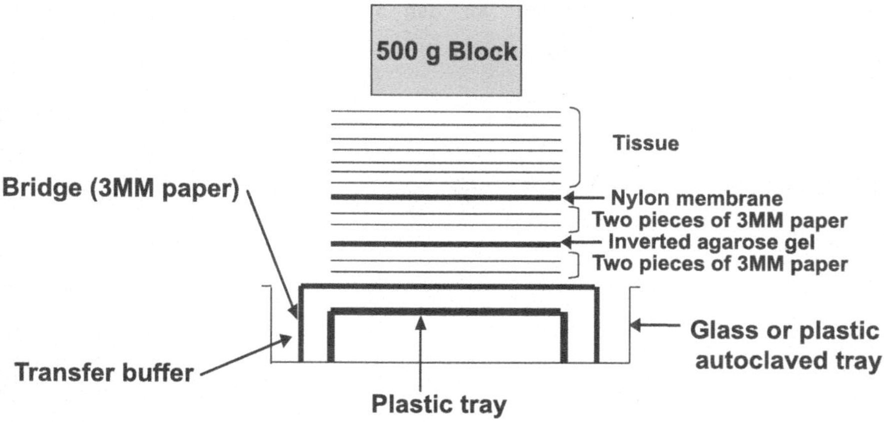

Fig. 7.1 Northern blotting set up

nally, add some filter papers and about 500 g of weight. Set up transfer as shown above and leave the RNA to transfer by capillary action overnight (Fig. 7.1).

After the transfer, carefully disassemble the transfer apparatus and remove the nylon membrane. Here, mark the right and left sides by cutting a small corner of the membrane with scissors. Neutralize the membrane in 100 mM Tris-HCl (pH 8) and fix the RNA onto nylon membranes using a UV crosslinker or by baking for 2 h. Now the blot is ready to be placed in the hybridization chamber. For prehybridization, put the blot in a hybridization bottle and soak it with at least 10–15 ml of pre-hybridization buffer (50% formamide, 5× SSPE, 5× Denhardt's reagent, 1% SDS). Let it rotate in the hybridization chamber at 56–60 °C for 4–5 h (by this time you can prepare the probe). Boil salmon sperm DNA for 5–10 min, snap-cool on ice, and add it to the hybridization buffer. In the hybridization step, the labeled probe is added to the pre-hybridization solution. Here, do not forget to boil the probe for 10 min and then chill quickly on ice. Open the hybridization chamber, and without disturbing the membrane, add the denatured (boiled) probe to the hybridization buffer and mix gently, then return it to the hybridization chamber and incubate overnight at 56 °C.

As a result of overnight hybridization, the complimentary sequences on the membrane should have hybridized with the probe. However, the membrane might have some background radioactivity due to non-specific binding of probe, which needs to be washed, in order to obtain a clean specific signal. For this purpose, we should perform at least three washings of the blot before developing it. Discard the hybridization solution in the designated container/sink. Keep the membrane in the hybridization chamber for washes. Add about 50 ml of 2× SSC with 0.1% SDS (room temperature) to the hybridization chamber, wash the membrane for 10 min, and then discard the washing solution. Repeat the above washing step. Discard the second wash solution. Add 50 ml of 0.1× SSC with 0.1% SDS, preheated to 56 °C, to the hybridization chamber. Wash for 20 min

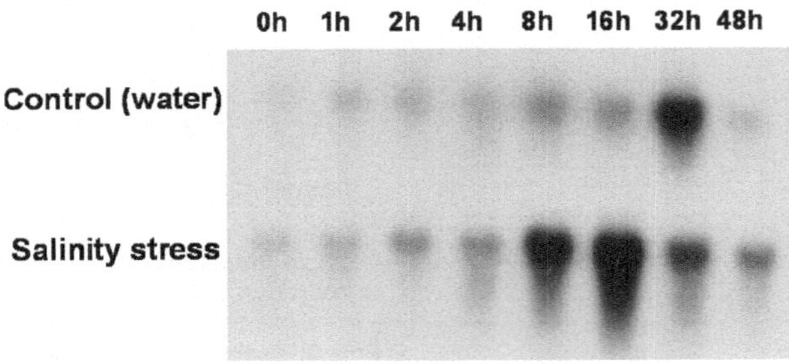

Fig. 7.2 Northern blot, after washing and developing the X-ray film. Blot shows the developmental regulation of gene A under control and salinity stress conditions

and discard the final wash solution. Remove the membrane from the hybridization chamber and put it on a dry piece of Whatman filter paper of similar size. Wrap it in Saran wrap and expose overnight to X-ray film after placing them in a cassette. X-ray film may be developed after 24 h, or can be left for longer periods if necessary (Fig. 7.2).

7.3.3 Applications

1. The Northern blotting technique is helpful in determining the status of a gene, i.e. whether the gene in question is active or not.
2. Depending on the inducibility of a novel gene by various signals, an idea can be obtained about the possible functions of the same.
3. Information from analysis as in point 2 above can also be extrapolated to describe the feature of the promoter of the gene being analyzed.
4. The tissue or stage-specific inducibility of a gene may also indicate its specific roles.
5. The data obtained from this technique can also be used for both qualitative and quantitative comparison of expression of gene(s).

7.4 In Situ Hybridization

7.4.1 Principle

In situ hybridization is a method of localizing and detecting specific mRNA sequences in tissue sections or cell preparations. In this technique, RNA or

DNA isolation is not required as in the other methods of RNA analysis. Specific mRNA is localized by hybridizing the complimentary strand of a nucleotide probe to the sequence of interest. The technique is appreciably sensitive as its detection limit range from ten to 20 copies of mRNA per cell. However, the technique also suffers from a major drawback associated with masking of low-copy signals due to associated protein or access of the probe to the target sequences, which are protected within complex cellular structures. Therefore, in order to detect the RNA in tissues or cells of interest, one has to increase the permeability within the cell without destroying its structural integrity. Theoretically speaking, there are as many protocols for carrying out in situ hybridization as there are different tissues that have been probed (Wilkinson 1994; Childs 1999). Here, we attempt to describe the basic steps of the process along with their underlying principle.

7.4.2 Procedure

7.4.2.1 Preparation of Sample

There are three ways to prepare tissues for in situ hybridization, as described below:

1. To make the sections of fresh tissue is not an easy job. Therefore, wherever possible, the tissue is snap-frozen (rapidly put into a –80 °C freezer) before sectioning. The completely frozen tissue is embedded in a special support medium for thin cryo-sectioning. The sections are lightly and rapidly fixed in 4% paraformaldehyde on a microscopic plate followed by their hybridization.
2. Large sections of the tissues are fixed in formalin and embedded in wax (paraffin sections) before being sectioned.
3. For special samples such as cell suspension from leaf tissues or callus, the cells are cytospun onto glass slides followed by fixing them onto the slides with methanol.

For permeabilization of the tissue, three commonly used reagents are HCl, detergents (Triton or SDS), and Proteinase K. HCl is thought to act by extraction of proteins and hydrolysis of the target sequence, which also help to decrease the level of background staining. Detergent is frequently used to permeabilize the tissue membranes by extracting the lipids. This is not usually required in tissue that has been embedded in wax, but may be more useful for intact cells or cryostat sections. Proteinase K is a non-specific endopeptidase attacking all peptide bonds and is active over a wide pH range. It is commonly used to remove proteins that surround the target sequence (Tautz and Pfeifle 1989; de Almeida-Engler et al. 1994).

7.4.2.2
Prehybridization

Prehybridization is generally carried out to reduce background staining. But in in situ hybridization, this step is very crucial than any other technique of transcript analysis described in this chapter. Many of the non-radioactive oligonucleotide probe detection methods utilize enzymes such as peroxidases or alkaline phosphatases to visualize the label. Therefore there should not be any interference of endogenous tissue enzymes, which could result in giving a very high background. If interference is expected, then we have to neutralize our tissue or tissue sections. This can be achieved as follows: (a) for peroxidases, treat the tissues with 1% H_2O_2 in methanol for 30 min, and (b) for alkaline phosphatases, the drug levamisole may be added to the substrate solution. This should be added in a very low concentration otherwise residual alkaline phosphatase activity is usually lost during hybridization. Finally, prehybridization involves incubating the tissue/section with a solution that is composed of all the elements of the hybridization solution, minus the probe.

7.4.2.3
Hybridization

Hybridization of the oligonucleotide to the target mRNA within the tissue depends on several factors, like temperature, solution pH, monovalent cation concentration, and presence of organic solvents. A typical hybridization solution which can be used at 37 °C temperature and with an overnight incubation period should have several essential components, as described here. Dextran sulfate is one of the most important components of a hybridization solution as it absorbs the maximum amount of water and thus reduces the amount of hydrating water for dissolving the nucleotides. Therefore, dextran sulfate effectively increases the probe concentration in solution. While formamide and dithiothreitol (DTT) reduce the thermal stability of bonds and allow hybridization at a lower temperature. The use of SSC (NaCl + sodium citrate) is also crucial as it dissociates into monovalent cations in solution, which interact with the phosphate groups of nucleic acids, and as a result, decrease the electrostatic interactions between the two strands and make them comparatively stable. Hybridization is generally reduced to a great extent in the presence of divalent ions. Therefore, EDTA is also added, which removes free divalent cations from the hybridization solution, thus increasing the efficiency of hybridization. In addition to this, salmon sperm DNA, tRNA, and Denhardt's solution are also added in the hybridization solution to decrease the chance of non-specific binding of the oligonucleotide probe.

Mainly radioactive probes are used for in situ hybridization. The advantage of a radiolabeled probe is its ability to detect very low levels of transcripts, while the major limitations with the use of radiolabeled probes are poor spatial resolution and the requirement of long exposure time for microautoradiography. However, exposure time depends on the radioisotope used and the amount of target molecules in the tissue under experiment. Recently, the application of non-radioactive labeled nucleotides [e.g. biotin-UTP, digoxigenin (DIG)-UTP] considerably improved the detection limits for in situ hybridization technique. Among the non-radioactive labeling methods developed so far, DIG-based detection has proven to be the most appropriate, due to its high specificity and sensitivity. Another advantage of the DIG method is the high signal to noise ratio, since no plant other than *Digitalis* has been shown to have this compound (O'Neill et al. 1994).

7.4.2.4 Washings

After overnight hybridization, the material is washed 2–3 times to remove unbound probe or probe which has loosely bound to partially homologous or mismatched sequences. Washing should be carried under stringency conditions similar to hybridization. However, the final wash should be carried out at low stringency, taking precautions not to dislocate the tissue.

7.4.3 Applications

1. Expression of a gene can be localized in a specific cell, and hence, can be correlated with its possible function(s).
2. The differential level of the same gene in different tissues of an organism can be analyzed.
3. Used for the identification of a microorganism in microbial ecosystem.
4. Detection of specific microbes can be done in plant tissue through mRNA analysis.
5. Diversity analysis of a microbes in a microbial ecology.
6. Identification of pathogenic microbes responsible for a disease in humans.

7.5 Dot Blot and Slot Blot

7.5.1 Principle

Specific transcript (mRNA) in an unfractionated preparation can be measured directly by immobilizing the sample in the form of a spot (*dot blot*) or in a manifold slot (*slot blot*). It is a relatively rapid technique for RNA detection and quantitation as compared with those described above (Yadetie et al. 2004). In the dot/slot blot, a desired RNA species is detected by using a labeled DNA/RNA probe. For the dot blot quantitation is usually visual, whereas the slot blot format is more easily and accurately quantitated by scanning with a densitometer.

7.5.2 Procedure

7.5.2.1 Sample Preparation

This technique can be equally optimized for quantitation of both DNA and RNA. For obtaining purified DNA for dot/slot blot, standard protocols of DNA isolation are followed. In the case of bacteria, we can use cell lysate directly as a DNA source. While using RNA for dot/slot blot, it is to be noted that all solutions and glassware involved with RNA work should be sterilized or treated to remove any RNase. Glassware should be washed in 0.2% diethylpyrocarbonate prior to use, followed by autoclaving.

7.5.2.2 Denaturation

Prior to the application of a DNA or RNA sample to the membrane, denaturation is required. For a DNA sample, Add 0.1 vol of 1 N NaOH and incubate for 5 min at 37 °C or heat the sample for 5 min in a boiling water bath and immediately put on ice. Add 1 vol of 2 M ammonium acetate, pH 7. Dilute the sample in a suitable buffer prior to its application to the membrane. In contrast, RNA is denatured by mixing with 100% formamide (50% final), 37% formaldehyde (7% final), and 20× SSC (1× final). Incubate the mixture at 68 °C for 15 min, followed by cooling on ice. Alternatively, RNA may also be denatured with glyoxal or with methyl mercuric hydroxide.

7.5.2.3 Membrane Preparation

Wet the membrane thoroughly in deionized water and then soak in 1 M ammonium acetate, pH 7.0, or in 6–10× SSC prior to use. A high salt buffer is necessary for retention of the DNA or RNA on nitrocellulose membranes. While a lower ionic strength buffer (2–5× final concentration) may be used for sample dilution with Nytran nylon membranes.

7.5.2.4 Sample Application

Apply sample aliquots to the membrane placed on the top of two sheets of dry filter paper (blot the membrane briefly to remove excess liquid before spotting sample). Allow sample area to dry prior to application of additional solution to the membrane.

7.5.2.5 Immobilization

For immobilization, bake the membrane at 80 °C in a vacuum oven for 20 min to 1 h or until dry. Alternatively, DNA or RNA on the membrane may also be immobilized by UV crosslinking. For this purpose, the membrane is exposed to a UV source of 254 nm.

7.5.2.6 Hybridization and Detection

Hybridization of the blot prepared above is carried out by using a labeled probe, as described for Northern blotting. Specific hybridized bands are detected through autoradiography, and for the densitometric scanning of nitrocellulose blots, the blot membrane is made clear by immersing the same in xylene, paraffin oil, or immersion oil.

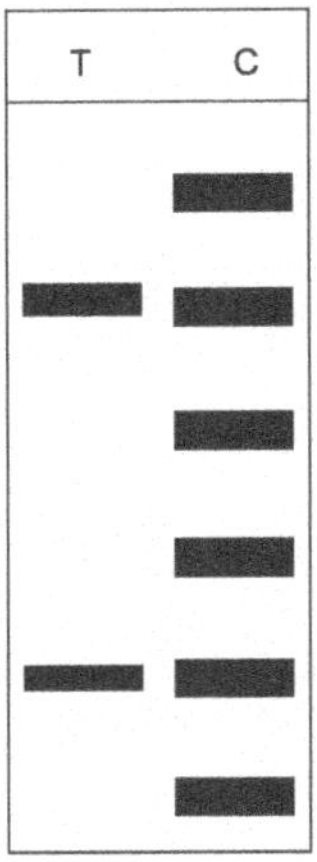

Fig. 7.3 Comparative profile expression on a slot of differentially expressed transcripts in control and treated samples. Total RNA isolated from control (*C*) and treated (*T*) microbes of six different species were subjected to slot blot analysis using the re-amplified PCR fragment as a probe. In the control conditions all the six probes used here are still expressing, while upon treatment four get down-regulated

Let us try to understand the technique with the help of a hypothetical example where cell lysate from six different microbial species has been used to make a slot blot. The objective of the analyis is to see the expression of a particular gene under control and treated conditions in these different species. After using the labeled sequence as probe, results as shown in Fig. 7.3 have been obtained. Based on these results, it can be safely concluded that the gene in question is strongly down-regulated in all species except the two which show only a marginal change in the level of the signal.

7.5.2.7 Dot Blot

For dot blot analysis, instead of making the slots, total RNA is loaded on the membrane in the form of dots. The rest of the procedure is exactly the same as described in slot blot analysis. We can describe dot blotting with an example that helps us in understanding the technique. In this experiment, six different microbial species have been analyzed. These species have been transformed with a specific stress-related gene, with the objective of improving their tolerance for a particular stress. Six non-transformed species have also been analyzed, along with their transformed counterparts. Through dot blot analysis, we can assess the degree of tolerance of these different species, as depicted in Fig. 7.4. It can be concluded from this experiment that expression of the gene in question is down-regulated in some species after just one week of growth under stress, while some tolerant ones still showed a high expression of the gene even after two or three weeks.

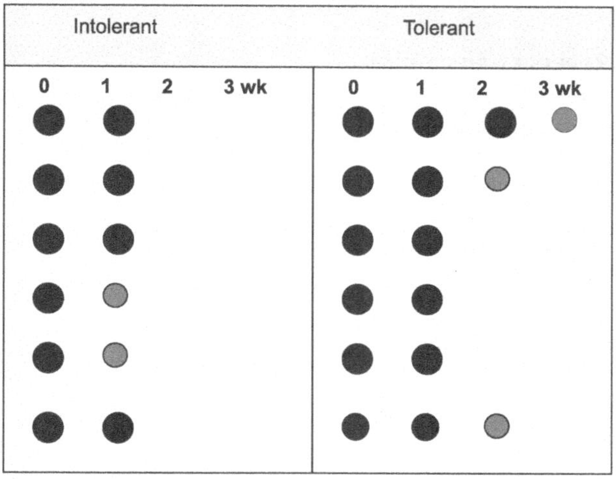

Fig. 7.4 Comparison of a transcript in tolerant and intolerant microbial species by Dot blot analysis. Total RNA used for dot blot analysis was isolated on the first day (*0*) and on weeks *1*, *2*, *3* of their growth

7.5.3 Applications

1. The dot/slot blot techniques, although they are crude methods in which isolated nucleic acid preparations in solution are applied directly to a transfer membrane, they skip the step requiring their analysis in agarose gel.
2. These techniques provide a rapid qualitative screening method for target sequences of RNA or DNA.
3. Assays can be performed with either purified nucleic acids or cell lysates as such.
4. They can be easily optimized for high-throughput analysis of samples such as identification of organisms and studies related to microbial community structures.

7.6 Reverse Transcriptase–Polymerase Chain Reaction

7.6.1 Principle

The routine PCR technique greatly amplifies the copy number of a given DNA for analysis, cloning, and storage. However, RT-PCR starts with mRNA or to-

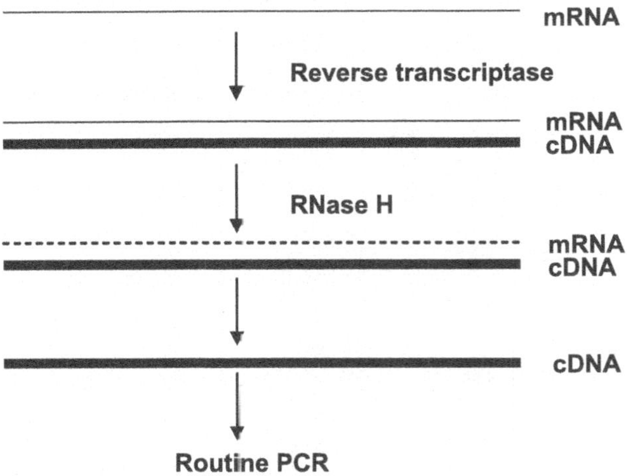

Fig. 7.5 Diagrammatic representation of the major steps in RT-PCR

tal RNA, makes a cDNA complimentary strand using reverse transcriptase, and then amplifies the product through routine PCR (Fig. 7.5). RT- PCR is most sensitive and accurate technique used for transcript analysis. It is used for both quantitation and comparision of RNA between two or more sources (Sperisen et al. 1992; Chelly and Kahn 1994).

There are several factors affecting the performance of RT-PCR as they interfere with amplification efficiency. These are Mg^{2+}/dNTPs/primer concentrations, efficiency of reverse transcription, enzyme activity, pH, annealing temperature, cycle number, temperature variation, tube-to-tube variation, etc. Since PCR results in a million-fold amplification within a short span of time, variation in any of the above factors during the amplification process significantly affects the final results. Therefore, routine RT-PCR cannot be used for the purpose of quantitative analysis. This major problem, however, can be solved using quantitative RT-PCR. This method uses an external template as the internal control for all the steps in RT-PCR process (Schneeberger et al. 1995).

7.6.2 Procedure

RT-PCR comprises of two steps, including the reverse transcription where mRNA is converted to first-strand cDNA and the amplification of first-strand cDNA through routine PCR. During the first step, oligo dT primers are generally used as mRNA and have the polyA tail at their 3' end Thus through this process, we can specifically make the first-strand cDNA of mRNA only. For the synthesis of cDNA from mRNA, reverse transcriptase is required. Reverse transcriptase can synthesize a DNA strand complementary to mRNA, and to carry out the routine PCR for amplification, mRNA is removed from the hybrid by using RNase H. For the purpose of reverse transcription, two types of reverse transcriptase are commercially available. One is from avian myeloblastosis virus (AMV) and other from Moloney muriene leukemia virus (MMLV). AMV-RTase has a powerful RNase activity, which can easily digest the RNA moiety of RNA–DNA hybrids but lacks proof-reading activity and shows maximum activity at 42 °C. In contrast, MMLV-RTase lacks RNase activity and shows maximum activity at 37 °C.

For carrying out the reaction, gene-specific primers along with enzyme and optimized RT buffer is required. Reverse transcription is carried out for 1.0–1.5 h and first-strand cDNA synthesis is followed by inactivation of enzyme at 70 °C for 5 min. After inactivation, tubes are immediately put on ice to avoid the formation of secondary structures. Next the sample is treated with RNase H to degrade the RNA, and cDNA is used for amplification through routine PCR. The routine PCR comprises of three steps: denaturation (carried out at 94–95 °C), annealing (annealing temperature depends on the T_m of the primes used in the amplification), and extension (carried out at 72 °C). These three steps are followed for 30–35 cycles.

Primer design is an extremely crucial step in RT-PCR. Following are the some important points, which should be kept in mind while designing primers:

1. GC content of primer sequence should be 40–60%.
2. There should be no continuous repeat of any one type of nucleotide.
3. Length of primers should be 18–25 nucleotides, with forward and reverse primers not differing in their length by more than three nucleotides.
4. There should be no inverted repeat sequence or self-complementary sequence. These types of sequences may form a hair-pin loop structure and interfere in the annealing process.
5. Two primers of a reaction should not be complementary to each other; else they would form primer-dimers, rendering them unavailable for amplification of the target sequences.
6. Primer pair should not differ in their T_m by more than 5 °C. A greater difference in T_m would not allow both the primers to anneal at same temperature.
7. Useful sequences such as restriction sites are usually added at 5' end, while the 3' end of primer is crucial; one should try to keep either G or C at the 3' end.
8. The T_m of a primer is calculated using the following formula: $T_m = 2(A+T) + 4(G+C)$.

7.6.3 Application of RT-PCR

1. RT-PCR is very useful technique for the detection of viruses containing RNA as the genetic material. Also, the amplified product can be used as a probe for the identification of virus species.

2. An amplified fragment through RT-PCR can be used for ligation into a suitable vector and transformed to a host where multiple copies can be generated.
3. By comparing the known quantity of RNA, we can measure the unknown quantity in our experimental sample.
4. Total RNA is converted to cDNA, which can be cloned into a suitable vector and transformed to a suitable host.
5. cDNA prepared through RT-PCR can be cloned into a suitable sequencing vector and used for sequencing.
6. To compare the differential expression of genes in an organism under two set of conditions, we isolate total mRNA and carry out the cDNA synthesis through RT-PCR.

7.7 DNA Microarray

7.7.1 Principle

Until the discovery of microarray technique, molecular biologists were working on the characterization of individual genes and monitoring its function. However, expression of the whole genome of an organism became possible with the advent of microchip technology. This technology could not be developed earlier as it needed the description of the complete genome sequence of an organism or the availability of large EST databases. After the completion of the genome sequences of various organisms and their data availability, the development of microarray technology became possible. This technology has made the simultaneous analysis of multiple genes very easy. The basic principle of this technology is described here. DNA molecules or oligonucleotides corresponding to the genes whose expressions have to be analyzed are used for making the probe. In this technique, oligonucleotides or cDNA molecules are attached in an ordered fashion to a solid support that can be a nylon membrane or a glass slide. With the availability of robotic spotters and automated liquid-handling stations, the technique has become fully automated and this has also made it possible to produce arrays with several thousand genes represented on a few square centimeters on the support. To determine the relative abundance of the corresponding transcripts in a total RNA preparation, the RNA species are converted into cDNA in the presence of fluorescent dyes. Commonly used labels are Cy3 and Cy5 available commercially. The labeled target sample is allowed to hybridize with spotted probes on the glass slide. Finally the intensity of the hybridization signal is a measure of the relative abundance of the corresponding mRNA in the sample. Thus comparing the intensities of hybridization signals for different mRNA samples allows the determination of changes in mRNA levels under the conditions tested for all of the genes represented on the arrays (Eisen et al. 1998; Lockhart and Winzeler 2000). Fodor (1997) proved that, by using this technology, one can display 409 000 spots in an area of 1.28 cm^2. Hence, all 20 000–25 000 genes of *Arabidopsis* can be displayed on a single slide. The microarray technique is highly sensitive as it can detect mRNAs at level of 1:100 000 or 1:500 000. Though the technique has its own limitations (such as high cost of operation, money-intensive set-up of facility, need for appropriate software and technical expertise), it recently became the technique of choice for researchers working with almost all organisms.

7.7.2 Procedure

The first step is designing the microarray itself. In principle, two different types of array techniques are used. In the first, fragments from either genomic clones or cDNA are amplified through PCR and spotted onto the appropriate solid support. In the second type, small oligonucleotides, designed based on available genetic information, are synthesized and spotted. Generally microscopic glass slides coated with polylysine are used for spotting the probe sample. The role of polylysine is to enhance the DNA/RNA binding to the plate through electrostatic interactions. For solid support, apart from glass slides, special coated plastic films are also used (Bertucci et al. 1999; Eisen and Brown 1999).

DNA samples are printed with a microarrayer (sometimes also called a spotter) onto microscope slides. The whole operation of automated spotting is performed inside a dust- and vibration-free chamber. Sometimes, evaporation of the sample may also take place at this stage, which can be avoided by maintaining good humidity in the chamber during operation. For efficient coupling of printed cDNA, slides are left at room temperature for 24 h. Finally, dried slides are put in a beaker for washing, followed by air-drying, and storage at room temperature until further use.

The basic principle of microarray technique remains the same, irrespective of the source of mRNA. Here, we describe the technique by taking an example of a microbial system. Microbes growing under natural environmental conditions are used as source 1 and microbes exposed to a kind of stress are used as source 2 for mRNA isolation. For qualitative and quantitative assay of the differentially expressed genes using a microarray, RNAs are extracted from both types of microbes. Messengers RNA are then converted into cDNA by reverse transcription. At this stage, cDNA from source 1 is labeled with a green dye (Cy3) and cDNA from source 2 is labeled with a red dye (Cy5). These labeled cDNA are used for hybridization with the probes spotted onto the solid support.

During hybridization, green- and red-labeled cDNAs are mixed together and put on the matrix of spotted single-strand DNA (probes). For hybridization, the chip is incubated overnight at 60 °C under high humidity conditions. At this temperature, DNA strands that encounter the complementary strands of the probes on the slide, create double-stranded DNA. The newly formed double-stranded DNA has one unlabeled and another labeled strand. After hybridization and washing, the microarrays are scanned at two different wavelengths corresponding to the absorbance of the red and green dyes and the signals are analyzed. A laser beam is passed through the microarray slide that excites each spot on the plate and the fluorescent emissions are gathered through a photomultiplicator (PMT) coupled to a confocal microscope. If the hybridization is stronger with one of the samples, the spot appears either red or green. If the intensities of binding of two dyes to target samples are same, then the spot on the microarray appears to be yellow. We have now two images from the same slide corresponding to the two dyes. For a given spot on the slide, we measure the signal intensities in the green dye emission wavelength and the signal intensities in the red dye emission wavelength. If the amount of fluorescent DNA fixed onto the plate is proportional to the mRNA amount used for hybridization, directly calculate the red/green fluorescence ratio. If this ratio is greater than 1 (red on the image), the gene expression is greater in the source 2; if this ratio is smaller than 1 (green on the image), the gene expression is greater in source 1. We can visualize and interpret these differences in expression using commercially available software (Fig. 7.6).

At the end of microarray analysis, we cluster the expression profiles obtained through arrays. Genes that share the same expression profile on several experiments gradually form clusters during phylogenetic analysis. Other techniques such as principal component analysis or neuronal networks are also now being used for microarray analysis and clustering. The final data is presented as hierarchical clustering, where each column represents the microarray data from one experiment and each row a specific gene (Chuaqui et al. 2002).

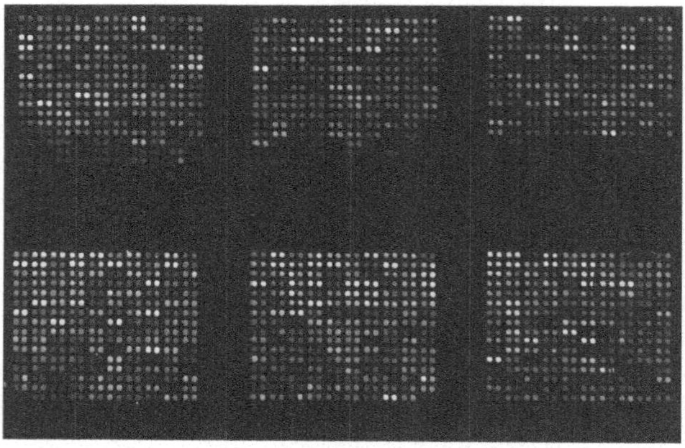

Fig. 7.6 A representative image of a cDNA microarray from barley. RNA from unstressed seedlings has been labeled with Cy3, while Cy5 has been used for labeling RNA extracted from seedlings which were osmotically stressed for 4 h (A. Pareek and H Bohnert, unpublished data)

7.7.3 Applications

1. A snapshot of the whole transcriptome of a system at a given time-point can be obtained since multiple genes can be analyzed simultaneously.
2. The whole genome can be used for expression analysis.
3. Gene expression studies can be performed for a subset of genes which are believed to be a part of the metabolic pathway.
4. Differential expression in the levels of gene(s) in the same organism under two different set of conditions or among two different organisms can be conducted.
5. Based on the coordinated expression profiles of a subset of genes, an inference can be obtained about their possible involvement in a metabolic pathway.

7.8 Conclusions

Transcriptome analysis is a very powerful tool in contemporary science. The above-mentioned techniques alone, or in combination with each other, can provide useful information for functional analysis of a gene or a group of genes. Additionally, high-throughput analysis of differential gene expression has proved to be a powerful tool for discovering novel genes through microarray analysis (Taniguchi et al. 2001). Changes in mRNA steady-state levels are mostly accomplished by changing the transcriptional rate of genes. Such fluctuations in relative mRNA amounts are indicative of changes in environment and developmental program or reflect responses to all kinds of stimuli. To properly understand a gene's function it is not only critical to know when, where, and to what extent a gene is expressed, it is also essential to discover other genes which are co-regulated with the gene of interest. Monitoring the transcriptome, i.e. the complement of all transcribed mRNAs of an organism, by measuring mRNA concentrations of defined genes in a multiparallel and quantitative way allows us to assign function to a multitude of unknown genes. Generally speaking, changes in mRNA abundance are related to changes in protein levels. Therefore, the information gathered from transcriptome analysis can easily be extrapolated to proteome analysis to gain additional information about certain biological processes at the physiological level. There are many examples in the literature where transcriptome analysis techniques have been applied and used in furthering our understanding of the complex aspects of molecular mechanisms of living organisms. In the present era of biotechnology, transcriptome analysis is at its peak to provide the solutions to all expected questions of bioscience.

Acknowledgements

The authors would like to gratefully acknowledge the funding received from the Department of Science and Technology, Department of Biotechnology, Government of India, The International Foundation for Science, Sweden, and The International Atomic Energy Agency, Austria.

References

Almeida Engler de J, Montagu Van M, Engler G (1994) Hybridization in situ of whole-mount messenger RNA in plants. Plant Mol Biol Rep 12:321–331

Arabidopsis Genome Initiative (2000) Analysis of the genome sequence of the flowering plant Arabidopsis thaliana. Nature 408:796–815

Bertucci F, Beranrd K, Loriod B, Chang YC, Granjeaud S, Birnbaum D, Ntuyen C, Peck K, Jordan B (1999) Sensitivity issues in DNA array-based expression measurements and performance of nylon microarrays for small samples. Hum Mol Genet 8:1715–1722

Butler D, Pockley P (2000) Monsanto makes rice genome public. Nature 404:534

Chelly J, Kahn A (1994) RT-PCR and mRNA quantitation. In: Mullis DB, Ferre F, Gibbs RA (eds) The polymerase chain reaction. Birkhauser, Boston, pp 97–109

Childs GV (1999) In situ hybridization with non-radioactive probes. In: Methods in molecular biology. Humana, Totowa, pp 131–141

Chuaqui RF, Bonner RF, Best CJ, Gillespie JW, Flaig MJ, Hewitt SM, Phillips JL, Krizman DB, Tangrea MA, Ahram M, Linehan WM, Knezevic V, Emmert-Buck MR (2002) Post-analysis follow-up and validation of microarray experiments. Nat Genet 32:509–514

Davenport RJ (2001) Rice genome. Syngenta finishes, consortium goes on. Science 291:807

Eisen MB, Brown PO (1999) DNA arrays for analysis of gene expression. Methods Enzymol 303:179–205

Eisen MB, Spellman PT, Brown PO, Botstein D (1998) Cluster analysis and display of genome-wide expression patterns. Proc Natl Acad Sci USA 95 :14863–14868

Fodor S (1997) Massively parallel genomics. Science 277:393

Jason W, O'Neill J W, Bier E (1994) Double-label in situ hybridization using biotin and digoxigenin-tagged RNA probes. BioTechniques 17:870–875

Lockhart DJ, Winzeler EA (2000) Genomics, gene expression and DNA arrays. Nature 405 :827–836

Sambrook J, Fritsch EF, Maniatis T (1989) Molecular cloning: a laboratory manual. Cold Spring Harbor Laboratory, Plainview

Schneeberger C, Speiser P, Kury F, Zeillinger R (1995) Quantitative detection of reverse transcriptase-PCR products by means of a novel and sensitive DNA stain. PCR Methods Appl 4:234–238

Sperisen P, Wang SM, Reichenbach, P, Nabholz M (1992) A PCR-based assay for reporter gene expression. PCR Methods Appl 1:164–170

Taniguchi M, Miura K, Iwao H, Yamanaka S (2001) Quantitative assessment of DNA microarrays – comparison with Northern blot analyses. Genomics 71:34–39

Tautz D, Pfeifle C (1989) A nonradioactive in situ hybridization method for the localization of specific RNAs in Drosophila embryos reveals translational control of the segmentation gene hunchback. Chromosoma 98:81–85

Trayhurn P, Duncan JS, Nestor A, Thomas MEA, Rayner DV (1994) Chemiluminescent detection of mRNAs on Northern blots with digoxigenin end-labeled oligonucleoides. Anal Biochem 222:224–230

Wilkinson DG (1994) In situ hybridization. A practical approach. IRL Press, Oxford

Yadetie F, Sandvik AK, Bergum H, Norsett K, Laegreid A (2004) Miniaturized fluorescent RNA dot blot method for rapid quantitation of gene expression. BMC Biotechnol 4:12

8 RNAi Technology: a Tool for Functional Validation of Novel Genes

R. Karan, S. Kumari, S.K. Yadav, and A. Pareek

8.1 Introduction

Until the past decade, mRNA was considered as a passive molecule serving only as a blueprint for the vast amount of information trapped in the highly significant biomolecule – DNA. But the more ebullient nature of RNA has come into the picture with recent discoveries about the possible role RNA can play in regulating gene expression. An emerging field of study involves the role of RNA in gene silencing. RNA mediates gene silencing either at the transcriptional level or post-transcriptional level. Recently, there has been a spurt of activity to study the intricacies of this phenomenon.

The first significant finding that paved the way for these studies came from a serendipitous discovery while attempting to enhance flower color in *Petunia*. In the year 1990, Napoli et al. were trying to engineer *Petunia* plants for increased anthocyanin production by the overexpression of chalcone synthase (chsA). Unexpectedly, instead of getting plants with higher anthocyanin content, variegated flowers with white patches were obtained (Napoli et al. 1990). It was later confirmed that the introduction of the chsA transgene led to the inhibition of endogenous gene expression. This phenomenon was termed "co-suppression" and several workers reported such instances independently. Further studies suggested that this phenomenon did not involve reduced transcription, rather degradation of the transcripts through a partial duplex mRNA was responsible for triggering this precise "gene silencing". Since this mechanism was involving post-transcriptional mRNA degradation, it was termed post-transcriptional

Ratna Karan, Sumita Kumari, Ashwani Pareek: School of Life Sciences,
Jawaharlal Nehru University, New Delhi, India,
email: Ashwani Pareek: ashwanip@mail.jnu.ac.in

Sudesh Kumar Yadav: Biotechnology Division,
Institute of Himalayan Bioresource Technology, Palampur-176061 (HP), India.

Soil Biology, Volume 11
Advanced Techniques in Soil Microbiology
A. Varma, R. Oelmüller (Eds.)

gene silencing in plants (in short, referred to as PTGS, Matzke and Matzke 1995). PTGS seems to serve a natural function of protecting the genome against mobile elements like viruses and transposons, orchestrating the functioning of the developmental programs of an organism. Entry of virus in a plant genome "pre-exposed" to a similar virus triggers the synthesis of dsRNA, which ultimately degrades the viral genome. Besides plants, homology-dependent gene silencing was found to occur commonly in fungal systems as well and these events were referred to as "quelling" (Cogoni et al. 1996).

In recent years, there has been an upsurge of information regarding the machinery involved in the RNAi mechanism and its potential role in functional genomics. In the present chapter, we provide details to understand how gene silencing works and how it can be employed as a tool of functional genomics to unravel the function of unknown genes. So far, three phenotypically different but mechanistically similar forms of RNAi have been reported which include "PTGS" or "co-suppression" in plants, "quelling" in fungi, and "RNAi" in the animal kingdom. For our purpose, we take the liberty of using all three terms interchangeably in the following text.

8.2 Machinery Involved in RNAi

Investigations forayed to decipher the outcome of RNAi have revealed its involvement in gene regulation. Both genetic and biochemical approaches have led to a greater understanding of the basis of silencing. The critical components involved in the processing of RNAi include inducer, Dicer, RNA-induced silencing complex (RISC), and RNA-dependent RNA polymerase (RdRp). All these components (details described in the following text) in coordination with other effector molecules work together in an organized fashion, resulting in silencing of the target gene (Fig. 8.1).

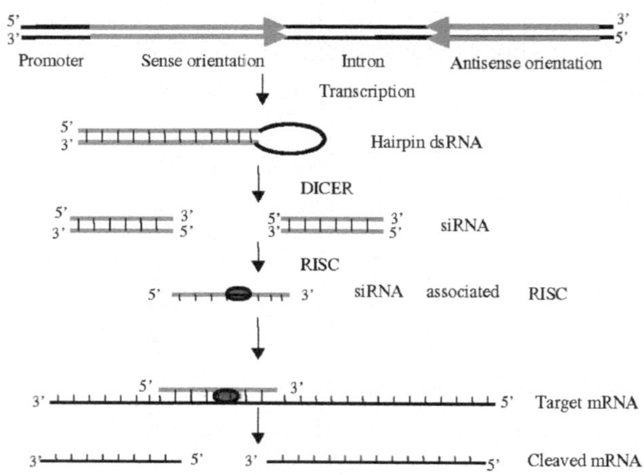

Fig. 8.1 Cartoon depicting the generalized RNAi-mediated gene silencing pathway

8.2.1 Inducer

A dsRNA produced either by a transgene introduced into an organism or by direct injection of dsRNA (as in the case of animal systems) homologous to target mRNA actually induces the onset of whole process. Such a RNA molecule is called an inducer (Palauqui and Balzergue 1999).

8.2.2 Dicer

Dicer is a double-stranded RNA-specific enzyme that belongs to the RNase III family of endonucleases. RNase III specifically cleaves double-stranded RNA (Robertson et al. 1967). RNase III-type enzymes – DROSHA (present in the nucleus), DICER (present in cytoplasm) in animals, and dicer-like (DCL) in plants – catalyze the processing of miRNA and siRNA precursors. Due to the ability to "dice" dsRNA molecule into equal pieces, it was termed "Dicer". Earlier, RNase III was regarded as unique to bacteria but biochemical studies on yeast RNA processing reactions and genome sequencing projects led to the identification of RNase III orthologs in fungi, plants, and animals, which established a RNase III superfamily (Rotondo and Frendewey 2001).

8.2.3 RNA-Dependent RNA Polymerase

RNA dependent RNA polymerase (RdRp) was first discovered in RNA viruses (Blumenthal and Carmichael 1979), which are responsible for transcription and replication of the viral genome. RdRp activity has also been detected in plants like Chinese cabbage (Astier-Manifacier and Cornuet 1971), tobacco (Duda et al. 1973), tomato (Boege and Sanger 1980), cucumber (Khan et al. 1986). RdRp amplifies the siRNA produced by the action of Dicer. This siRNA then systemically moves from one cell to another cell, so it has systemic inheritance from one part of a plant to another part of the same plant.

8.2.4 RNA-Induced Silencing Complex

Small RNAs associate with factors such as ARGONAUTE (AGO) proteins in effector complexes to guide target RNA cleavage, translational repression, or chromatin modification. This RNA-induced silencing complex (RISC) recognizes siRNA produced by the Dicer. It has both unwinding and endonuclease activity. The siRNA duplex containing ribonucleoprotein (RNPs) particles is subsequently rearranged into the RISC (Hammond et al. 2000). It mostly remains associated with the antisense strand of RNA (Nykanen et al. 2001). Finally, the RISC complex associated with the single-stranded antisense RNA identifies and pairs with its complementary homologous target mRNA which is to be degraded. Endonucleolytic cleavage of the target mRNA occurs at the center of the siRNA–mRNA hybrid (Elbashir et al. 2001).

8.2.5 miRNA and siRNA

According to their origin or function, broadly two types of naturally occurring small RNA have been described: short interfering RNAs (siRNAs), and micro RNAs (miRNAs). RNA-templated RNA polymerization, e.g. from viruses or hybridization of overlapping transcripts from repetitive sequences such as transgene arrays or transposons, gives rise to siRNAs which guide mRNA degradation or chromatin modification. In addition, endogenous transcripts that contain complementary or near-complementary 20- to 50-base pair inverted repeats fold back on them to form dsRNA hairpins. These dsRNAs are processed into miRNAs that mediate translational repression or mRNA degradation. This class of small RNAs, sharing mechanistic similarity to siRNA, but with characteristic differences, is called microRNA (miRNA) and was known long before the term siRNA was coined.

8.3 RNAi as a Tool of Functional Genomics

Much before the discovery of PTGS, antisense RNA technology was being used to achieve gene silencing. However, loss of function was never achieved; the expression of the targeted gene could be reduced to almost 70% of the original levels. With the available platform of discoveries made so far, a group of workers from the Carnegie Institution of Washington made an attempt to use double-stranded RNA to achieve gene silencing (Fire et al. 1998). Interestingly, the dsRNA was found to be substantially more effective at producing interference than either sense or antisense strands individually. This was the first ever endeavor to obtain gene silencing through artificially induced dsRNA and this phenomena was termed "RNA interference" (RNAi).

In the post-genome-sequencing era, the ever-burgeoning repository of genome sequences has now focused the impetus to validate the functions of all of these predicted genes. "Loss-of-function" mutants have been the most preferred tools to study functional aspects of genes but the conventional methods to obtain such mutants, viz. homologous recombination and random mutagenesis, are tedious and have met little success so far. Currently, PTGS is the most favored technique available for large-scale functional assays of genes (Baulcombe 1999; Vauchret et al. 2001; Waterhouse et al. 2001).

RNAi can be defined in simple terms as "homology-dependent sequence-specific degradation of mRNA that leads to gene silencing". Unlike antisense RNA technology, it has been found to be more potent and efficacious in complete knockdown of a particular gene against which a double-stranded RNA is produced. A double-stranded RNA is introduced or induced in an organism.

8.3.1 Production of dsRNA

Besides the naturally occurring dsRNAs, a dsRNA targeted against a specific gene can also be induced artificially which is recognized by the enzymatic machinery present inside the cell and finally leads to the breakdown of the mRNA complementary to the antisense strand of siRNA associated with RISC complex. dsRNA can be synthesized in vitro and then introduced into the cell; vector-based dsRNA production can also be achieved in the plant cell in vivo. ihRNA (hairpin RNA) is the common choice in the plants for PTGS. Based on its highest potency, dsRNAs can be produced in cells by one of the possible mechanisms as described below:

8.3.1.1 Two Independent Complementary Transcripts

In this method, the selected fragment of gene is cloned in sense orientation and antisense orientation as independent expression cassettes in separate vectors. Both cassettes are transfected in cells simultaneously where cloned fragments in the two cassettes are expressed separately and lead to the production of long dsRNA, which initiates the RNAi pathway inside the transfected cell (Fig. 8.2). This strategy is mainly used in animal cells that lack the interferon response, such as embryonic cells and somatic cells.

8.3.1.2 Single Transcript with Inverted Repeat

In this case, the selected DNA fragment of gene is cloned in sense and antisense orientation in a single construct flanking an intron that expresses and produces

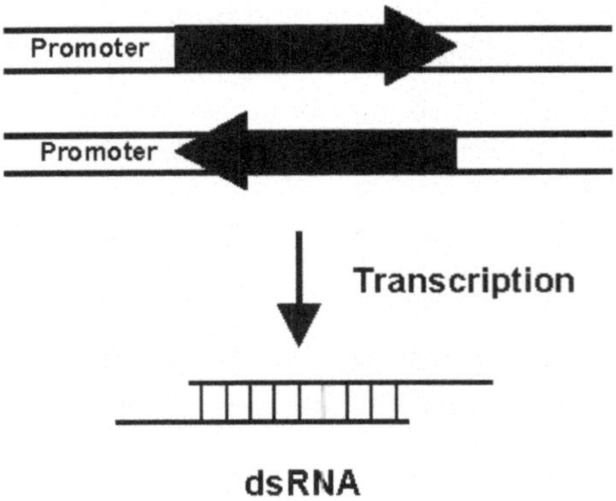

Fig. 8.2 Production of dsRNA from two independent complementary transcripts. *Large arrows* depict the direction of functional mRNA synthesis

hairpin dsRNA (Fig. 8.3) that are efficiently processed by Dicer. This strategy is employed in both plant and animal cells.

8.3.2 Constitutive and Inducible RNAi

Constitutive expression of dsRNA may cause deleterious effects on the developmental stages of organism if the product of that particular gene is required for metabolism during development. Generally, CaMV35S promoter-driven dsRNA production is utilized for unraveling the functions of genes and also to check the efficiency of newly designed RNAi vectors. Constitutive gene silencing cannot be used with genes involved in fundamental processes such as embryo viability. To overcome the deleterious effects of constitutive dsRNA production, there is a need to develop the construct, which could only produce dsRNA during specific developmental stages, in a tissue-specific manner. Now, inducible expression of RNAi cassettes are available in which the inhibition of expression of the desired gene can be achieved at the desired period during growth and development. The induction of RNAi cassettes is being done by applying ethanol (Caddick et al. 1998), estradiol (Guo et al. 2003), and dexamethasone (Wielopolska et al. 2005). Several vector systems have been developed so far which have enabled both constitutive as well as inducible gene silencing in plant systems (Table 8.1).

An inducible RNAi system should work when there is a need to silence the gene so that unwanted gene silencing could be avoided. The inducible RNAi

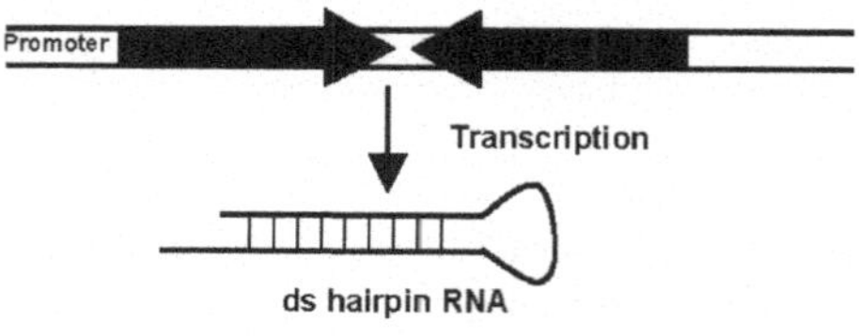

Fig. 8.3 Production of dsRNA from single transcripts with inverted repeat. *Large arrows* depict the direction of functional mRNA synthesis

Table 8.1 Different types of RNAi vectors

RNAi Vector	Promoter	Reference
pHELLESGATE 12 (gateway vector)	Constitutive (CaMV35S)	CSIRO, Australia
pANDA (gateway vector)	Constitutive (ubiquitin)	Miki and Shimamoto (2004)
PX7-RNAi	Estradiol (irreversibly inducible)	Guo et al. (2003)
POpOff1	Dexamethasone (reversibly inducible)	Wielopolska et al. (2005)

system should be reversible in nature to "switch on" and "switch off" the dsRNA production whenever the silencing of a gene is desired. Inducible promoters provide an alternative approach for temporal and spatial gene expression control (Table 8.2).

Thereafter, RNA interference can be employed successfully for gene functional analysis by reverse genetic approaches.

Table 8.2 Constitutive/inducible promoters used for regulating RNAi/antisense gene constructs

Gene	Constitutive/inducible promoter	Transgenic plant	RNAi/antisense gene	Reference
Arabidopsis beta-amylase (*BMY8*) gene	CaMV35S	*Arabidopsis*	RNAi	Kaplan and Guy (2005)
A. thaliana chromatin-remodeling protein 11 (CHR11), CaMV35S promoter	CaMV35S	*Arabidopsis*	RNAi	Huanca-Mamani et al. (2005)
Meloidogyne incognita dual oxidases (peroxidase and NADPH oxidase) gene	CaMV35S	*Meloidogyne incognita*	RNAi	Bakhetia et al. (2005)
A. thaliana phytoene desaturase gene, driven by heat-shock gene promoter (*HSP18.2*)	Heat inducible	*Arabidopsis*	RNAi	Masclaux et al. (2004)

Table 8.2 *(continued)* Constitutive/inducible promoters used for regulating RNAi/antisense gene constructs

Gene	Constitutive/inducible promoter	Transgenic plant	RNAi/antisense gene	Reference
Torenia hybrida, chalcone synthase (*CHS*)	CaMV35S	*Torenia hybrida*	RNAi	Fukusaki et al. (2004)
Arabidopsis putative vacuolar sorting receptor (atbp80)	CaMV35S	*Arabidopsis*	Antisense	Laval et al. (2003)
Rice, metallothionein gene (*OsMT2b*)	CaMV35S	Rice	RNAi	Wong et al. (2004)

8.3.3 Antisense RNA and RNAi

Antisense RNA is a type of RNA molecule, which is complementary to a specific mRNA. Antisense RNA can be produced by cloning the gene of interest in antisense orientation relative to the promoter. It is relatively easy and inexpensive to produce antisense RNA rather than dsRNA. However, antisense RNA has variable efficacy and specificity in comparison with RNAi because antisense RNA hybridizes with its corresponding mRNA and inhibits protein synthesis transiently whereas in the case of RNAi the corresponding mRNA is cleaved which leads to relatively more intense gene silencing effects rather than by antisense RNA.

8.4 Potential Areas of Application

As the repository of information about thousands of genes has increased over the years, unraveling processes unknown, the quest for knowing more of it has certainly increased. In this post-genomic era, a plethora of sequence information provides a platform to expedite the process of ascribing functions to genes. Functional genomics has come to rescue to satiate this quest. Until now homologous recombination was used for underexpression studies, which unfortunately claimed valuable time and money. Chemical mutagenesis and T-DNA insertions have also been the method of choice for study of loss of function in plant systems. However, associated shortcomings limit their successful applications.

These methods require a large population to screen the mutants and often more than one generation to select a suitable mutant. A mutation may not even show up for the gene of interest and the function of gene under study may remain unknown. There may also be a combination of mutations, thus making it difficult to decipher the effect of knockdown. Thereafter, antisense technology came up as a new promising area but overtime experiments carried out to achieve silencing have posed questions on its efficiency. Complete knockdown is often difficult to obtain through antisense technology, limiting the success to 60–70% only. At this point in time, discovery of RNAi emerged as a savior to achieve efficient gene silencing, paving the way for delineating the functions of unknown genes. Ongoing experiments to dig out more of this mechanism have postulated several biological roles for this process, the most evident being the ability to elicit a defense response in plant systems against viruses (Voinnet 2001). This homology-dependent sequence-specific phenomenon lowers the titer of invading viruses through an endogenous RNase-inducible mechanism leading to viral RNA degradation. (Goldbach et al. 2003). Interestingly, this natural biological phenomenon can be tamed effectively to generate a transient loss of function assays to assess gene function as a rapid alternative to stable transformation. By introducing host cDNA fragments within the viral genome, it is possible to redirect this mechanism to corresponding endogenous host mRNAs, thereby allowing down-regulation of host gene expression (Hein et al. 2005; Scofield et al. 2005). Studies in virus-induced PTGS have also revealed the involvement of viral suppression that interferes with PTGS. This provided a basis to look for such suppressors in other organisms. Studies in *Caenorhabditis elegans* indicate an increased activity of transposable elements in RNAi-defective mutants. RNAi may have a role in maintaining the genome stability, although not much has been worked out in this respect (Hannon 2002). dsRNA has also been observed to induce DNA methylation and chromatin remodeling (Wassenegger 2000). This provides further evidence for its active role in genome organization. Induction of dsRNAs to incite RNAi was used successfully in plants. However, applications in mammals did not yield favorable results as long as dsRNAs (>30 nt) induced a sequence non-specific interferon response, which in turn resulted in global inhibition of mRNA translation (Elbashir et al. 2001). To overcome hindrance, transfection of siRNA into mammalian cell lines was attempted and it was found to be efficient in silencing the endogenous genes (Dykxhoorn et al. 2003). Using siRNAs, a number of disease–related genes have been targeted efficiently, thus unveiling the therapeutic potential of this technique. A mutated allele for spinobulbular muscular atrophy (SBMA) was targeted through siRNA in human kidney 293T cells which resulted in decreased levels of mutated transcript along with reduced polyglutamine toxicity (Caplen et al. 2002). This opened a new area to be exposed for the treatment of diseases caused by mutated alleles. The successful specific inhibition of K-RAS V12 expression, an oncogene in human tumor cells, resulted in loss of anchorage-dependent growth and tumorogenicity through virus-mediated siRNA delivery. This unraveled the possibilities of tumor-specific gene therapy (Brummelkamp et al. 2002). The siRNA construct directed against HIV-1 rev mRNA (Lee et al. 2002) and HIV-1 co-receptor CCR5 (Qin et al. 2002) was found to be effective in reducing HIV-infected cells. These findings indicate that siRNA could be useful in antiviral strategies. siRNA can also be applied to whole animals by hydrodynamic delivery, resulting in gene silencing in various tissues (Lewis et al. 2002; McCaffrey et al. 2002). These findings offer a mere highlight of the tremendous potential that this technique holds in itself for the benefit of mankind and stills need to be explored in depth.

8.5 Conclusions

RNAi has come up as a major breakthrough in the field of molecular biology, providing altogether a new face to the unexplored nature of the enigmatic molecule "RNA". Beyond providing a better understanding of the interplay of several factors in gene regulation, this technique offers immense potential in the field of therapeutics, functional genomics, and molecular breeding.

Although this technique offers many credentials awaiting to be tapped, any decision regarding its application must not be impetuous, particularly when intended for human therapeutics. We must ascertain its existing limitations. Do humans really lack RdRp that can induce transitive silencing to exert potential side-effects and can we avoid the saturation of RISC and possibilities of site-directed mutagenesis? There are several questions that need to be addressed before the technique can be actually put to use.

Acknowledgements

Authors would like to thank UGC (R.K.) and CSIR (S.K.) for their research fellowships, and thank the International Foundation for Science, Sweden, the International Atomic Energy Agency, Vienna, and DBT, Government of India, for supporting the RNAi-related research in the laboratory.

References

Astier-Manifacier S, Cornuet P (1971) RNA-dependent RNA polymerase in Chinese cabbage. Biochim Biophys Acta 232:484–493

Bakhetia M, Charlton W, Atkinson HJ, McPherson MJ (2005) RNA interference of dual oxidase in the plant nematode *Meloidogyne incognita*. Mol Plant Microbe Interact 18:1099–1106

Baulcombe DC (1999) Fast forward genetics based on virus-induced gene silencing. Curr Opin Plant Biol 2:109–113

Blumenthal T, Carmichael GG (1979) RNA replication: function and structure of Qbeta-replicase. Annu Rev Biochem 48:525–548

Boege F, Sanger HL (1980) RNA dependent RNA polymerase from healthy tomato leaf tissue. FEBS Lett 12:91–96

Brummelkamp TR, Bernards R, Agami R (2002) A system for stable expression of short interfering RNAs in mammalian cells. Science 296:550–553

Caddick MX, Greenland AJ, Jepson I, Krause KP, Qu N, Ridell KV, Salter MG, Suhuch W, Sonnewald U, Tomsette AB (1998) An ethanol inducible gene switch for plants: manipulation of carbon metabolism. Nat Biotechnol 16:177–180

Caplen NJ, Taylor JP, Statham VS, Tanaka F, Fire A, Morgan RA (2002). Rescue of polyglutamine-mediated cytotoxicity by double stranded RNA mediated RNA interference. Hum Mol Genet 11:175–184

Cogoni C, Irelan JT, Schumacher M, Schmidhauser TJ, Selker EU, Macino G (1996) Transgene silencing of the al-1 gene in vegetative cells of *Neurospora* is mediated by a cytoplasmic effector and does not depend on DNA–DNA interactions or DNA methylation. EMBO J 15:3153–3163

Duda CT, Zaitlin M, Siegel A (1973) In vitro synthesis of double stranded RNA by an enzyme system isolated from tobacco leaves. Biochim Biophys Acta 319:62–71

Dykxhoorn DM, Novina CD, Sharp PA (2003) Killing the messenger: short RNAs that silence gene expression. Nat Rev Mol Cell Biol 4:457–467

Elbashir SM, Lendeckel W, Tuschl T (2001) RNA interference is mediated by 21- and 22-nucleotide RNAs. Genes Dev 15:188–200

Fire A, Xu S, Montgomery MK, Kostas SA, Driver SE, Mello CC (1998) Potent and specific genetic interference by double-stranded RNA in *C. elegans*. Nature 391:806–810

Fukusaki E, Kawasaki K, Kajiyama S, An CI, Suzuki K, Tanaka Y, Kobayashi A (2004) Flower color modulations of *Torenia hybrida* by downregulation of chalcone synthase genes with RNA interference. J Biotechnol 111:229–240

Goldbach R, Bucher E, Prins M (2003) Resistance mechanisms to plant viruses: an overview. Virus Res 92:207–212

Guo HS, Fei JF, Xie Q, Chua NH (2003) A chemical regulated inducible RNAi system in plants. Plant J 34:383–392

Hammond SM, Bernstein E, Beach D, Hannon GJ (2000) An RNA-directed nuclease mediates post-transcriptional gene silencing in *Drosophila* cells. Nature 404:293–296

Hannon GJ (2002) RNA interference. Nature 418:244–251

Hein I, Barciszewska-Pacak M, Hrubikova K, Williamson S, Dinesen M, Soenderby IE, Sundar S, Jarmolowski A, Shirasu K, Lacomme C (2005) Virus-induced gene silencing-based functional characterization of genes associated with powdery mildew resistance in barley. Plant Physiol 138:2155–2164

Huanca-Mamani W, Garcia-Aguilar M, Leon-Martinez G, Grossniklaus U, Vielle-Calzada JP (2005) CHR11, a chromatin remodeling factor essential for nuclear proliferation during female gametogenesis in *Arabidopsis thaliana*. Proc Natl Acad Sci USA 102:17231–17236

Kaplan F, Guy CL (2005) RNA interference of *Arabidopsis* beta-amylase8 prevents maltose accumulation upon cold shock and increases sensitivity of PSII photochemical efficiency to freezing stress. Plant J 44:730–743

Khan ZA, Hiriyanna KT, Chavez F, Fraenkel-Conrat H (1986) RNA-directed RNA polymerases from healthy and from virus-infected cucumber. Proc Natl Acad Sci USA 83:2383–2386

Laval V, Masclaux F, Serin A, Carriere M, Roldan C, Devic M, Pont-Lezica RF, Galaud JP (2003) Seed germination is blocked in *Arabidopsis* putative vacuolar sorting receptor (atbp80) antisense transformants. J Exp Bot 54:213–221

Lee NS, Dohjima T, Bauer G, Li H, Li MJ, Ehsani A, Salvaterra P, Rossi J (2002) Expression of small interfering RNAs targeted against HIV 1 rev transcripts in human cells. Nat Biotechnol 20:500–505

Lewis DL, Hagstrom JE, Loomis AG, Wolff JA, Herweijer H (2002) Efficient delivery of siRNA for inhibition of gene expression in postnatal mice. Nat Genet 32:107–108

Masclaux F, Charpenteau M, Takahashi T, Pont-Lezica R, Galaud JP (2004) Gene silencing using a heat-inducible RNAi system in *Arabidopsis*. Biochem Biophys Res Commun 321:364–369

Matzke MA, Matzke AJM (1995) How and why do plants inactivate homologous transgenes? Plant Physiol 107:679–685

McCaffrey AP, Meuse L, Pham TT, Conklin DS, Hannon GJ, Kay MA (2002) RNA interference in adult mice. Nature 418:38–39

Miki D, Shimamoto K (2004) Simple RNAi vectors for stable and transient suppression of gene function in rice. Plant Cell Physiol 45L:490–495

Napoli C, Lemieux C, Jorgensen R (1990) Introduction of a chimeric chalcone synthase gene into petunia results in reversible co-suppression of homologous genes in trans. Plant Cell 2:279–289

Nykanen A, Haley B, Zamore PD (2001) ATP requirements and small interfering RNA structure in the RNA interference pathway. Cell 107:309–321

Palauqui JC, Balzergue S (1999) Activation of systemic acquired silencing by localised introduction of DNA. Curr Biol 9:59–66

Qin XF, An DS, Chen IS, Baltimore D (2002) Inhibiting HIV 1 infection in human T cells by antiviral mediated delivery of small interfering RNA against CCR 5. Proc Natl Acad Sci USA 100:183–188

Robertson HD, Webster RE, Zinder ND (1967) A nuclease specific for double stranded RNA. Virology 12:718–719

Rotondo G, Frendewey D (2001) Pac I ribonucleases of *Schizosaccharomyces pombe*. Methods Enzymol 342:168–193

Scofield SR, Huang L, Brand AS, Gill BS (2005) Development of a virus-induced gene-silencing system for hexaploid wheat and its use in functional analysis of the Lr21-mediated leaf rust resistance pathway. Plant Physiol 138:2165–2173

Vauchret H, Beclin C, Fagard M (2001) Post-transcriptional gene silencing in plants. J Cell Sci 114:3083–3091

Voinnet O (2001) RNA silencing as a plant immune system against viruses. Trends Genet 17:449–459

Wassenegger M (2000) RNA-directed DNA methylation. Plant Mol Biol 42:203–220

Waterhouse PM, Graham MW, Wang MB (2001) Virus resistance and gene silencing in plants is induced by double stranded RNA. Proc Natl Acad Sci USA 95:13959–13964

Wielopolska A, Townley H, Moore I, Waterhouse P, Helliwell C (2005) A high-throughput inducible RNAi vector for plants. Plant Biotechnol J 3:583–590

Wong HL, Sakamoto T, Kawasaki T, Umemura K, Shimamoto K (2004) Down-regulation of metallothionein, a reactive oxygen scavenger, by the small GTPase OsRac1 in rice. Plant Physiol 135:1447–1456

9 Molecular Matchmaking: Techniques for Biomolecular Interactions

R. Oberoi, P. Kumar, and S.K. Lal

9.1 Introduction

Protein interactions play pivotal roles in virtually all the cellular processes. They are intrinsic to every cellular process, ranging from DNA replication, transcription, splicing, and translation, to secretion, cell cycle control, signal transduction, metabolism, formation of cellular macrostructures, and enzymatic complexes. Thus the identification of protein–protein interactions remains fascinating and very helpful in understanding biological phenomena.

9.2 Tools for the Study of Protein–Protein Interactions

In recent years, the convergence of biochemistry, cellular, and molecular biology has made available a number of powerful techniques for studying such interactions. Together, these constitute an impressive collection of tools for studying interactions among proteins. These techniques vary in their sensitivity, efficiency, and rapidity, but judicial deployment of a combination of them has proved to be effective and reliable.

Two broad approaches are generally applied to the study of protein–protein interactions: experimental and computational. Computational methods (Valencia and Pazos 2002) are used to infer protein interaction networks and predict the function of proteins. When the molecular structure of two proteins is known, the molecular prediction (or docking problem) of protein interactions

Virology Group, International Centre for Genetic Engineering & Biotechnology (ICGEB), New Delhi, India, Tel: 91-11-26177357, Fax: 91-11-26162316, email: sunillal@icgeb.res.in

Soil Biology, Volume 11
Advanced Techniques in Soil Microbiology
A. Varma, R. Oelmüller (Eds.)

can be analyzed. Therefore, as more genomic, structural and protein interaction data become available, the ability to predict protein interactions in silico is strengthened. The experimental approaches include physical/biochemical, genetic and biophysical methods to select and detect proteins that bind another protein. Traditionally, the tools available to analyze protein–protein interactions in multicellular organisms have been restricted to biochemical (also referred to as physical methods) approaches. However, despite obvious advantages, biochemical approaches can be time-consuming. Biochemical methods that detect proteins that bind to other proteins generally result in the appearance of a band on a polyacrylamide gel. Under this category, protein affinity chromatography, affinity blotting, co-immunoprecipitation, far-westerns, cross-linking are popular techniques to detect proteins that interact with a known protein (Phizicky and Fields 1995). Certain spectroscopic techniques, including fluorescence polarization spectroscopy (FPS), surface plasmon resonance, and mass spectroscopy, are used for several cases of protein interactions. Biacore's surface plasmon resonance technology has become widely popular. This is a label-free technology for monitoring biomolecular interactions as they occur. It also uses spectroscopy to measure changes in molecular size. The instrument monitors changes in refractive index that occur at a liquid/metal interface when biomolecules interact. Several new fluorescent imaging-based biophysical techniques are also available for studying protein–protein interactions, such as fluorescence resonance energy transfer (FRET), bioluminescence resonance energy transfer (BRET), fluorescence correlation spectroscopy, and biomolecular fluorescence complementation (Boute et al. 2002). Other widely applicable methods are library-based methods. A variety of methods have been developed to screen large libraries for genes or fragments of genes whose products may interact with a protein of interest. As these methods are by their nature highly qualitative, the interactions identified must be subsequently confirmed by biochemical approaches. Library screens are generally performed in bacteria or yeasts, organisms with rapid doubling times. Thus, these procedures can be completed rapidly. Protein probing and phage display are common library screening techniques. Protein probing uses a labeled protein as a probe to screen an expression library in order to identify genes encoding interacting proteins. Since all combinations of protein–protein interactions are assayed, including those that might never occur in vivo, the possibility of identifying artifactual partners exists and is a typical disadvantage of most exhaustive screening procedures. A second drawback derives from the use of a bacterial host, where not all post-translational modifications needed for the interaction might occur. Despite obvious advantages, biochemical approaches can be tedious and time-consuming. Also coming along the pike is the application of microarrays and protein chips to protein–protein interactions (MacBeath and Schreiber 2000). All in vitro methods suffer from one common drawback, i.e., the genes encoding the interacting proteins are not readily available. An answer to this problem was the introduction of the yeast two-hybrid system by Fields and Song in 1989.

Currently, the yeast two-hybrid system is the most widely used genetic assay for the detection of protein–protein interactions (Fields and Sternglanz 1994; Fashena et al. 2000; Bartel and Fields, 1995). The yeast two-hybrid system has become popular because it requires little individual optimization and because, compared with conventional biochemical methods, the identification and characterization of protein–protein interactions can be completed in a relatively short time-span and is inexpensive. Most importantly, novel protein–protein interactions can be easily selected from a pool of potential interaction partners (e.g., a cDNA expression library; Gyuris et al. 1991; Chevray and Nathans 1992) and genetic systems not only yield information on the interaction itself but also directly provide the cDNA encoding the novel interaction partner. Furthermore, no previous knowledge about the interacting proteins is necessary for a screen to be performed. Since its conception, the two-hybrid system has become one of the most widely used experimental methods. The basic method is constantly being improved and widely used with a range of improvements and modifications to overcome drawbacks and limitations. It is no longer applicable to study only protein–protein interactions but has been extended to allow screening for DNA and RNA interactions, assaying interactions in the cytosol rather than being limited to the nucleus, and screening in bacterial or mammalian hosts.

9.2.1 The Two-Hybrid System

The classic two-hybrid assay exploits the modular nature of the yeast *Saccharomyces cerevisiae* transcriptional activator, GAL4, required for the expression of genes encoding enzymes for galactose utilization (Johnson 1987). GAL4 consists of two separable and functionally distinct essential domains: (a) the DNA binding domain (DBD; Keegan et al. 1986) which binds to specific DNA sequences [upstream activation sequences (UAS; Giniger et al. 1985)] in GAL4 responsive promoters, and (b) a transcription activation domain (TAD; Ma and Ptashne 1987) required for the transcriptional activation of the GAL4 responsive genes. Theoretically the two-hybrid principle is very straightforward. To study interaction between two proteins X and Y, protein X (the bait) is fused in-frame to DBD and protein Y (the prey) is fused to the TAD, where either hybrid protein alone fails to activate the transcription. The bait and prey fusions are co-expressed in yeast, where the interaction of proteins X and Y reconstitutes the proximity of GAL4 domains, reconstituting a functional transcription factor, and transcription of downstream reporter occurs. Commonly, auxotrophic markers that can be selected for are used in combination with the *lacZ* gene encoding the bacterial β-galactosidase. The common auxotrophic markers HIS3 and LEU2 allow the selection of interactions by monitoring growth on selective plates lacking histidine or leucine, respectively, whereas lacZ can be easily measured using a colorimetric assay.

9.2.2 The Split-Ubiquitin System

This is a genetic technique, based on the split-ubiquitin system (Johnsson and Varshavsky 1994a, b; Stagljar et al. 1998), which offers the advantage that it can be used to detect interactions between virtually any type of protein in the cell – that is, between two integral membrane proteins, between a membrane protein and a cytoplasmic protein, or between two cytoplasmic proteins, provided that one of them is artificially anchored to the membrane. To date, this system is the most widely used of the alternative yeast-based two-hybrid systems.

The split-ubiquitin system is an alternative assay for the in vivo analysis of protein interactions. The system pioneered/proposed by Johnsson and Varshavsky (1994a) was originally developed to detect interactions between soluble proteins and later modified to work with membrane proteins.

9.2.3 Reverse Two-Hybrid System

In this system, the conventional yeast two-hybrid system has been modified to allow genetic selection of events responsible for the dissociation of particular interactions, e.g., mutations, drugs, or competing proteins. For the reverse two-hybrid system, yeast strains are generated such that the expression of interacting hybrid proteins increases the expression of a counter-selectable marker that is toxic under particular conditions (negative selection; Vidal et al. 1996a). Under these conditions, dissociation of the interaction provides a selective advantage (as the counter-selectable marker is no longer expressed), thereby facilitating detection: a few growing yeast colonies in which hybrid proteins fail to interact can be identified among millions of non-growing colonies expressing interacting hybrid proteins. This system has a variety of uses. For example, mutations that prevent an interaction can be selected from large libraries of randomly generated alleles (Vidal et al. 1996b). Similarly, molecules that dissociate or prevent an interaction can be selected from large libraries of peptides or compounds.

9.2.4 Sos Recruitment System (Cyto Trap Yeast Two-Hybrid System)

This system was developed by Aronheim et al. (1994, 1997). It is another modification of the yeast two-hybrid system to bypass the reconstitution of transcription factor and takes advantage of a cell proliferation signaling pathway. In this system, the protein–protein interactions are artificially tethered to yeast cell membranes. Interaction is detected by activation of the Ras signal transduction cascade by localizing a signal pathway component, human Sos (h-Sos), to its site of activation in the yeast plasma membrane.

9.2.5 Yeast One-Hybrid System

The one-hybrid system is an extension, by simplification, of the two-hybrid concept. The yeast one-hybrid or single hybrid system is a genetic system to identify DNA binding proteins. It provides a genetic screen to identify cDNAs encoding polypeptides that bind short sequences (motifs) of DNA, usually *cis*-acting regulatory elements of expressed genes (Li and Herskowitz 1993; Inouye et al. 1994). In this method also, the bipartite structure of the yeast transcription factor GAL4 is exploited. Each cDNA in the library being explored is expressed as a fusion protein with the activation domain of the GAL4 protein. This fusion protein interacts directly with a DNA binding site/target element and transactivates reporter genes (*HIS3*, *lacZ*). The usual upstream activating sequences (within the promoters of these reporter genes) in the yeast two-hybrid systems are replaced by the target DNA motif. This motif is introduced in multiple copies to provide increased sensitivity to the screen.

9.2.6 Double Interaction Screen

Yu et al. (1999) developed the double interaction screen (DIS) to identify partners of DNA binding transcription factors. DIS is a modification that combines yeast two-hybrid and one-hybrid screens, used to identify partners of DNA binding transcription factors. As in the one-hybrid screen, a *cis*-acting regulatory element is cloned upstream of reporter genes *lacZ* and *HIS3*. This DNA motif is known to be a direct target of the transcription factor (TF) in question, i.e., protein X, and also contains binding sites for other transcription factors whose activities are independent of protein X. Thus, two baits are available in the screen, the *cis*-regulatory element itself, [which is used in the first screen to "anchor" a native full length TF (protein X) to DNA upstream of reporter gene] and X anchored to the regulatory element via native binding sites. Next, screening of the cDNA library allows identification of three types of proteins: (a) DNA binding proteins that interact directly with the regulatory element, (b) protein bait partners that also bind to specific DNA sequences, and (c) protein bait partners that interact only at the protein level.

9.2.7 Yeast Three-Hybrid or Tri-Hybrid System

Different cellular mechanisms often involve interactions between more than two proteins. The three-hybrid system is based on the reconstitution of a transcriptional activator complex either to search for or to study a protein that interacts with two others, providing information about ternary complexes. The technique detects either direct or mediated interactions between two fusion proteins. As in the yeast two-hybrid system, one protein is a fusion with DBD (that is DBD-X) and the other with the AD (that is AD-Y) of the GAL4 proteins. Different variations that involve third partners as native proteins, in the absence of any fused domains, are referred to as "tribrid" systems. The third protein can act either as a bridging factor (it interacts with both X and Y, which alone do not interact with each other), a stabilizing factor (it promotes/induces/strengthens the weakly interacting proteins X and Y), or as a regulating factor (it post-translationally modifies X and/or Y in order for them to interact, and in this case it may not necessarily be part of the reconstituted transcriptional activator). In either case, the third partner allows transcriptional activator formation and stimulates reporter gene transcription by the reconstituted transcription factor. Hence, the interaction between X and Y is mediated by the third protein. Another utility of the three-hybrid system is that, if X and Y interact and reconstitute the transcription factor, the system can be used to search for inhibitors. The three-hybrid system actually encompasses a range of different systems to study RNA–protein, small organic ligand–receptor or protein–protein interactions, which all have in common the basic principle of the two-hybrid systems but are mediated by a third partner. These third partners are quite diverse, from proteins to small molecules and nucleic acids.

9.3 Procedure

1. Take 50 µl of freshly grown appropriate yeast reporter strain. Inoculate into a 250-ml baffled flask containing 100 ml of YPD. Place on shaker at 30 °C with shaking (150 rpm) overnight.
2. Check cell density of $1–4\times10^7$ using a spectrophotometer ($OD_{600} = 1.00$).
3. Transfer cells into two 50-ml sterile falcon tubes and centrifuge at 3000 rpm for 2 min at room temperature.
4. Resuspend the cell pellet with 10 ml of Lithium acetate (LiAc) solution, centrifuge at 3000 rpm for 5 min, and discard the supernatant.
5. Resuspend cells in 500 µl of LiAc solution with gentle shaking and store tubes in ice until further use.

6. Take 100 μl of cells in a sterile micro centrifuge tube, add 10 μl of plasmid DNA, mix well, and incubate at room temperature for 5 min.
7. Add 280 μl of PEG 3350 solution and mix by inverting the tube 4–6 times.
8. Incubate at 30 °C for 45 min.
9. Add 43 μl of DMSO and mix by inverting the tube 4–6 times.
10. Heat shock at 42 °C for 5 min, chill on ice for 1–2 min.
11. Centrifuge at 4000 rpm for 1 min at room temperature and resuspend cells in 0.1 ml of sterile H_2O.
12. Spread plate transformation mix on selective media plates and incubate at 30 °C for 3 nights.
13. Pick the largest colonies and restreak them on the same selection medium for master plates. Plates sealed with parafilm may be stored at 4 °C for 3–4 weeks.

9.3.1 Reagents, Materials, and Equipment

Regular molecular biology laboratory equipment, like microcentrifuge, incubator, water bath, and a laminar hood.

9.3.1.1 Reagents and Materials

YPD or the appropriate SD liquid medium, sterile 1xTE/LiAc (prepare immediately prior to use from 10× stocks), sterile 1.5-ml micro centrifuge tubes for the transformation, appropriate SD agar plates (100-mm plates), appropriate plasmid DNA in solution, appropriate yeast reporter strain for making competent cells, Herring Testes carrier DNA (10 mg/ml; denature the carrier DNA by placing it in boiling water for 20 min and immediately cool it on ice), sterile 40–50% PEG-LiAc solution (make PEG solution in 1× 0.1 M LiAc), 10× TE buffer (0.1 M Tris-HCl, 10 mM EDTA, pH 7.5, autoclaved), 0.1 M LiAc, 100% DMSO, glass spreader to spread cells on plates.

9.3.1.2 Composition of Reagents

1. YPD medium: yeast extract (1 g/100 ml), peptone (2 g/100 ml), dextrose (2 g/100 ml).

2. YPD plates: yeast extract (1 g/100 ml), peptone (2 g/100 ml), dextrose (2 g/100 ml), agar (2 g/100 ml).
3. LiAc solution: 0.1 M LiAc (0.1 g/10 ml), 10 mM Tris-HCl (pH 8.0), 1 mM EDTA (50 μl/10 ml).
4. 50% PEG 3350 solution: 50% PEG 3350 in LiAc solution.
5. 10× dropout (SD) LT^-: YNB (1.87 g/250 ml), dextrose (5.0 g/250 ml), agar 5.0 g/250 ml, amino acid mixture* (25 ml/250 ml), H_2O (225 ml), histidine (500 μl).
6. 10× dropout (SD) LTH^-: YNB (1.87 g/250 ml), dextrose (5.0 g/250 ml), agar (5.0 g/250 ml), amino acid mixture* (25 ml/250 ml), H_2O (225 ml).
7. 10× TE pH 8.0: 10 mM Tris-HCl (6.0578 g), 1 mM EDTA (1.8612 g).
   *Amino acid mixture: L-isoleucine (300 mg/l), L-valine (1500 mg/l), L-adenine hemisulfate (200 mg/l), L-arginine HCl (200 mg/l), L-lysine HCl (300 mg/l), L-methionine (200 mg/l), L-phenylalanine (500 mg/l), L-threonine (2000 mg/l), L-tyrosine (300 mg/l), L-uracil (200 mg/l).

9.3.2 Notes and Points to Watch

- For the highest transformation efficiency, use the competent cells within 1 h of their preparation.
- Prepare the media plates in advance and allow them to dry at room temperature for 2–3 days.
- To obtain even growth on plates, continue to spread the transformation mix over the agar surface until all liquid has been absorbed.
- Calf thymus DNA is not recommended as carrier DNA.

References

Aronheim A, Engelberg D, Li N, al-Alawi N, Schlessinger J, Karin M (1994) Membrane targeting of the nucleotide exchange factor Sos is sufficient for activating the Ras signaling pathway. Cell 78:949–961

Aronheim A, Zandi E, Hennemann H, Elledge SJ, Karin M (1997) Isolation of an AP-1 repressor by a novel method for detecting protein–protein interactions. Mol Cell Biol 17:3094–3102

Bartel PLS, Fields S (1995) Analyzing protein–protein interactions using two-hybrid system. Methods Enzymol 254:241–263

Boute N, Pernet K, Issad T (2002) The use of resonance energy transfer in high throughput screening: BRET versus FRET. Trends Pharmacol Sci 23:351–354

Chevray PM, Nathans D (1992) Protein interaction cloning in yeast: identification of mammalian proteins that react with the leucine zipper of Jun. Proc Natl Acad Sci USA 89:5789–5793

Fashena SJ, Serebriiskii I, Golemis EA (2000) The continued evolution of two-hybrid screening approaches in yeast: how to outwit different preys with different baits. Gene 250:1–14

Fields S, Song OK (1989) A novel genetic system to detect protein–protein interactions. Nature 340:245–246

Fields S, Sternglanz R (1994) The two-hybrid system: an assay for protein–protein interactions. Trends Genet 10:286–292

Giniger E, Varnum SM, Ptashne M (1985) Specific DNA binding of GAL4, a positive regulatory protein of yeast. Cell 40:767–774

Gyuris J, Dencso L, Polyak K, Duda E (1991) Complex interaction of yeast nuclear proteins with the enhancer/promoter region of SV40. Curr Genet 20:359–363

Inouye C, Remondelli P, Karin M, Elledge S (1994) Isolation of cDNA encoding a metal response element binding protein using a novel expression cloning procedure: the one hybrid system. DNA Cell Biol 13:731–742

Johnson M (1987) A model fungal gene regulatory mechanism: the GAL genes of *Saccharomyces cerevisiae*. Microbiol Rev 51:458–476

Johnsson N, Varshavsky A (1994a) Split ubiquitin as a sensor of protein interactions in vivo. Proc Natl Acad Sci USA 91:10340–10344

Johnsson N, Varshavsky A (1994b) Ubiquitin-assisted dissection of protein transport across membranes. EMBO J 13:2686–2698

Keegan L, Gill G, Ptashne M (1986) Separation of the DNA binding from the transcriptional activation function of a eukaryotic regulatory protein. Science 231:699–704

Li JJ, Herskowitz I (1993) Isolation of ORC6, a component of the yeast origin of recognition complex by a one-hybrid system. Science 262:1870–1874

Ma J, Ptashne M (1987) A new class of yeast transcriptional activators. Cell 51:113–119

MacBeath G, Schreiber SL (2000) Printing proteins as microarrays for high-throughput function determination. Science 289:1760–1763

Phizicky EM, Fields S (1995) Protein–protein interactions: methods for detection and analysis. Microbiol Rev 59:94–123

Stagljar I, Korostensky C, Johnsson N, te Heesen S (1998) A genetic system based on split-ubiquitin for the analysis of interactions between membrane proteins in vivo. Proc Natl Acad Sci USA 95:5187–5192

Valencia A, Pazos F (2002) Computational methods for prediction of protein interactions. Curr Opin Struct Biol 12:368–373

Vidal M, Brachmann RK, Fattaey A, Harlow E, Boeke JD (1996a) Reverse two-hybrid and one-hybrid systems to detect dissociation of protein–protein and protein–DNA interactions. Proc Natl Acad Sci USA 93:10315–10326

Vidal M, Braun P, Chen E, Boeke JD, Harlow E (1996b) Genetic characterization of a mammalian protein–protein interaction domain by using a yeast reverse two-hybrid system. Proc Natl Acad Sci USA 93:10321–10326

Yu Y, Yussa M, Song J, Hirsch J, Pick L (1999) A double interaction screen identifies positive and negative *ftz* regulators and Ftz-interacting proteins. Mech Dev 83:95–105

10 Environmental Proteomics: Extraction and Identification of Protein in Soil

Z. Solaiman, M. A. Kashem and I. Matsumoto

10.1 Introduction

Proteomics involves the systematic study of proteins in order to provide a comprehensive view of the structure, function and regulation of biological systems. Advances in instrumentation and methodologies have fueled an expansion of the scope of biological studies from simple biochemical analysis of single proteins to measurements of complex protein mixtures. Proteomics is rapidly becoming an essential component of biological research such as health, environmental and agricultural sciences. Environmental proteomics concerns the study of proteins and peptides found in water, sediment, soils, etc. Coupled with advances in bioinformatics, the proteomics approach to comprehensively describing biological systems will undoubtedly have a major impact on our understanding of the microbes, soil and protein interactions. It has the potential to improve our knowledge further on function, cellular localization, post-translation modification and the source of proteins found in environmental samples. Proteomics complements genomics (i.e. nucleic acid-based) approaches to study microbial diversity and functions.

Initially, proteomics focused on the generation of protein maps using two-dimensional polyacrylamide gel electrophoresis. The field has since expanded to include not only protein expression profiling, but also the analysis of post-translational modifications and protein–protein interactions. Protein expression, or the quantitative measurement of the global levels of proteins, may still be done with two-dimensional gels; however, mass spectrometry has been incorporated

Zakaria Solaiman: School of Earth and Environmental Sciences, The University of Adelaide, SA 5005, Australia; Present address: School of Earth and Geographical Sciences, The University of Western Australia, Crawley, WA 6009, Australia, email: solaiman@cyllene.uwa.edu.au

Mohammed Abul Kashem and Izuru Matsumoto: Proteomics Laboratory of Pathology, School of Medical Science, Blackburn Building (D06), University of Sydney, NSW 2006, Australia

Soil Biology, Volume 11
Advanced Techniques in Soil Microbiology
A. Varma, R. Oelmüller (Eds.)

to increase sensitivity, specificity and to provide results in a high-throughput format. A variety of platforms are available to conduct protein expression studies and this site provides links to these resources. In order to identification and characterization the protein component of a given sample, a number of technologies could be utilized. In this chapter we highlight some critical points to give an outline of this technique for soil proteomics studies.

Proteins are released into the soil environment after the death and disruption of the cells of organisms, or as extracellular enzymes, which are excreted by a number of microorganisms (Skujinš 1976). These are also exuded from plant roots (Brenner et al. 1998). Although the extracellular protein present in soil is quickly become decomposed into small polypeptide fragments by indigenous soil microbes, some portion is considered to be resistant to microbial decomposition by binding with clay mineral and humic substances (Boyd and Mortland 1990). Soils are known to contain a wide variety of cell-free enzymes (Skujinš 1976) that display considerable stability (Zantua and Bremner 1977). These enzymes have been recognized by indirect enzyme assay of soil solution or soil extract, but there is a scarcity of research on the extracellular enzyme/protein molecules measured rather than enzyme activity (Murase et al. 2003). In earlier research it was shown that mineralizable organic nitrogen can be extracted from soils by a neutral phosphate buffer solution (Matsumoto et al. 2000). Recently, Ogunseitan (2006) outlined soil proteomics study in detail.

10.2 Sample Preparations

Soil samples should be fresh or stored at −80 °C for a short period of time on extraction. Protein extraction can be performed directly or indirectly from soil samples. Direct extraction can be performed by bead beating, sonication, vortex or chemical lysis. The indirect method involves isolation of microbial cells before protein extraction.

Protein solubilization and cell lysis are key factors for effective analysis. In general, samples should be lysed before submission. Precipitated samples must be solubilized in resuspension solution and clarified by centrifugation if necessary. The solution or supernatant is then analyzed. Contaminants such as polysaccharide, phenolic compounds, nucleic acid, lipids and insoluble material should be removed from the sample prior to submission.

Precautions should be taken to reduce the keratin contamination because this is the primary limitation to the sensitivity of protein identification. Protein precipitation with trichloroacetic acid may interfere with isoelectric focusing. Proteins should never be heated in the presence of urea. The reason is that cyanates, which accumulate in urea solutions, carbamylate the primary amino groups on proteins at elevated temperatures, producing species with altered isoelectric point, changed susceptibility to proteases and altered mass.

10.3 Protocols for Protein Extraction from Soil

Very few extraction methods have been developed to extract protein from soil.

10.3.1 Extraction of Extracellular Protein

This extraction method is after Murase et al. (2003).

1. Mix 100 g soil with 300 ml of 67 mM phosphate buffer (pH 6.0), consisting of $Na_2HPO_4.12H_2O$ (2.38 g/l) and KH_2PO_4 (8.17 g/l) and mildly shaken at 25 °C for 1 h.
2. Filter soil extract through no. 6 filter paper (Advantec Toyo, Tokyo) and then passed through 0.2 μm pore size cellulose acetate membrane filter to remove bacterial cells in 50 ml centrifuge tube.
3. Add TCA to a final concentration of 5%, keep filtrate at 4 °C for at least 12 h.
4. Centrifuge at 3400 rpm (2100 *g*) for 30 min to remove the TCA soluble component.
5. Discard supernatant, wash TCA insoluble fraction with ethanol into a 1.5-ml microtube and centrifuge at 1200 rpm (11 000 g) for 2 min.
6. Resuspend pellet in ethanol using sonication, centrifuge and discard supernatant.
7. Resuspend pellet in diethyl ether, centrifuge and discard supernatant.
8. Dry pellet in a vacuum desicator.
9. Dissolve pellet in a 20 μl of sample buffer for direct analysis or store at –20 °C.

10.3.2 Extraction of Whole-Cell Protein

The method of total protein extraction from soil outlined here comes from Singleton et al. (2003); and it was originally based on modifications and developments on the method proposed by Ogunseitan (1993).

1. Take 1 g of soil (50% WHC) in an Eppendorf cetrifuge tube (1.5 ml).
2. Add 100 μl of protease inhibitor cocktail (Sigma P 2714).
3. Add 1 ml of extraction buffer.
4. Mix for 10 s on a vortex mixer.
5. Do 4 cycles of snap freezing in liquid nitrogen and thawing to 25 °C.
6. Centrifuge at 20 000 rcf for 15 min at 4 °C.
7. Take about 600 μl clear supernatant.
8. Measure protein concentration in extract by Bradford dye protein reagent (Biorad).

9. Perform SDS-PAGE (polyacrylamide gel electrophoresis) analysis after adding size marker.

10.4 Protein Loading

The protein concentration in the sample is important for effective loading. Several methods are used for protein assay. The Bradford method is more convenient but has a few limitations. However, for doing final experiments, it should be better to optimize the protein loading through gel running. If you are "fishing" for proteins or working with low-abundance protein, it is better to load more protein. You can rehydrated a maximum of 200 μl of solution per strip, therefore, a high concentration of protein in the sample is better. Precipitated protein should be re-dissolved in buffer to gain as high a concentration as possible.

10.5 Protein Expression Analyses

To characterize protein expression differences among species, you can run one dimension polyacrylamide gel electrophoresis (PAGE) or high-resolution 2D-PAGE of whole-cell protein extract may be necessary.

10.5.1 SDS-PAGE

Dissolve protein in 10% acrylamide for separation in SDS-PAGE with a size marker and then stain with Coomassie brilliant blue R250 (CBB-R) or with silver staining, after electrophoresis (Fig. 10.1).

10.5.2 Two-Dimension SDS-PAGE Analysis

Two-dimension SDS-PAGE analysis is performed in steps. The procedures stated elsewhere (Biorad; Proteome System Ltd.; Kashem et al. 2007) are described in detail below:

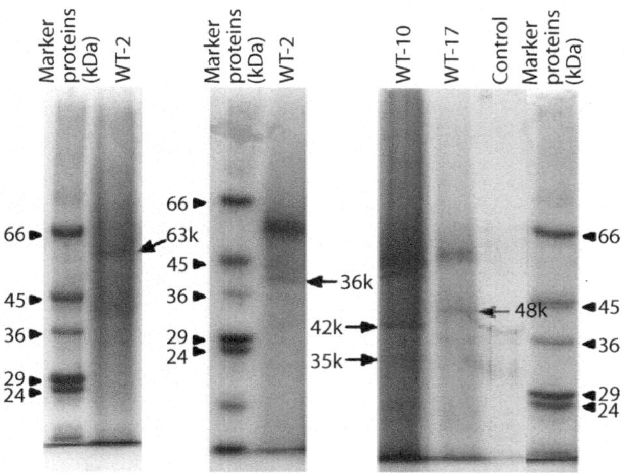

Fig. 10.1 Electrophoresis of extracellular proteins extracted from greenhouse soil. Proteins marked with *arrows* were subjected to N-terminal amino acid sequencing. The amount of each sample applied was equivalent to 75 g of soil, except WT-2 (equivalent to 7.5 g of soil). The control was prepared from 67 mM phosphate buffer. Protein markers (MW-SDS-200; Sigma) were loaded at about 9 mg in all. WT-1, WT-2, WT-10 and WT-17 are different soils. This figure was reproduced from Murase et al. (2003) with permission from the author as well as from Elsevier Science Ltd

10.5.2.1 First Dimension

Isoelectric focusing (IEF) represents the first dimension. Each sample protein applied to an IPG strip migrates to its isoelectric point (pI), the point at which its net charge is zero. Different pH ranges and sizes of IPG are available in the market.

Features and benefits of IPG strips:

1. Narrow- and wide-range strips with overlap options, in three lengths, allow optimal resolution of most protein samples.
2. Control in manufacturing ensures reproducible performance.
3. IPG strips reduce preparation time and reduce reagent waste.
4. Strips are labeled for polarity to ensure proper orientation.

If you are "fishing" for proteins, then it is best to start with a 3–10 strip. The problem with the 3–10 strip is that pretty much all proteins fall in that range, so you have a lot of spots overlapping. Once you know the range you are interested in, you could use a strip with a higher resolution (lower range), i.e. the gel size remains the same (7, 11, or 18 cm) but the resolution is much greater, so you can focus on just the proteins which fall in a 4–7 range or 5–6 range. In our experience, most of the metabolic proteins are found within pH 4–8 and polymer-containing samples present some difficulty for separation of protein in the IPG gel. Strip rehydration is allowed to take place for at least 6 h and air bubbles trapped beneath the strip should be removed. Electro focusing is carried out for 100–10 000 V for 8 h and further run for 9 h at a constant 10 000 V, depending on strip length and the presence of salts and detergent in the original sample.

Protocol for isoelectric focusing:

1. Rehydrate IPGs in a disposable Dry Strip tray with 200 μl of extract and 2 μl of orange tracking dye for 6 h – gel-side down, remove backing tape and any air bubbles beneath the gel.
2. Assemble IEF – damp the wicks, center the IPG strips under a covering fluid (paraffin oil). Ensure no air bubbles are trapped under the strips. Make sure +pH end is near the anode and that strip gel is in contact with the wicks.
3. Run 1D on ElectrophoretIQ3 – first phase with increasing voltage protocol, second phase with maximum voltage (i.e. 100–10 000 V for 8 h, then 10 000 V constant for further 9 h).
4. Next morning, drain paraffin oil into waste bottle, blot underside of strips and place in Dry Strip tray channels.
5. Equilibrate the IPGs for 2× 10 min on shaker in equilibration buffer (6 M urea, 2% SDS, 50 mM Tris-acetate pH 7, bromophenol blue).

10.5.2.2
Second Dimension

This technique has become a core technology in proteomics applications since the introduction of 2D electrophoresis 30 years ago. Currently, 2D electrophoresis is one of the preferred analytical techniques used to resolve and separate hundreds to thousands of proteins and protein isoforms. The first dimension separates proteins based on their inherent isoelectric point (pI). The second dimension is mass-driven, separating the focused proteins on the basis of molecular weight through the use of a denaturing polyacrylamide gel electrophoresis (Fig. 10.2).

When preparing protein extracts for isoelectric focusing, it is best to avoid solutions with high ionic strength and ionic detergents such as SDS. High salt and detergent content interfere with the initial phase of 2D electrophoresis and proteins do not separate or focus properly. Also, during sample preparation, the removal of nucleic acids and/or cellular debris improves protein separation and decrease background interference for visualization.

Protocol after Kashem et al. (2007) for 2D SDS-PAGE:

1. Wash 6–15% GelChips with MilliQ water and then with running buffer.
2. Fill top of gel with running buffer to aid placement of IPG.
3. Fill bottom tank with running buffer.
4. Place gels in tank. Equilibrated IPGs are slotted into the recess of 6–15% GelChips and pressed firmly against the top of the SDS gel with a thin spatula (ensure plastic backing is against long glass plate).

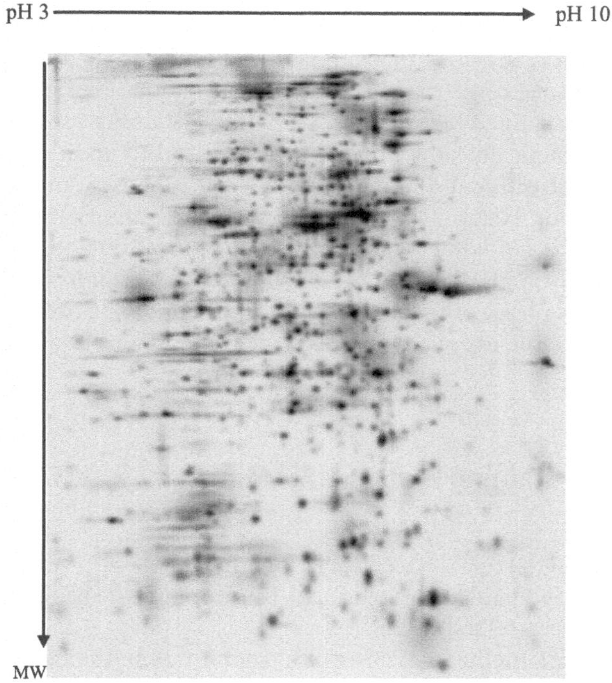

Fig. 10.2 Two-dimensional (2D) SDS-PAGE obtained from environmental samples. Protein bands were separated based on their iso-electric point as well as molecular weight (*MW*)

5. Place molecular weight marker equidistant from the end of the +pH strip and the end of the gel.
6. Ensure rubber gasket has not slipped (stop voltage leaking).
7. Place blanks in any unused slots (normally on side slots).
8. Fill top tank with buffer up to line.
9. Run gels according to default setting on ElectrophoretIQ3 – until BPB front has migrated to the bottom of the gel. One voltage phase should be used.

10.6. Gel Staining

Dyes are used to detect proteins following electrophoresis, and the intensity of staining provides a measure of protein abundance. Precautions should be taken after running the gel, such as using clean dishes and freshly made stain solutions for staining the gels to prevent contamination from keratin, dust, saliva or any other proteins you are using in your laboratory. In general, we use a Coomassie blue stain. If you see a faint protein band using this stain, then you can use silver staining. Silver stain is more sensitive than Coomassie blue but may render proteins impervious to mass spectrometry (MS). However, if you use silver stain, the gels have to be totally de-stained before digestion. The de-staining process for silver stained gels is a kinetically slow process and may lead to additional protein loss during repeated treatment. If staining with silver is chosen, please do not use any cross-linking for fixation (such as glutaraldehyde fixation). If you are preparing the samples, it is far better to pool your sample together and run them on a single lane to get the highest concentration effect and to get it to stain by a colloidal blue stain. Also, please note that extreme caution has to be used to avoid contamination with keratin, especially for low-level protein samples.

10.6.1 Coomassie Brilliant Blue Staining Protocol (For Mini Gels)

1. Fix gel in 100 ml of 46% methanol, 7% acetic acid for 1 h.
2. Stain gel in 100 ml of 46% methanol, 7% acetic acid, 0.1% Coomassie brilliant blue R-250 (filter this before use) for 1 h.
3. Destain gel in 100 ml of 5% methanol, 7.5% acetic acid for 24 h. Replace if needed.
4. Store the gel in 1–2% acetic acid in clean sealed sample tubes at 4 °C.

10.6.2 Silver Staining Protocol

Silver staining is a procedure used to detect low levels of biological compounds such as DNA and proteins in an immobilized medium, e.g. polyacrylamide gel (Okaley et al. 1980). Silver stain can be used with both DNA and proteins. It is generally used to detect levels of compounds that are present in very small quantities. Silver staining is more sensitive (0.1 ng protein per band) than traditional Coomassie blue method (50–100 ng protein per band).

The reducers, potassium ferricyanide (30 mM) and sodium thiosulfate (100 mM), should be made fresh. Mix the two reducers in a 1:1 ratio and immediately add the reducers to cover the gel pieces.

Once the silver brown color disappears, remove the reducers and wash with water until the gel piece is clear [note: incubation after washing with water in ammonium bicarbonate (100 mM) will speed this process].

Silver-stained gels are usually stored in 1% acetic acid at 4 °C. The residual acetic acid should be removed by thoroughly rinsing the gel with water before destaining. Make sure that the gel piece is clear before proceeding with digestion.

10.7 Image Analysis

After scanning the gel, the images are analyzed by computer-based software program. The analysis of sets of 2DE images currently forms a bottleneck in the proteomics research pipeline. A single wide-range pH gel can resolve over 3000 separate spots, many of which correspond to individual protein species. The number of identifiable spots from the same sample can be analyzed by the software. Many commercial image analysis packages have been developed to analyze 2D images. We have used Phoretix TM 2D expression software. These programs facilitate the generation of statistical data concerning proteins that have been identified as differentially expressed. Image analysis programs are employed with the view to ascertaining differential protein expression in the visualized proteome for comparative samples.

10.8 Spot Cut

When you acquire an image of your gel, please take special care not to allow your gel to contact any contaminated surfaces during the process. When you cut out the bands of interest, be sure to use extremely clean surfaces and new scalpels for band excision. Ideally this should be done in a laminar flow hood to minimize contamination from dust, hair, skin flakes, dirt, etc. Even trace amounts of such contaminants usually contain keratins in much larger amounts than the proteins present in the gel bands of interest. Therefore, such contaminants can cause the failure of attempts to characterize the proteins. Once cut, gel bands can be stored frozen in water or 1% acetic acid in clean, sealed sample tubes. Blank bands from the same gel are very helpful for measuring the background and trypsin peak.

10.9 Protein Digestion

Proteins of interest are excised either manually, or with a Spot Picker. Proteins are denatured, reduced and alkylated before digestion with trypsin overnight. In-gel digestion is performed with sequencing-grade, modified trypsin supplied frozen by Promega Corp. Trypsin is made up by dissolving in 20 μg of trypsin in 200 μl of bicarbonate buffer. This 0.10 μg/μl solution can be used according the protein content in the sample in a 1:30 ratio (enzyme:substrate by weight). This is an approximate value for the trypsin catalyst. Peptides released from gel plugs are then extracted, purified using C18 ZipTips from Millipore Corp. and spotted onto MALDI targets for mass spectrometry. These operations may be performed either manually or with a spotting robot. If there is sufficient amount of protein, it can be spotted directly from the dilute solution onto the MALDI target with matrix. If there is not enough sample, preconcentrate by drying the sample down in a Speedvac to a smaller volume. However, this also increases the salt and/or urea concentration and make it difficult to see the ions directly by MALDI. You may have urea crystals crashing out at the bottom of the tube. You can stop the drying down process when there is still some (~50 μl of liquid) left in the centrifuge tube. This sample can be taken (containing the peptides) and zip-tipped to get rid of the extra urea. Most of the urea is probably crashed out at the bottom of the tube. Care should be taken about the contamination of karatin or other proteins.

10.10 Mass Spectrometry Analysis

There are several methods for submitting proteins to identification, but the most powerful to date is mass spectrometry (MS). Proteins are sent into a pair of tandem MS devices (MS/MS). The proteins are sorted and groups of proteins of similar mass to charge ratio (m/z) are sent to be ionized and characterized to determine the identiy of each protein. This process is automated so that thousands of proteins can be identified for each experiment. Several mass spectrometers that can be used for proteomics including the Agilent MSD ion trap SL, Thermo Finnigan LTQ, Thermo Finnigan Deca, Applied Biosystems Voyager-STR-DE MALDI-TOF MS, Micromass MALDI and Micromass Q-TOF. We use Qstar XL Excell Hybride MS system (AB applied Biosystem). The several thousand tandem mass spectra obtained from a sample also contain the trypsin as well as gel spectra. The trypsin/gel spectra should be removed from the samples and all the sample's spectra calibrated, using at least two major spectra of trypsin.

10.11 Spectral Analysis

The application of certain constraints, such as mass accuracy limits of the instrument, narrowing down taxonomic category (such as microbes, human, plant, rat, etc.), specifying modifications on residues (oxidation, propionamide, biotin k, phosphorylation), peptide tolerance (low tolerance is better), spectra area (monoisotopic peak >2000), etc., helps make the search more efficient. There are many spectra that do not result in a successful identification due to the poor quality of the fragmentation pattern achieved. Sometimes a poor fragmentation is due to the charge state of the ion (>3+ or 1+), specific sequence of an ion or simply poor sensitivity. The spectra obtained from MALDI-TOF are searched against the predicted fragment ions from the trypsin digestion of proteins contained in a database such as NCBI, using mascot (http://www.matrixscience.com/).

10.12 N-Terminal Amino Acid Sequencing

Alternative techniques can be used according to Murase et al. (2003) after electrophoresis in 1D SDS-PAGE as follows:

1. For sequencing of N-terminal amino acids, separate proteins by SDS-PAGE before elctroblotting onto a polyvinylidene difluoride (PVDF membrane) in a blotting apparatus.
2. After electrophoresis, place the gel between a sheet of PVDF membrane and several sheets of filter paper (CB-09A type; Atto), all of which are soaked with blotting buffer (0.38 g/l SDS, 2.92 g/l glycine, 5.82 gl Tris), in a blotting apparatus and electroblot proteins at a constant current of 100 mA for 1 h.
3. Wash PVDF membrane with deionized water, stain with 0.1% CBB-R in 50% methanol and 10% acetic acid for 2 min, then destain in a solution of 45% methanol and 7% acetic acid until the protein bands become clear. After washing with deionized water, dry the membrane in air and store at –20 °C until use.
4. Cut out the protein band on the PVDF membrane with a clean razor and then analyze by a sequenator.
5. Search homology using obtained sequence from the database.

10.13 Conclusions

Proteomics is a method can be used to investigate the functions of microbes indigenous to soils. It is a culture-independent technique and it could explore the ways for analysis of microbial community and functional relationships in studying soil microbiology.

Acknowledgements

We thank Dr. Murase as well as Elsevier Science Ltd. for giving us permission to use some of the figures and text.

References

Boyd SA, Mortland MM (1990) Enzyme interaction with clays and clay-organic matter complexes. In: Bollag JM, Stotzky G (eds) Soil biochemistry, vol 6. Dekker, New York, pp 1–28

Brenner ED, Lambert KN, Kaloshian I, Williamson VM (1998) Characterization of LeMir, a root-knot nematode-induced gene in tomato with an encoded product secreted from the root. Plant Physiol 118:237–247

Kashem MA, James G, Harper C, Wilce P, Matsumoto I (2007) Differential protein expression in the corpus callosum (splenium) of human alcoholics: A proteomics study. Neurochem Int 50: 450-459

Matsumoto S, Ae N, Yamagata M (2000) Extraction of mineralizable organic nitrogen from soils by a neutral phosphate buffer solution. Soil Biol Biochem 32:1293–1299

Murase A, Yoneda M, Ueno R, Yonebashi K (2003) Isolation of extracellular protein from greenhouse soil. Soil Biol Biochem 35:733–736

Ogunseitan OA (1993) Direct extraction of proteins from environmental samples. J Microbiol Methods 17:272–281

Ogunseitan OA (2006) Soil proteomics: extraction and analysis of proteins from soils. In: Nannipieri P, Smalla K (eds) Nucleic acids and proteins in soil. Springer, Berlin Heidelberg New York, pp 95–116

Okaley BR, Kirsch DR, Morris NR (1980) A simplified ultrasensitive silver stain for detecting proteins in polyacrylamide gels. Anal Biochem 105:361–363

Singleton I, Merrington G, Colvan S, Delahunty JS (2003) The potential of soil protein-based methods to indicate metal contamination. Appl Soil Ecol 23:25–32

Skujinš J (1976) Extracellular enzymes in soil. CRC Critical Reviews in Microbiology 4:383–421

Zantua MI, Bremner JM (1977) Stability of urease in soils. Soil Biol Biochem 9:135–140

11 DGGE and RISA Protocols for Microbial Community Analysis in Soil

Z. Solaiman and P. Marschner

11.1 Introduction

Soil microorganisms are pivotal for nutrient cycling and maintenance of soil health. Interactions between different species of the microbial community are important for ecosystem functioning. Traditional microbiological techniques have often failed to describe these interactions and may therefore be inadequate in detecting perturbations within soil microbial communities because 99% of soil microorganisms are not culturable (Schwieger and Tebbe 1997). Several culture-independent methods have been developed for the assessment of microbial community structure and identification of species within the community. Most common are methods that rely on extraction of DNA from soil and subsequent characterization of DNA sequences. Several protocols for extraction of soil DNA suitable for further molecular analysis have been devloped, among them those which are described in this chapter: direct extraction of DNA from soil, PCR amplification of rRNA genes, followed by DNA sequence analysis by denaturant gradient gel electrophoresis (DGGE) or ribosomal intergenic spacer analysis (RISA; Borneman 1999; Van Elsas and Wolters 1995). We also use fatty acid methyl esters (FAME) techniques to study microbial community analysis. The protocols of DNA-based techniques are described in detail in this chapter, the FAME method is described in Chapter 12 in this book.

Zakaria Solaiman: School of Earth and Environmental Sciences, The University of Adelaide, SA 5005, Australia; Present address: School of Earth and Geographical Sciences, The University of Western Australia, Crawley, WA 6009, Australia, email: solaiman@cyllene.uwa.edu.au

Petra Marschner: School of Earth and Environmental Sciences, The University of Adelaide, DP 636, SA 5005, Australia

Soil Biology, Volume 11
Advanced Techniques in Soil Microbiology
A. Varma, R. Oelmüller (Eds.)

11.2 Soil DNA Extraction

11.2.1 Equipment

Fast Prep cell disruptor/bead-beater (BIORAD) and centrifuge

11.2.2 Chemicals

1. Phosphate buffer (PB) at pH 7.2 (autoclaved) preparation 200 mM (Table 11.1). Dissolve either $NaH_2PO_4.H_2O$, or $Na_2HPO_4.H_2O$, or Na_2HPO_4 in deionized water and add slowly NaH_2PO_4 solution to adjust the pH to 7.2.
2. Polyvinyl pyrrolodine (PVP) solution (autoclaved) or polyvinyl polypyrrolodine (PVPP) powder. Dissolve 100 mg PVP per milliliter of phosphate buffer or use PVPP as powder.
3. 3 M $CaCl_2$ (autoclaved). Dissolve 333 g $CaCl_2$ per liter of Milli-Q water.
4. 20% SDS. Dissolve 20 g SDS in 100 ml Milli-Q by slowly heating the suspension. If crystalized, reheat carefully for a few minutes before use to dissolve SDS.
5. Binding matrix Q-BIOgene (BIO 101)
6. Guanidine thiocyanide (2.75 M) solution. To 200 ml of MilliQ water slowly add 322.25 g of guanidine thiocyanate while mixing continuously. Mix until the salt is completely dissolved. Add 3.35 g sodium citrate and mix until completely dissolved. Add water to bring to final volume of 1 l. Filter through a Whatman No. 42 filter paper.
7. Wash buffer. Mix Tris-HCl (10 mM), EDTA (0.5 mM) and NaCl (5 mM). Then dilute this mixture with ethanol (>95%) at 1:1 ratio.

11.2.3 DNA Extraction Protocol

This protocol is modified after Wechter et al. (2003). The DNA is liberated from the microbial cells by homogenization with glass beads in presence of a phosphate buffer and SDS (surfactant). After centrifugation the supernatant containing DNA and proteins is transferred into a fresh tube. Proteins are then removed by precipitation. The DNA is bound to a silica matrix and is washed twice with an ethanol–salt buffer to remove humic substances and other contaminants. For extracts from some soils, additional washing with guanidine thiocyanide solu-

Table 11.1 Phosphate buffer (PB) preparation

Reagent	mg/mmol	g/100 ml
NaH_2PO_4	137.99	2.75 (for pH adjustment)
$NaH_2PO_4.H_20$	156.01	3.12
$Na_2HPO_4.H_20$	177.99	3.57
Na_2HPO_4	141.9	2.84

tion may be required to remove substances which inhibit the polymerase during PCR. Because of the high salt concentration the DNA remains bound to the silica matrix. The remaining ethanol must be removed completely, as it may inhibit the polymerase. The DNA is liberated from the silica matrix by adding ultrapure water to the dried silica matrix pellet.

1. Fill 2-ml screw cap tube with (0.1 mm) glass beads up to the first line (approx. 10 mm from the bottom). Then add 5–8 (2 mm) glass beads and 1 (40 mm) glass bead.
2. Weigh out soil (500 mg; or roots with adhering soil) and place into the tube.
3. Add 450 µl PB and 450 µl PVP and 2 µl of 3 M $CaCl_2$; tightly close cap and process in First Prep (30 s at speed 5.5); then centrifuge tubes at 14 000 rpm for 10 min.
4. Transfer supernatant in a 1.5-ml microcentrifuge tube, then add 400 µl PVP or 0.1 g PVPP and 30 µl 20% SDS.
5. Vortex for 5 s and then incubate at 4 °C for 5 min.
6. Centrifuge tubes at 14 000 rpm for 10 min.
7. Pour out supernatant in a 2.0-ml microcentrifuge tube and add 300 µl binding matrix (shake binding matrix before use).
8. Invert tubes by hand or place on shaker for 5 min.
9. Centrifuge 1 min at 14 000 rpm and discard supernatant.
10. Add 500 µl of 5.5 M guanidine thiocyanate, vortex and spin for 20 s and discard supernatant.
11. Add 500 µl of 5.5 M guanidine thiocyanate, vortex and spin for 20 s and discard supernatant.
12. Add 500 µl wash buffer, resuspend pellet by vortexing (bound DNA is washed).
13. Centrifuge at 14 000 rpm for 1 min and discard supernatant.
14. Add 500 µl SEWS (wash) buffer, resuspend pellet by vortexing.
15. Centrifuge at 14 000 rpm for 1 min and discard supernatant.
16. Centrifuge again at 14 000 rpm for 2 min (remove as much ethanol from the pellet as possible) and remove as much of the liquid as possible using a pipette.
17. Dry pellet for 10 min with the lid open (ethanol evaporates).
18. Add 100 µl sterile ultrapure water, resuspend pellet by vortexing (DNA is liberated from the binding matrix) and then centrifuge at 14 000 rpm for 2 min (DNA is in the supernatant).

19. Carefully remove as much of the supernatant as possible without getting any of the binding matrix into the DNA extract. Transfer DNA extract in new 1.5-ml tube (can be stored at −20 °C for several months). If the solution is not completely clear, centrifuge again and transfer clear supernatant to fresh tube. The extract should not contain any binding matrix as this may interfere with the PCR.

11.3 Polymerase Chain Reaction Protocol for DGGE

This polymerase chain reaction (PCR) protocol for DGGE (modified after Clark and Atkins 2004) is a process by which a target sequence is amplified one million-fold. The principle of PCR is that short oligonucleotide primers bind to the DNA flanking the target region. The primers serve as starting points for the polymerase enzyme, which amplifies the DNA region between the primers and increases the concentration of the target DNA region exponentially. DNA amplification is performed in a thermocycler.

For DGGE, PCR can be carried out using a wide range of primers, for example universal primers for bacteria (Muyzer et al. 1993), fungi (Vainio and Hantula 2000), actinomycetes (Heuer et al. 1997). Here we outline a DGGE protocol only for bacterial community structure. In our laboratory, PCR is routinely performed in 20–25 μl total volume containing 1 μl template DNA from 10× diluted samples (10–20 ng soil DNA), 2.0–2.5 μl of 10× PCR reaction buffer (with final $MgCl_2$ concentration 2.5–3.15 mM). But larger volumes can also be used. Diluted DNA extracts, e.g. 10×, 20×, or 50× may give better amplification than the undiluted extract because the concentration of inhibiting substances is reduced. Different dilutions should be tested and selection of the correct dilution should be based on the band intensity in an agarose gel or a DGGE gel. Our studies showed that dilutions up to 50× did not result in loss of bands in DGGE. Annealing temperatures and $MgCl_2$ concentration may have to be adjusted for each new primer set used (Clark and Atkins 2004). We outline nested PCR here, but this may not be necessary and it should be tested whether just the second round PCR yields sufficient product for DGGE.

11.3.1 First Round PCR

The bacterial primer set for the first round PCR (Weisburg et al. 1991):
fD1 (8–27): 5' CCGAATCGTCGACAACAGAGTTTGATCCTGGCTCAG 3'
rD1 (1361–1378): 5' CCCGGGATCCAAGCTTAAGGAGGTGATCCAGCC 3'

The PCR master mix (20 μl) is described in Table 11.2.

The cycling conditions are as follows: 93 °C for 5 min, then 95 °C for 30 s, then 55 °C for 30 s, for 34 cycles, followed by 72 °C for 2.30 min, then 72 °C for 5 min and finally hold at 4 °C constant.

11.3.2 GC Clamp 16S PCR (Second Round PCR)

The second round PCR is performed in 25 µl volume.

The primer set for the GC clamp 16S PCR (Muyzer et al. 1993):

GC clamped F 341: 5' CGCCCGCCGCGCGGCGGGCGGGGCGGGGGC-ACGGGGGGCCTACGGGAGGCAGCAG 3'

R 534: 5' ATTACCGCGGGTGCTGG 3'

The PCR master mix (25 µl) is described in Table 11.3.

The cycling conditions are as follows: 93 °C for 5 min, then 94 °C for 30 s, then 53 °C for 30 s, for 29 cycles, followed by 72 °C for 2 min, then 72 °C for 10 min and finally hold at 4 °C constant.

Table 11.2 First round PCR master mix

Reagent	Quantity
Water	11.3 µl
PCR buffer 10×	2.0 µl
dNTPs	2.0 µl (2 mM dNTPs)
$MgCl_2$	2.0 µl of 25 mM $MgCl_2$
Reverse primer rD1	0.8 µl of 5 pm/µl
Forward primer fD1	0.8 µl 5 pm/µl
Taq polymerase	0.08 µl of 5U/µl
Template DNA	1.0 µl (10–50 ng DNA)

Table 11.3 Second round PCR master mix

Reagent	Quantity
Water	11.9 µl
PCR buffer	2.5 µl
dNTPs	2.5 µl (2 mM dNTPs)
$MgCl_2$	2.5 µl of 25 mM $MgCl_2$
GC F 341	1.0 µl of 5 pm/µl
GC R 534	1.0 µl of 5 pm/µl
Taq polymerase	0.1µl of 5 units/µl
Template DNA	1.0 µl from fD1/rD1 PCR

The amplification success of DNA should be tested in an agarose gel after the first and the second PCR.

11.4 DGGE Techniques

DGGE was originally designed to detect single base mutations in DNA sequences. Muyzer et al. (1993) expanded the use of DGGE to assess the microbial community composition in environmental samples. A given section of the DNA or RNA is amplified; the resulting DNA sequence has the same length for all species, but the species differ in sequence. The DGGE gel has a linear gradient of increasing concentrations of denaturants (urea, formamide) which separates the sequences based on their stability towards the denaturants. The separation of different sequences during electrophoresis is related to the guanidine and cytosine(GC) content of the fragment. The stability of double-stranded DNA increases with GC content because there are three bonds between G and C compared with only two bonds between adenine and thymine (AT). Thus, sequences with a high GC content migrate further through the gel than low-GC sequences, resulting in a banding pattern that reflects the structure of the community being assessed. To prevent smearing of the bands, the 5' end of the forward primer contains a 35–40 base pair GC clamp to ensure that the two strands are not completely separated (Marschner et al. 2005; Kirk et al. 2004).

DGGE has advantages of being reliable, reproducible and rapid. After electrophoresis, bands can be excised and sequenced (e.g. Yang and Crowley 2000). However there are several limitations to the identification of species from DGGE bands: (i) one band may contain the sequences of several species and sequencing may or may not reveal the presence of all species in a band, (ii) fragments used in DGGE are typically relatively short (300–1000 bp) to give good band resolution and these fragments are often not long enough to identify organisms at a species level, and (iii) only abundant species can generate a band. In DGGE the detection limit for bacteria may be as high as 10^6 cells per gram of soil (Gelsomino et al. 1999).

The optimal gradient in DGGE needs to be tested for a given soil. The first DGGE should have a wide gradient, e.g. 15–70%. If most bands are found in a small proportion of the gel, e.g. over 35–55%, then this narrower gradient can be used for subsequent gels for better band resolution.

11.4.1 Equipment

Biorad DGGE kit.

11.4.2 Chemicals

1. Acrylamide solution: acrylamide-bis-acrylamide 37.5:1, 40%. Preferably use purchased acrylamide solution. If not available, make up own stock solution: acrylamide 38.93 g, bis-acrylamide 1.07 g, H_2O (ultra pure) to 100 ml.
2. Formamide: deionized formamide.
3. Denaturing solution, 0%: acrylamide 20 ml, 50× TAE buffer 2 ml, H_2O (ultra pure) to 100 ml.
4. Denaturing solution, 100%: urea 42 g, acrylamide 20 ml, 50× TAE buffer 2 ml, formamide 40 ml, H_2O (ultra pure) to 100 ml. This solution precipitates in the cold. To re-suspend, place bottle with a stirring bar in a beaker with water and put on stirrer with a hot plate. Stir slowly and heat up to max. 40 °C. Both 0% and 100% solutions can be stored at 4 °C for several weeks, but 1- to 2-week-old solutions work best.
5. TAE buffer, 50×: Tris base 242 g, concentrated acetic acid 57.1 ml, 0.5 M EDTA (pH 8.0) 100 ml, H_2O to 1000 ml. EDTA only dissolves in solutions with pH 8.0. Weigh EDTA then prepare 4 M NaOH and add 50 ml to the EDTA and stir. When it is dissolved, check pH and fill up to the final volume with deionized water. To make TAE buffer, weigh Tris into bottle, add concentrated acetic acid and EDTA solution and add approximately 600 ml deionized water. Stir until dissolved (takes about 1 h). Then adjust to final volume with deionized water.
6. TAE runnning buffer, 1×, buffer used in the DGGE tank: 50× TAE buffer 140 ml, H_2O (deionized) to 7000 ml. Needs to be freshly prepared for each run.
7. Dye preparation
   a. DGGE gel dye: mix bromophenol blue 50 mg, xylene cyanol 50 mg and 1× TAE buffer 10 ml.
   b. Sample loading dye: mix bromophenol blue solution (2%) 0.25 ml, xylene cyanol solution (2%) 0.25 ml, glycerol (99%) 7.0 ml and milli-Q water 2.5 ml.

11.4.3 Assembling the Gel Chamber

1. Clean glass plates and spacer with ethanol and wipe dry with a tissue paper.
2. Place the glass plates and spacer together with the sandwich clamps in the casting stand in which a rubber strip should be placed at the bottom to prevent leakage. Make sure that the glass plates are perfectly aligned at the bottom.

11.4.4 Casting the Gel

11.4.4.1 Mixing Solutions

Place all solutions except the 100% denaturing solution on ice to avoid premature polymerization, and then mix solutions in tubes also placed on ice.

Final volume of the solutions in each tube: 18 ml for 1 mm thick gel.

Volume of the 0% and 100% denaturing solution depends on desired gradient. For example:

18 ml of 35% solution = 11.5 ml of the 0% denaturing solution + 6.5 ml of the 100% denaturing solution.

18 ml of 55% solution = 8.1 ml of the 0% denaturing solution + 9.9 ml of the 100% denaturing solution

1. Add 160 μl of the DGGE gel dye to the solution with the higher denaturant concentration. Then mix. The gel dye is not necessary for the separation but allows visual verification of the gradient (dark blue where the denaturant concentration is highest, becoming lighter as the concentration of the denaturant decreases).
2. Add 160 μl of ammonium persulfate (10%; stored at –20 °C and thawed before use) to both solutions.
3. Add 16 μl of TEMED to both solutions.
4. Mix well.

11.4.4.2 Gel Casting Procedures

1. Gels should be cast immediately after preparation of the solutions because the solutions polymerize within 30–45 min after addition of ammonium persulfate and TEMED.
2. Load solutions into the syringes. Join the short tubings from the two syringes with the long tubing. Add a narrow gauge needle to the end of the long tubing and place needle between the two glass plates. Insert syringes into the gradient former. Then slowly turn the wheel of the gradient former so that there is a steady flow of solution but not too fast, otherwise the gradient will not be even.
3. Put in comb.
4. Cover gels with damp paper towels and leave gels to polymerize for at least 2 h or overnight at room temperature before loading sample.
5. Fill the DGGE tank with 1× TAE buffer (7 l). Turn on the heater of the DGGE tank and check that set temperature is 60 °C. It will take about 1 h for the buffer to reach that temperature.

11.4.5 Loading of the Samples

1. Mix by using 20 µl of PCR product (5 µl were used to verfiy the amplification in an agarose gel) samples with 8 µl of loading dye.
2. Pipette the entire sample in a pocket of the gel.
3. When all samples are loaded, put the lid back on the tank. Turn on the heater and the pump, wait for 1–2 min. Connect lid to the power supply and adjust to 150 V for 5 h or 70 V overnight (about 16 h). Shorter run times with a high voltage are also possible, but the resolution is often better when low voltage and longer run times are used.

11.4.6 Staining and Imaging of the Gels

There are several procedures for DNA staining, such as ethidium bromide, silver and Sybr gold/green. Ethidium bromide is the most commonly used staining procedure in molecular biology because it is inexpensive and the resolution is good. Silver staining is the most sensitive but it is more time-consuming than the other two staining methods and results in large amounts of toxic liquid waste. In our laboratory, we use Sybr gold for staining and it is relatively quick and reliable. The outline procedure is as follows (Marschner et al. 2005):

1. After running gel, remove the gel holder with the gels. Let the gels cool for about 5 min.
2. Remove gel with longer glass plate attached and put the gel facing up in a plastic staining tray.
3. Spread the staining solution (10 ml of 1× TAE buffer with 2 µl of SYBR gold) on the gel and use alignment card to distribute the solution evenly over the gel. Close the staining tray and incubate for 30–45 min in the dark.
4. After the incubation, visualize the banding under UV light and take a picture of the gel. Store the image for further processing.

11.5 Ribosomal Intergenic Spacer Analysis

RISA is also a DNA-based method for microbial community analysis. It involves PCR amplification of a region of the rRNA gene operon between the small (16S) and large (23S) subunits termed the intergenic spacer region. By using oligonucleotide primers targeted to conserved regions in the 16S and 23S genes, RISA fragments can be generated from the dominant bacteria in an environmental sample. While the majority of the rRNA operon serves a structural function, the taxonomic value of the intergenic spacer region lies in the significant heterogeneity in both length and nucleotide sequence. In RISA, we exploit the length of heterogeneity of the intergenic spacer region between 150 bp and 1500 bp with the majority of the intergenic spacer region lengths being between 150 bp and 500 bp (Fisher and Triplett 1999).

The RISA protocol for bacterial community analysis described here is based on Borneman and Triplett (1997) and Yin et al. (2000). RISA has also been used for fungal community analysis (Ranjard et al. 2001).

RISA has following features compared to DGGE:

1. The amplified region is longer (positions 23R to 1406F, i.e. 1383 bp).
2. No GC tail is attached to the forward primer.
3. The species differ in the length of the amplified region.
4. The species are separated according to the length of the region in an agarose or an acrylamide gel. The acrylamide gel may give a better resolution than the agarose gel.

In some cases, samples that cannot be amplified for DGGE can be amplified for RISA. This may be due to the fact that the primers for RISA do not contain a GC tail, which is more difficult to amplify.

11.5.1 Equipment

Polyacrylamide gel casting system

11.5.2 Chemicals

1. Denaturing loading buffer (10 ml): 95% formamide 9.5 ml, 20 mM EDTA (0.4 ml of 0.5 M stock solution), 0.05% bromphenol blue 5 mg, 0.05% xylene cyanol 5 mg.
2. TBE buffer (5×): Tris 54.0 g/l, boric acid 27.5 g/l, EDTA 3.72 g/l.

11.5.3 PCR Protocol

1. Primer set:
   1406F: TGY ACA CAC CGC CCG T

23R: GGG TTB CCC CAT TCR G
Note that Y, B, R and B stand for a random mix of the following bases. Y= C/T; B= G/C/T; R= A/G; B= G/C/T.

2. PCR master mix (in 25 µl total volume): H_2O 12.8 µl, PCR buffer 2.5 µl, dNTPs (2 nmol/µl) 3.0 µl, $MgCl_2$ 2.5 mM 2.5 µl ($MgCl_2$ solution is only necessary if polymerase buffer does not contain Mg), primer 1 (5 pmol/µl) 1.0 µl, primer 2 (5 pmol/µl) 1.0 µl, polymerase (5 U/µL) 0.25 µl, template DNA 2.0 µl.
3. PCR cycle: 94 °C for 5 min and then 34 cycles of 94 °C for 1 min, 52 °C for 1 min, 72 °C for 2 min, 72 °C for 10 min, then hold at 4 °C constant.

11.5.4 Gel Preparation and Loading

11.5.4.1 Preparation of the Acrylamide Gel

1. 5% polyacrylamide with a 37.5:1 acrylamide:bisacrylamide ratio containing 7 M urea: 40% solution of 37.5:1 acrylamide:bisacrylamide 12.5 ml, urea 42 g, 5× TBE buffer 20 ml, Milli-Q water to 100 ml.
2. For one gel (1 mm thick): 5% acrylamide solution (see above) 35 ml, Temed 30 µl, 10% APS 300 µl.

Gels can be cast by loading the acrylamide solution into a syringe and then dispensing the solution through a needle between the two glass plates. When the gel sandwich is filled, insert the comb. Allow to polymerize for a least 2 h. After removing the comb, rinse out the wells to remove precipitated urea. Pre-electrophorese gel at 60 °C for 30–45 min at 40 V.

11.5.4.2 Loading of the Samples

For RISA, the double-stranded DNA product from the PCR has to be denatured before the PCR product is loaded on the RISA gel. The denaturation is carried out by adding a denaturing buffer and heating the samples to 80 °C for 10 min. After denaturation, the samples should be placed on ice and loaded quickly on the gel to prevent re-annealing of the two DNA strands.

1. Precipitate PCR products with ethanol and resuspend in 25 µl TE buffer for purification of DNA. This step may not be necessary if the gels are not stained with silver.
2. Add 15 µl denaturing loading buffer.

3. Heat samples at 80 °C for 10 min, then place on ice for <10–15 min.

Load samples immediately onto a pre-heated and pre-run gel at 60 °C for 45 min at 40 V.

11.5.5 Gel Running

Run gel with 200 V for 5 h at 60 °C in a 1× TBE buffer.

11.5.6 Staining and Imaging of the Gels

As like as DGGE gel, RISA gel can be stained with ethidium bromide, silver and Sybr gold/green. We use Sybr gold for staining and it is relatively quick and reliable. The outline procedure is as follows (Marschner et al. 2005):

1. After running gel, remove the gel holder with the gels. Let the gels cool for about 5 min.
2. Remove gel with longer glass plate attached with and put the gel facing up in a plastic staining tray.
3. Spread the staining solution (10 ml of 1× TAE buffer with 2 µl of SYBR gold) on the gel and use alignment card to distribute the solution evenly over the gel. Close the staining tray and incubate for 30–45 min in the dark.
4. After the incubation, visualize the banding under UV light and take a picture of the gel. Store the image for further processing.

11.6 Data Analysis

Gel banding patterns generated by DGGE or RISA can be compared visually, but due to the often high number of bands per sample (between 10 and >30) this is unsatisfactory, particularly when many samples are compared. Digitization of the banding patterns allows subsequent analysis by a range of multivariate analysis techniques, such as principal component analysis or cluster analysis. When samples from different gels are to be compared, it is important to normalize the banding pattern both for band intensity and band position. Due to differences in staining intensity, and for DGGE, slight differences in gradient between gels, variation in banding pattern between samples only due to the fact that they are placed on different gels ("gel effect"). To normalize the band position, a standard should be included in every gel to be compared. This standard (containing no more than 5–10 bands) may be a mix of DNA of pure cultures or one sample with a few bright bands. The positions of the bands in the samples are then expressed relative to 3–4 bands in the standard. Normalization of band intensity can be carried out by dividing the intensity of each band in a given sample by the average intensity of the gel. We found that this normalization is quite effective, but does not completely remove the "gel effect". It is therefore important to: (a) place the samples randomly on the different gels, or (b) place the replicates of a given treatment on different gels.

11.7 Conclusions

With culture-independent methods we are starting to get a better picture of the enormous biodiversity of microorganisms in soils. New primers are constantly being developed, allowing the amplification of specific microbial groups or functional genes. However, there are a number of limitations to DNA-based methods: (a) due to amplification bias during PCR, the abundance of a DNA sequence in a sample and band intensity may not be directly related; (b) most methods can only detect the most abundant species/DNA sequences present and may therefore underestimate biodiversity. Despite these limitations, DNA-based methods such as DGGE and RISA are powerful tools that can provide insight into microbial community composition. Despite our better understanding of microbial diversity in the environment, it is intriguing that we currently do not know whether the most abundant species/DNA sequences are also the functionally most important ones. Moreover, the link between diversity and ecosystem function or sustainability is far from clear. Clearly, more research is needed and DNA-based methods will play an important role in such studies.

References

Borneman J (1999) Culture-independent identification of microorganisms that respond to specified stimuli. Appl Environ Microbiol 65:3398–3400

Borneman J, Triplett EW (1997) Molecular microbial diversity in soils from eastern Amazonia: evidence for unusual microorganisms and microbial population shifts associated with deforestation. Appl Environ Microbiol 63:2647–2653

Clark I, Atkins S (2004) Microbial environmental surveillance – a molecular biology manual. IACR, Rothamsted, pp 50

Fisher MM, Triplett EW (1999) Automated approach to ribosomal intergenic spacer analysis of microbial diversity and its application to freshwater bacterial communities. Appl Environ Microbiol 65:4630–4636

Gelsomino A, Keijzer WAC, Cacco G, Van Elsas JD (1999) Assessment of bacterial community structure in soil by polymerase chain reaction and denaturing gradient gel electrophoresis. J Microbiol Methods 38:1–15

Heuer H, Krsek M, Baker P, Smalla K, Wellington EMH (1997) Analysis of actinomycete communities by specific amplification of genes encoding 16S rRNA and gel electrophoretic separation in denaturing gradients. Appl Environ Microbiol 63:3233–3244

Kirk JL, Beaudette LA, Hart M, Moutoglis P, Klironomos JN, Lee H, Trevors JT (2004) Methods of studying soil microbial diversity. J Microbiol Methods 58:169–188

Marschner P, Baumann K, Solaiman Z (2005) Molecular approaches to study the microbial community structure and function in the rhizosphere. In: Podila GK, Varma A (eds) Biotechnological applications of microbes. I.K. International, New Delhi, pp 311–334

Muyzer G, De Waal EC, Uitterlinden AG (1993) Profiling of complex microbial populations by denaturing gradient gel electrophoresis analysis of polymerase chain reaction-amplified genes encoding for 16S rRNA. Appl Environ Microbiol 59:695–700

Ranjard L, Poly F, Lata J-C, Mougel C, Thioulous J, Nazaret S (2001) Characterization of bacterial and fungal soil communities by automated ribosomal intergenic spacer analysis fingerprints: biological and methodological variability. Appl Environ Microbiol 67:4479–4487

Schwieger F, Tebbe CC (1997) Efficient and accurate PCR amplification and detection of a recombinant gene in DNA directly extracted from soil using the expand™ high fidelity PCR system and T4 gene 32 protein. Biochemica 2:21–23

Vainio EJ, Hantula J (2000) Direct analysis of wood-inhabiting fungi using denaturing gradient gel electrophoresis of amplified ribosomal DNA. Mycol Res 104:927–936

Van Elsas JD, Wolters A (1995) Polymerase chain reaction (PCR) analysis of soil microbial DNA. In: Akkermans ADL, Elsas Van JD, De Bruin (eds) Molecular and microbial ecology mannual. Kluwer, Dordrecht, pp 1–10

Wechter P, Williamson J, Robertson A, Kluepfel D (2003) A rapid, cost-effective procedure for the extraction of microbial DNA from soil. World J Microbiol Biotechnol 19:85–91

Weisburg WG, Barns SM, Pelletier DA, Lane DJ (1991) 16S ribosomal DNA amplification for phylogenetic study. J Bacteriol 173:697–703

Yang CH, Crowley DE (2000) Rhizosphere microbial community structure in relation to root location and plant iron nutritional status. Appl Environ Microbiol 66:345–351

Yin B, Crowley DE, Sparovek G, De Melo WJ, Borneman J (2000) Bacterial functional redundancy along a soil reclamation gradient. Appl Environ Microbiol 66:4361–4365

12 Soil Microbial Community Structure and Function Assessed by FAME, PLFA and DGGE – Advantages and Limitations

P. Marschner

12.1 Introduction

Microorganisms play a pivotal role in nutrient availability, plant growth and plant health. It has been estimated that one gram of soil contains approximately 10^9 prokaryotes and more than 2000 genome types, with an average genome type representing less than 0.05% of the soil community (Torsvik et al. 1990). Until a few decades ago, soil microorganisms could only be studied by direct microscopic observation or culture-dependent methods. It is now recognized that culture-dependent methods such as dilution plating on standard media assess less than 5% of soil microorganisms (Bakken 1985). Although some previously unculturable bacterial species can now be isolated using special media and incubation methods (Janssen et al. 2002), most soil bacteria remain non-culturable. The advent of culture-independent methods has revolutionized soil microbial ecology because they make it possible to study a much greater fraction of soil microorganisms. With these new techniques, many new microbial species and even families have been discovered (Hugenholtz et al. 1998) and new insights obtained into microbial community composition, the interactions between microorganisms and the factors influencing microbial communities in soil.

In this chapter, three of the most widely used methods, fatty acid methylester (FAME), phospholipid fatty acid (PLFA) analyses and denaturing gradient gel electrophoresis (DGGE) are briefly described and their advantages and limitations outlined. Details on DGGE and FAME methodology can be found in Chapter 11. It should be noted that there are many variations to the methods described, more information can be found in the cited literature and methodological manuals.

Soil and Land Systems, School of Earth and Environmental Sciences,
The University of Adelaide, SA 5005 DP 636, Australia, Tel: +61 08 8303 7379
Fax: +61 08 8303 6511, email: Petra.marschner@adelaide.edu.au

Soil Biology, Volume 11
Advanced Techniques in Soil Microbiology
A. Varma, R. Oelmüller (Eds.)

12.2 Microbial Community Structure Based on Fatty Acid Patterns

Phospholipid fatty acids are components of the membranes of all organisms and each species has a characteristic fatty acid pattern. In soil microbial ecology, fatty acid patterns have been initially used to identify isolated microorganisms, but are now very common in studies of soil microbial communities.

To obtain fatty acid profiles, the fatty acids are extracted from soil (Frostegård et al. 1993a, b). FAME patterns are based on all fatty acids extracted (polar and non-polar fatty acids), whereas for phospholipid fatty acids (PLFA) patterns, the polar phospholipids are separated from the non-polar lipids by exchange columns (Bååth et al. 1995; Frostegård et al. 1993a). For PLFAs and FAMEs the fatty acids are methylated and then detected by gas chromatography (GC).

Phospholipids are dephosphorylated within minutes in soil (White 1993). Thus, PLFA profiles are derived from active microorganisms only, whereas FAME profiles may also include fatty acids from microorganisms that died recently. Nevertheless, shorter-chain fatty acids (C<20, which predominate in microbial cell membranes) are rapidly decomposed in soil, thus the majority of FAMEs are also derived from the living biomass (Jandl et al. 2005). Fatty acids extracted from soil may also be derived from plant residues, roots or soil animals, however the fatty acids from these organisms are usually longer-chain (C>20) than those of microorganisms (C10–C20; Jandl et al. 2005).

The fatty acid pattern is used to determine community composition. The biomass of groups such as gram-negative bacteria, gram-positive bacteria, actinomycetes, fungi and other soil organisms can be estimated by determining the concentration of so-called signature fatty acids (White 1993; White et al. 1996), which are specific for a given group (Table 12.1). While there are 10–15 bacterial signature fatty acids, only 2–3 fatty acids are characteristic for fungi, which may lead to an underestimation of fungal biomass. However, Klamer and Bååth (2004) found a good correlation between the concentration of the fungal signature fatty acid 18:1ω9c and ergosterol (a sterol that is only found in fungi; Klamer and Bååth 2004).

Under stress such as nutrient limitation or drought, many microorganisms produce specific fatty acids. These have been used to assess the physiological status of microbial communities (Bååth and Anderson 2003; White 1993). However, it should be noted that the relationship between stress and these fatty acids was determined in laboratory cultures and it is not clear whether such a relationship also exists in diverse soil communities.

PLFA patterns have been used to study the effect of a range of factors on soil microbial communities, e.g. pH or acid rain in forest soils (Bååth and Anderson 2003; Bååth et al. 1995; Frostegård et al. 1993a; Pennanen et al. 1998), heavy metal addition (Frostegård et al. 1993b; Kandeler et al. 2000) or soil amendments (Marschner et al. 2003; Zelles et al. 1995).

Many studies report the effect of environmental factors on microbial community structure assessed by FAME. Examples include effects of management and

Table 12.1 Signature fatty acids, based on Zak et al. (2000) and Olsson et al. (1997). Nomenclature is based on the ratio of number of carbon atoms:number of double bonds in the fatty acid, followed by the position of the double bond from the methyl end of the molecule. *Cis-* and *trans-* configurations are indicated by c and t, respectively; prefixes a and i indicate *anteiso-* and *iso-* branching; 10Me indicates a methyl group on the tenth C atom from the carboxyl end of the molecule; cy refers to cyclopropane fatty acids (Frostegård et al. 1993a). *AM* Arbuscular mycorrhizal fungi

	Signature fatty acid		Signature fatty acid
Gram+ bacteria	10me16:0	Bacteria	14:0
	i15:0		15:0
	a15:0		17:0
	i16:0		a17:0/17:1ω8c
	i17:0		
Gram− bacteria	16:1ω7c	Actinomycetes	10me18:0
	16:1ω5c		
	cy17:0	Fungi	18:1ω9c
	18:1ω7c		18:2ω6
	18:1ω7t		18:3ω3
	18:1ω5c		
	cy19:0	AM fungi	16:1ω5

crop rotation (Buyer and Drinkwater 1997; Buyer and Kaufman 1996), the introduction of foreign bacterial strains (Gagliardi et al. 2001), heavy metal pollution (Brim et al. 1999), biosolid application (Sullivan et al. 2006), plant species (Ibekwe and Kennedy 1998; Marschner et al. 2005) or salinity (Pankhurst et al. 2001).

Fatty acids specific for arbuscular mycorrhizal (AM) fungi and certain storage structures have been used to measure AM root colonization and hyphal biomass in soil (Olsson et al. 1997, 1998).

12.2.1 FAME Extraction and Data Analysis

The method described below for FAME extraction is based on Pankhurst et al. (2001) and Hawke (personal communication) and uses 6 g soil, but smaller amounts of soil can also be used. If the amount of soil available varies considerably between samples, a pre-test should be performed to determine the smallest possible sample size. This is done by comparing the FAME patterns derived from different amounts of soil. The lower limit of soil would then be the amount of soil that still gives a similar pattern as the greatest amount. If small amounts of soil are used, it may be advantageous to also use smaller volumes of the extractants. After sampling, the soils should be kept on ice for transport and then stored at −20 °C or −80 °C to prevent decomposition of the fatty acids.

12.2.1.1
Equipment

Culture tubes with Teflon-lined screw-cap, water bath, centrifuge, end-over-end shaker, gas chromatograph. All glassware should be washed with deionized water and twice with chloroform to avoid fatty acid contamination.

12.2.1.2
Reagents

Reagent 1: NaOH (45 g), methanol (HPLC grade; 150 ml), deionized distilled water (150 ml).
Reagent 2: 6.0N HCl (150 ml), methanol (HPLC grade; 275 ml).
Reagent 3: hexane (HPLC grade; 200 ml), methyl-*tert*-butyl ether (HPLC grade; 200 ml).
Reagent 4: NaOH (10.8 g), deionized distilled water (900 ml).

12.2.1.3
Extraction Procedure

1. Place 6 g fresh soil in a culture tube with a Teflon-lined screw-cap.
2. Add 6 ml of reagent 1 for saponification.
3. Seal with a Teflon-lined screw cap and vortex to ensure good soil/liquid contact.
4. Place tubes in a 100 °C water bath for 30 min and then cool in ice-water.
5. Centrifuge tubes for 3 min at 3000 g.
6. Transfer 3 g of the supernatant in a glass tube with screw-cap.
7. Add 6 ml of reagent 2 and vortex.
8. Incubate at 80 °C for 10 min, then allow to cool.
9. Add 1.5 ml of reagent 3 and close tube with screw-cap.
10. Place end-over-end shaker for 10 min.
11. Centrifuge the extract at 3000 g for 3 min.
12. Transfer top phase to a new tube.
13. Add 4 ml of reagent 4 and close tube with screw-cap.
14. Place on an end-over-end shaker for 5 min.
15. Centrifuge for 3 min at 3000 g.
16. Collect the top phase into a GC vial.
17. Evaporate completely under a nitrogen gas stream.
18. Add 10 µl of 19:0 methyl ester standard, then add 200 µl of reagent 3 and mix carefully.
19. The samples can be analyzed directly or stored at –20 °C for later analysis

To estimate recovery of fatty acids during extraction, a known concentration of fatty acid 13:0 can be added after step 6 (13:0 is usually not found in soils or is present at very low concentrations only). The peak height of 13:0 is then be compared with that of 19:0. For example, if the concentration of 13:0 is 5.0 times higher than that of 19:0, the peak of 13:0 should be 5.0 times greater than that of 19:0. If it is only 2.5 times greater, that suggests an extraction efficiency of 50%.

12.2.1.4 Data Analysis

The fatty acids are separated by GC. MIDI is a GC software package originally used for the identification of microorganisms from pure cultures by FAME patterns. It can also be used to generate FAME patterns from soil. However FAMEs can also be separated by GC without the MIDI system. Standard fatty acid mixes are commercially available (e.g. from Supelco). They may contain up to 37 fatty acids and include the most common fatty acids found in soils. Peaks in the samples are then identified by matching their retention time with those of the standard mix. Commercially available standard FAME mixes also come with a recommendation of the column and the temperature program to be used. The temperature program may need to be adjusted to obtain optimal separation of the peaks.

During sample preparation, the internal standard [nonadecanoic acid (19:0)] is added to all samples. This fatty acid is not found in soils and can therefore be used to normalize the fatty acid patterns. Normalization is performed by dividing the peak of each individual fatty acid by the peak of the internal standard (19:0). Dividing this value by the dry weight of the soil gives the fatty acid concentration in μg per g soil.

Fatty acid concentrations can be expressed in micrograms or millimoles per gram soil or in weight% (wt%). The weight% of fatty acids is calculated by dividing the peak of each fatty acid by the sum of fatty acids of the sample and multiplying this value by 100.

12.2.2 PLFA Analysis

The method described here is based on a procedure described by Bardgett et al. (1996), which was based on Blight and Dyer (1959) as modified by White et al. (1979).

Principle: fatty acids are extracted with a reagent (Blight and Dyer 1959) containing chloroform, methanol and citrate buffer. Lipid extracts are separated into neutral, glyco- and phospholipids via passage through an exchange column. The phospholipids are converted into fatty acid methylesters which can be determined by GC.

12.2.2.1
Materials and Reagents

Citrate buffer (0.15 M), pH 4: 31.52 g citrate dissolved in 1 l water, pH adjusted pH 4.
Extraction reagent (Blight and Dyer 1959): chloroform, methanol and citrate buffer mixed in a ratio of 1:2:0.8 (by vol.).
KOH-MeOH (0.2 M): 5.61 g KOH dissolved in 500 ml methanol gelöst (can be stored at 5 °C for up to 2 months).
Acetic acid (1 M): 5.72 ml of 100% acetic acid diluted with deionized water to a final volume of 100 ml.
Internal standard: methylnondecanoate (C19:0), 230.8 µg/ml: 5.770 mg methylnondecanoate dissolved in 25 ml isooctane.
Rinsing solution for columns: chloroform and methanol 1:1 (v/v).
Acetone, hexane chloroform solution (4:1, v/v), methanol toluol solution (1:1, v/v).
Silica-bonded columns for fractionation (e.g. 500 mg, 3 ml; from Varian).
Glass centrifuge tubes (25 ml, 10 ml).
Heating block to concentrate samples and for methanolysis.
Optional: Baker system with vacuum pump for rapid elution from columns.
All glassware should be washed with deionized water and twice with chloroform to avoid fatty acid contamination.

12.2.2.2
PLFA Procedure

1. Lipid Extraction
   Weigh out 0.5 g soil (organic matter-rich soils) to 2.0 g soil (organic matter-poor soils) in 25-ml glass centrifuge tubes. Negative controls (without soil) should be subjected to the same treatment as the samples.
   Add 1.5 ml citrate buffer, 1.9 ml chloroform, 3.8 ml methanol and 2 ml Blight and Dyer reagent, vortex.
   Shake tubes for 2 h, followed by centrifuging at 2500 g for 10 min.
   Transfer supernatant into a new 25-ml glass centrifuge tube. Wash soil pellet again with 2.5 ml Blight and Dyer reagent (vortex, centrifuge) and combine with first supernatant.
2. Phase Separation
   Add 3.1 ml chloroform and 3.1 ml citrate buffer to the supernatant and vortex.
   Centrifuge at 2500 g for 10 min.
   Transfer 1–3 ml from the lower (organic) phase into a new 10-ml glass tube.
   Dry at 40 °C under a stream of N_2. The tubes can be stored at –20 °C until the following steps are carried out.

3. Lipid Fractionation
   Note: elution can be speeded up by applying negative pressure to the bottom of the columns, e.g. Baker system.
   Condition silica-bonded columns with 2× 1 ml chloroform.
   Dissolve dried sample in 300 μl chloroform and add to the column using a Pasteur pipette. Rinse the pipette with 2× 300 μl chlororform.
   Elute neutral lipids with 5 ml chloroform and discard eluate.
   Elute glycolipids with 20 ml acetone and discard eluate.
   Elute phospholipids with 5 ml methanol and transfer eluate into a 10-ml centrifuge tube.
   Dry at 40 °C under a stream of N_2.
4. Alkaline Methanolysis
   Add 30 μl internal standard (C19:0) to sample and then dissolve sample in 1 ml methanol toluol solution (1:1, v/v).
   Add 1 ml of 0.2 M KOH-MeOH and incubate for 15 min at 37 °C in a water bath.
   Add 2 ml hexane–chloroform solution, 0.3 ml of 1 M acetic acid and 2 ml deionized water, vortex.
   Centrifuge for 5 min at 2500 g.
   Transfer upper (organic) phase into a 10-ml glass tube using a Pasteur pipette.
   Add 2 ml hexane–chloroform solution to the lower phase, vortex and combine the supernatant with the previous one.
   Dry at 40 °C under a stream of N_2. The dried extract can be stored at –20 °C until analysis.
5. Fatty Acid Determination
   Dissolve dried extract in 100 μl isooctane and transfer into GC vial.
   Fatty acids can be separated by a 50 m capillary column (HP-5, Agilent) and detected by a flame ionization detector.

12.2.2.3
Calculations and Data Analysis

$$c[nmol/g] = \frac{A_{FM} \cdot c_{IS} \cdot 2 \cdot 1000 \cdot 100}{SW \cdot DW \cdot A_{IS} \cdot MW_{FM}}$$

Correction for negative control:

$$c[nmol/g] = \frac{\left(\left(\frac{A_{FM} \cdot c_{IS} \cdot 2 \cdot 1000}{A_{IS} \cdot MW_{FM}}\right) - c_{NC}\right) \cdot 100}{SW \cdot DW}$$

c [$nmol/g$]: concentration of fatty acid methylester
A_{FM}: area of the fatty acid methylester
MW_{FM}: molecular weight of the fatty acid methylester (μg/μmol)
C_{IS}: concentration of internal standard (μg)
A_{IS}: area of internal standard
SW: weight of soil sample (g)
DW: dry weight of soil (%)
2: corrects for the use of 3 ml from 6 ml organic phase in phase separation
1000: factor for expression (nmol)
c_{NC}: Concentration in negative control (nmol)

FAME and PLFA patterns can be analyzed in a number of ways: (a) concentration of individual fatty acids, (b) concentration of signature fatty acids (see Table 12.1) and their ratios, e.g. ratio of bacterial to fungal fatty acids, (c) diversity using the Shannon–Weaver index (H') = $-\Sigma(n!/N) \times \ln(n!/N)$, where n = concentration of each fatty acid and N = sum of concentration of all bands of the sample (Zak et al. 1994), and (d) multivariate analyses such as cluster analysis, principle component analysis, etc.

12.2.3 Advantages and Limitations of Fatty Acid Patterns

1. Advantages
- Extraction procedure is relatively simple and quick if only FAMEs are measured, more laborious for PLFAs.
- Phospholipids are rapidly dephosphorylated (White 1993) and fatty acids decomposed in soil (Jandl et al. 2005). Therefore PLFAs are considered to be derived mainly from living organisms.
- Signature fatty acids can be used as indicators for biomass of certain microbial groups and the ratios between them. The sum of microbial signature fatty acids in a given sample can be used as a measure of microbial biomass.
- Peak patterns represent community composition and can be compared using multivariate analyses. With some multivariate analysis methods, such as canonical correspondence analysis or multidimensional scaling, it is possible to relate community composition with environmental factors.
- Certain fatty acids have been used to assess the physiological status of microbial communities (Bååth and Anderson 2003; White 1993). However the relationship between stress and these fatty acids was determined in laboratory cultures and it is not clear whether such a relationship also exists in diverse soil communities.

2. Limitations
- Only a small number of fatty acids are truly characteristic for certain groups, many are ubiquitous and may be derived from other soil organisms (Jandl et

al. 2005). Hence the background of unspecific fatty acids may mask differences in microbial community structure.

- It is often assumed that, due to the limited number of signature fatty acids for fungi, fungal biomass may be underestimated. Nevertheless, the correlation between fungal fatty acid content and other measures of fungal biomass, e.g. ergosterol, is quite high (Klamer and Bååth 2004).
- FAME patterns may also contain fatty acids from dead microorganisms and plant residues.
- No information about species composition.

12.3 Denaturing Gradient Gel Electrophoresis

Denaturing gradient gel electrophoresis (DGGE) is a method that relies on separating species according to differences in sequence of the target DNA or RNA region. DGGE involves several steps which are described below: (1) DNA/RNA extraction from soil, (2) polymerase chain reaction (PCR) of the target region, and (3) DGGE itself.

12.3.1 DNA Extraction from Soil

DNA extraction from soil is often more difficult than from many other ecological samples (Wintzigerode et al. 1997) because:

1. Microbial cells can be located within soil aggregates and adhere tightly to soil organic matter and minerals. Thus efficient DNA/RNA extraction is only possible by mechanical breakdown of aggregates and release of the cells adhering to soil particles.
2. Spores of fungi and gram-positive bacteria have thick and tough walls; DNA will only be released from spores after rupture of the spore walls.
3. Many soils contain large amounts of humic substances and phenolic compounds, which inhibit the polymerase enzyme in the PCR. For successful PCR, the inhibiting substances have to be removed or the extract needs to be diluted.

In the indirect DNA extraction methods, the cells are first separated from the soil followed by DNA extraction from the cell suspension. Direct methods extract the DNA from the soil sample directly without prior separation of the cells. An example of an indirect method is described by Duarte et al. (1998). They separated the cells from soil aggregates by shaking the soil with a buffer solution and then subjecting the cell pellet to bead beating. Subsequently the DNA was purified by isopropanol precipitation (Duarte et al. 1998). The advantage of separating the cells from the soil prior to bead beating is that the concentration of humic acids is minimized, which may be important in soils with very high organic matter content.

The method described in Chapter 11 is an example of a direct method. Soil samples are homogenized in a bead beater. After removal of proteins and humic acids by precipitation, DNA is bound to a silica matrix, purified by several washing steps and finally desorbed into water.

Kozdroj and Van Elsas (2000) compared four methods: two direct methods (bead beating, grinding in liquid N) and two indirect methods. They found that DNA yield decreased in the following order: direct bead beating > grinding in liquid N > indirect methods, whereas DNA purity was greater in the indirect than in the direct methods. However, they found that the cell pellet in the indirect methods only contained a subset of the microbial community in the soil, namely cells that were easily dislodged from aggregate surfaces. Bead beating methods extracted substantial amounts of humic substances and PCR was only possible after purification (Kozdroj and Van Elsas 2000).

There are a number of commercial kits for the extraction of soil DNA or RNA, which are often as efficient and less labor-intensive than "home-made" methods. Two of the most widely used kits are the soil DNA or RNA kits from Qbiogene and MoBio. It should be noted that, due to the difficulties in DNA extraction from soils mentioned above, DNA extraction kits developed for plant tissues or microbial pure cultures will often not efficiently extract DNA from soils.

Humic substances which can inhibit the polymerase in PCR may be removed by:

1. Addition of 200 µl of 100 mM $AlNH_4(SO_4)_2$ solution to tubes before bead beating [$AlNH_4(SO_4)_2$ should be filter-sterilized (2 µm) before use; Braid et al. (2003)].
2. Addition of PVPP or PVP (0.1 g/0.2 g soil) before bead beating. Autoclave PVP before use (Wechter et al. 2003).
3. Washing the DNA bound to silica matrix with 2× 500 µl of 5.5 M guanidine thiocyanate, then vortexing, spinning for 20 s and discarding supernatant.
4. Using exchange resins (Kuske et al. 1998).

12.3.2 Polymerase Chain Reaction

Polymerase chain reaction (PCR) is a process by which a target sequence is amplified one million-fold. It has revolutionized molecular studies because it allows the detection of sequences that have very low initial concentrations. Briefly, PCR involves binding of short oligonucleotide primers to the DNA flanking the target region. The primers serve as starting points for the polymerase enzyme, which amplifies the DNA region between the primers and increases the concentration of the target DNA region exponentially because the sequences amplified in one cycle act as templates in the next cycle. Amplification is carried out in thermocyclers which can change the temperature of the sample within seconds. Thermocyclers typically undergo programs with the following temperatures: 95 °C for separation of the double strands, 45–55 °C (depending on the primers used) for primer annealing and 75 °C for extension during which the polymerase enzyme creates a matching strand of the target DNA sequence between the primers. PCR programs usually consist of 20–40 such cycles and are designed to maximize the yield of the target DNA. Thus, the final DNA yield is considered to be a poor indicator of the initial target DNA concentration. Success of the PCR can be visualized under UV light by running the products in an agarose gel (1–2%) and staining with ethidium bromide or Sybr green or Sybr gold. For quality control it is important to include in every PCR run at least one negative control (water instead of sample) and at least one postive control. The positive control may consist of: (a) solution containing target DNA sequence (b) DNA from target organism, or (c) extract of the environmental sample that contains target DNA.

Primer design is critical for the specificity of the PCR reaction. Primers can be designed from existing databases or from sequences derived from own studies. In each case the specificity has to be checked carefully. Universal primers are designed to amplify the DNA of a large group of organisms such as bacteria or fungi. However, Watanabe et al. (2001) showed that many so-called universal eubacterial primers were in fact not universal because they did not amplify the DNA of certain bacterial species or genera (Watanabe et al. 2001). Fungal primers may also amplify DNA from other eukaryotes such as soil animals and plants (Borneman and Hartin 2000).

Primers can be family-, genus- or kingdom-specific or have a functional gene as target. Some examples are given below:

1. Bacteria (Muyzer and Smalla 1998; Watanabe et al. 2001)
2. Actinomycetes (Heuer et al. 1997)
3. Fungi (Vainio and Hantula 2000)
4. *Pseudomonas* (Widmer et al. 1998)
5. *Bacillus* (Garbeva et al. 2003)
6. Ammonia oxidizers (Avrahami et al. 2003)
7. Nitrogen-fixing (*nif*) genes (Chelius and Lepo 1999).

A PCR protocol for bacteria is described in Chapter 11. PCR conditions vary with primer. The reader is referred to the appropriate conditions in the relevant publication.

When using soil extracts, it is important to perform PCR pre-tests. Often, undiluted DNA extracts contain high concentrations of substances inhibiting the polymerase. Therefore DNA extracts may have to be diluted. Dilutions between 1:10 to 1:500 or even 1:1000 have been used. It is recommended to try several dilutions of the DNA extracts (using between three and five samples) and select the dilution that gives the most consistent amplification.

There are a number of pitfalls of PCR (Wintzigerode et al. 1997):

1. PCR amplification of DNA in soil extracts may be inhibited by humic acids or humic substances. Purification of extracts to remove inhibiting substances may lead to loss of DNA.
2. Not all sequences are amplified to the same extent due to PCR bias or differential PCR amplification.
3. Possible PCR artifacts: (a) formation of chimeric molecules (two single strands which differ slightly in sequence form a double strand) and (b) deletion or point mutations during PCR (polymerase reading errors).
4. A given organism may contain multiple copies of the same gene with similar or slightly different sequence leading to overestimation of the abundance of this organism compared to organisms with single gene copies.
5. Primer design is based on known sequences, thus the DNA of some species of target organisms may not be amplified because their DNA sequence is slightly different.

Consequently, the frequency of genes or species determined after PCR may not truly reflect their frequency in the sample. Primers targetting the same genes may differ in amplification bias, thus the frequency of genes or species will also depend on primer choice.

12.3.3 DGGE Procedures

Denaturing gradient gel electrophoresis (DGGE) was originally designed to detect single base mutations in isolates but is now often used to assess microbial community composition in environmental samples (Muyzer et al. 1993).

After DNA extraction and PCR, the amplified DNA region from different species has approximately the same length (which, depending on the primers used, may be 300–1000 base pairs long) but each differs in sequence. The separation of different sequences during electrophoresis is based on the guanidine and cytosine (GC) content of the fragment. For details see Chapter 11. Denaturation, the partial separation of the two strands, is achieved in polyacrylamide gels by denaturing chemicals (urea, formamide) with the concentration of the denaturants increasing from the top to the bottom of the gel. Thus sequences with a high GC content, which are more stable, migrate further through the gel than GC-poor sequences. When the PCR product of a microbial community is electrophoresed in a DGGE gel, the fragments migrate through the gel and form a band when they reach the concentration of denaturants (DGGE) at which they denature (Fig. 12.1). The result is a banding pattern that varies with community composition. A GC tail (clamp) is added to one primer in the PCR to avoid complete separation of bands, which would result in smeared bands (Muyzer and Smalla 1998).

Depending on the primers used, the community composition of different microbial groups can be assessed. Patterns generated from DNA reflect the community composition of the total community of the target organisms or genes, whereas RNA patterns include only the active fraction, and in case of mRNA of a functional gene, the gene-expressing fraction of the community.

For interpretation of banding patterns it should be noted that species often contain several copies of the same gene that differ slightly in sequence. Hence one species may produce several bands in DGGE (Muyzer and Smalla 1998). In contrast, a given band may contain the sequence of more than one species because they differ only by a few base pairs within the amplified DNA region (Yang and Crowley 2000).

After electrophoresis, bands can be excised and sequenced (e.g. Yang and Crowley 2000). However there are several limitations to the identification of species from DGGE bands: (a) one band may contain the sequences of several species and sequencing may or may not reveal the presence of all species in a band, (b) fragments used in DGGE are typically relatively short (300–1000 bp) to give good band resolution and these fragments are often not long enough to identify organisms at a species level, and (c) only abundant species generate a band because the detection limit for bacteria in DGGE may be as high as 10^6 cells/g soil (Gelsomino et al. 1999).

Community composition of eubacteria has been studied by DGGE in the rhizosphere of different plant species (Duineveld et al. 1998; Marschner and Baumann 2003; Marschner et al. 2001a, b; Yang and Crowley 2000), soils from different ecosystems and geographical regions (Gelsomino et al. 1999), heavy metal-polluted soils (Kandeler et al. 2000) and hot springs (Ferris et al. 1996). DGGE has also been used to assess the community composition of bacterial groups such as ammonia oxidizers (Avrahami et al. 2003; Baeckman et al. 2003), sulfate reducers

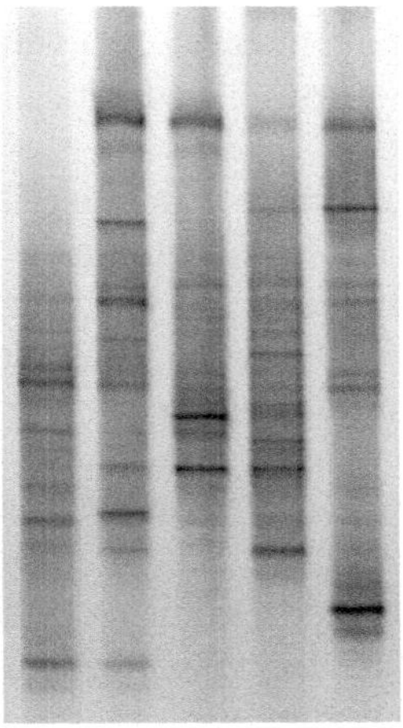

- Denaturant gradient concentration increases from top to bottom of the gel.
- DNA moves downwards through the gel until it reaches denaturant concentration at which the double strands separate ⇒ formation of a band.
- GC-rich sequences are more stable than GC-poor sequences ⇒ GC-rich sequences migrate further down in the gel.
- Banding pattern represents community composition of target organisms.
- Each band represents a species or group of species with similar melting behaviour.

Fig. 12.1 Example of DGGE profiles of five bacterial communities

(Sass et al. 1998), *Desulfovibrio* sp. (Wawer and Muyzer 1997) and bacteria with *nif* genes (Piceno and Lovell 2000; Rosado et al. 1998). The community composition of fungi (Pennanen et al. 2001; Vainio and Hantula 2000) and actinomycetes (Heuer et al. 1997) in soils was also determined by DGGE.

12.3.3.2
Comments on DGGE

For assembling the gel sandwiches and pouring the gel, the reader is referred to Chapter 11 or the manual of the DGGE equipment. Here only a few additional comments are presented.

- Given the large variability in community structure, at least four replicates of a given treatment/site should be run in the DGGE.
- It is crucial to ensure that glass plates and spacers are carefully aligned to avoid leaking of the gel.
- Gels should be poured slowly to avoid "smiling" of the gels.
- For better resolution, it is recommended to let the gels polymerize overnight or at least 5–6 h at room temperature. During this period, the gels should be kept moist by placing wet paper towel on the top of the gels and covering the gels with a plastic bag.
- Combs with different number of "teeth" are available. Combs with e.g. 20 "teeth" result in larger wells which are easier to load than small wells (from combs with 25 or more "teeth").
- For the initial DGGE, a wide gradient (e.g. 25–75%) should be used. This selection can be based on the gradient used in the reference from which the primers were taken. However, it is recommended to use a wider gradient initially because each soil will differ in community structure and hence in range of GC content of the amplified region. In the wide gradient the bands will usually be concentrated in a certain part of the gel. A narrower gradient can then be selected for better separation of the bands.
- With larger sample numbers it is usually necessary to use more than one gel. Since gradient and staining intensity are not identical in different gels, the influence of gel on banding patterns can be quite strong and may lead to wrong conclusions about treatment effects. To avoid this "gel effect", the different replicates of a given treatment/site should be placed on different gels. This can be done randomly (when the sample number is very large), or by placing only one or two replicates of each treatment on a gel and the other replicates on other gels.
- When more than one gel is used, the band position should be normalized with respect to position and intensity (see below). In this case it is important to run at least one "standard" in each gel and expressing the position of the bands of the sample relative to 3–4 bands in the standard. This standard can be a sample, a commercially available base pair ladder or a mix of DNA from

pure cultures. Ideally, the standard should contain at least 3–4 strong bands that can be easily identified.
- Electrophoresis time ranges between 3 h and 18 h, with a higher voltage at shorter run times. Band resolution is often better at longer run times and lower voltage, but it is recommended to try short and long electrophoresis times initially.
- Gels may be stained with ethidium bromide, Sybr green/gold or silver. Silver staining usually results in better band resolution. However, it is more time-consuming than the other two staining methods and silver-stained bands cannot be sequenced. In order to avoid damage, the gels should always be submerged when handling them during staining.

12.3.3.3
Data Analysis

If only a few samples are compared, visual comparison of the samples may be sufficient. However, band patterns may be very complex (between 10 and 40 bands per sample) and visual comparison may be subjective. Therefore it is recommended to digitize the band patterns and compare them statistically.

Digitization involves expressing band position and band intensity in numerical values. There is a range of commercially available digitization software. They usually allow both manual and automatic band detection and should also include the possibility of normalization of the band position relative to that of bands in the standard.

If more than one gel is used, the band position should be normalized with respect to position and intensity.

Normalization of intensity can be done by dividing the intensity of a given band by either the average intensity of the sample or the average intensity of the gel. In both cases the values may be multiplied by 100, to express them as %.

Digitized banding patterns can be analyzed by a wide range of statistical analysis (e.g. Marschner and Baumann 2003; Marschner et al. 2001a; Yang and Crowley 2000), similar to those mentioned above for fatty acid patterns.

12.3.3.4
Advantages and Limitations of DGGE

1. Advantages
- Rapid assessment of complex communities of target organisms.
- Banding patterns can be compared visually or after digitization by multivariate analyses.
- Bands can be excised and sequenced.

- Wide range of primers allows studying microbial communities with different resolution: kingdoms (e.g. bacteria, fungi), genera (e.g. *Pseudomonas*, *Bacillus*), genes (e.g. *nif* genes).
- Patterns of expression of certain genes can be compared if mRNA is used as template.

2. Limitations
- High detection limit (10^6 cells/g soil), thus only abundant species are detected.
- Community complexity can be underestimated as one band may contain several species, or overestimated as one species may form several bands.
- Artifacts due to chimeric double strands (DNA double strands formed by two single DNA strands that differ slightly in sequence).
- Length of fragment may not be sufficient to allow species identification by sequencing.
- Due to PCR step, DGGE is not truly quantitative (see section on PCR above).
- DNA can be bound to soil particles (Cai et al. 2006) where it is protected against microbial decomposition. Therefore DGGE profiles may also include bands from dead organisms.
- Community composition of target group only, no information about general community (as in fatty acid-based methods), unless a large number of different primers are used.

12.4 Conclusions

Fatty acid-based methods such as FAME and PLFA and DNA-based methods such as DGGE can provide insights into microbial community composition at different scales. Fatty acid profiles reflect the general microbial community composition and are quantitative but provide no information on species composition. DNA-based methods have the advantage that the target community (kingdom, species, genes) can be selected. However they are not quantitative because they rely on gene amplification in PCR. Ideally, fatty acid-based methods should be combined with DNA-based methods to provide a comprehensive picture of the microbial community.

Both DGGE and FAME/PLFA can be used to compare soil microbial communities as affected by, e.g. management or soil type. Undoubtedly, one goal in soil microbial ecology is linking community structure and function. However, we are still a long way from this goal. To move towards this goal, assessment of microbial community structure should be combined with other methods such as nutrient analysis, enzyme activity determination and real-time PCR of functional genes.

References

Avrahami S, Liesack W, Conrad R (2003) Effects of temperature and fertilizer on activity and community structure of soil ammonia oxidizers. Environ Microbiol 5:691–705

Bååth E, Anderson TH (2003) Comparison of soil fungal/bacterial ratios in a pH gradient using physiological and PLFA-based techniques. Soil Biol Biochem 35:955–963

Bååth E, Frostegård A, Pennanen T, Fritze H (1995) Microbial community structure and pH response in relaton to soil organic matter qualityin wood ash fertilized, clear-cut or burned coniferous forest soils. Soil Biol Biochem 27:229–240

Baeckman JSK, Hermansson A, Tebbe CC, Lindgren PE (2003) Liming induces growth of a diverse flora of ammonia-oxidizing bacteria in acid spruce forest soil as determined by SSCP and DGGE. Soil Biol Biochem 35:1337–1347

Bakken LR (1985) Separation and purification of bacteria from soil. Appl Environ Microbiol 49:1482–1487

Bardgett RD, Hobbs PJ, Frostegård Å (1996) Changes in soil fungal:bacterial biomass ratios following reductions in the intensity of management of an upland grassland. Biol Fertil Soils 22:261–264

Blight EG, Dyer WJ (1959) A rapid method of total lipid extraction and purification. Can J Biochem Physiol 37:911–917

Borneman J, Hartin RJ (2000) PCR primers that amplify fungal rRNA genes from environmental samples. Appl Environ Microbiol 66:4356–4360

Braid MD, Daniels LM, Kitts CL (2003) Removal of PCR inhibitors from soil DNA by chemical flocculation. J Microbiol Methods 52:389–393

Brim H, Heuer H, Kroegerrecklenfort E, Mergeay M, Smalla K (1999) Characterization of the bacterial community of a zinc-polluted soil. Can J Microbiol 45:326–338

Buyer JS, Drinkwater LE (1997) Comparison of substrate utilisation assay and fatty acid analysis of soil microbial communities. J Microbiol Methods 30:3–11

Buyer JS, Kaufman DD (1996) Microbial diversity in the rhizosphere of corn grown under conventional and low-input systems. Appl Soil Ecol 5:21–27

Cai P, Huang Q, Zhang X, Chen H (2006) Adsorption of DNA on clay minerals and various colloidal particles from an alfisol. Soil Biol Biochem 38:471–476

Chelius MK, Lepo JE (1999) Restriction fragment length polymorphism analysis of PCR-amplified nifH sequences from wetland plant rhizosphere communities. Environ Technol 20:883–889

Duarte GF, Rosado AS, Seldin L, Cook WA, Van Elsas JD (1998) Extraction of ribosomal RNA and genomic DNA from soil for studying the diversity of the indigenous bacterial community. J Microbiol Methods 32:21–29

Duineveld BM, Rosado AS, Van Elsas JD, Van Veen JA (1998) Analysis of the dynamics of bacterial communities in the rhizosphere of the chrysanthemum via denaturing gradient gel electrophoresis and substrate utilization patterns. Appl Environ Microbiol 64:4950–4957

Ferris MJ, Muyzer G, Ward DM (1996) Denaturing gradient gel electrophoresis profiles of 16S rRNA-defined populations inhabiting a hot spring microbial mat community. Appl Environ Microbiol 62:340–346

Frostegård A, Bååth E, Tunlid A (1993a) Shifts in the structure of soil microbial communities in limed forests as revealed by phospholipid fatty acid analysis. Soil Biol Biochem 25:723–730

Frostegård A, Tunlid A, Bååth E (1993b) Phospholipid fatty acid compostion, biomass, and activity of microbial communities from two soil types experimentally exposed to differing heavy metals. Appl Environ Microbiol 59:3605–3617

Gagliardi JV, Buyer JS, Angle JS, Russek CE (2001) Structural and functional analysis of whole-soil microbial communities for risk and efficiency testing following microbial inoculation of wheat roots in diverse soils. Soil Biol Biochem 33:25–40

Garbeva P, Van Veen JA, Van Elsas JD (2003) Predominant *Bacillus* spp. in agricultural soil under different management regimes detected via PCR-DGGE. Microb Ecol 45:302–316

Gelsomino A, Keijzer WAC, Cacco G, Van Elsas JD (1999) Assessment of bacterial community structure in soil by polymerase chain reaction and denaturing gradient gel electrophoresis. J Microbiol Methods 38:1–15

Heuer H, Krsek M, Baker P, Smalla K, Wellington EMH (1997) Analysis of actinomycete communities by specific amplification of genes encoding 16S rRNA and gel electrophoretic separation in denaturing gradients. Appl Environ Microbiol 63:3233–3244

Hugenholtz P, Goebel BM, Pace NR (1998) Impact of culture-independent studies on the emerging phylogenetic view of bacterial diversity. J Bacteriol 180:4765–4774

Ibekwe AM, Kennedy AC (1998) Fatty acid methyl ester (FAME) profiles as a tool to investigate community structure of two agricultural soils. Plant Soil 206:151–161

Jandl G, Leinweber P, Schulten HR, Ekschmitt K (2005) Contribution of primary organic matter to the fatty acid pool in agricultural soils. Soil Biol Biochem 37:1033–1041

Janssen PH, Yates PS, Grinton BE, Taylor PM, Sait M (2002) Improved culturability of soil bacteria and isolation in pure culture of novel members of the divisions *Acidobacteria, Actinobacteria, Proteobacteria* and *Verrumicrobia*. Appl Environ Microbiol 68:2391–2396

Kandeler E, Tscherko D, Bruce KD, Stemmer M, Hobbs PJ, Bardgett RD, Amelung W (2000) Structure and function of the soil microbial community in microhabitats of a heavy metal polluted soil. Biol Fertil Soils 32:390–400

Klamer M, Bååth E (2004) Estimation of conversion factors for fungal biomass determination in compost using ergosterol and PLFA 18:2w6,9. Soil Biol Biochem 36:57–65

Kozdroj J, Van Elsas JD (2000) Application of polymerase chain reaction – denaturing gradient gel electrophoresis for comparison of direct and indirect methods of soil DNA used for microbial community fingerprinting. Biol Fertil Soils 31:372–378

Kuske CR, Banton KL, Adorada DL, Stark PC, Hill KK, Jackson PJ (1998) Small-scale DNA sample preparation method for field PCR detection of microbial cells and spores in soil. Appl Environ Microbiol 64:2463–2472

Marschner P, Baumann K (2003) Changes in bacterial community structure induced by mycorrhizal colonisation in split-root maize. Plant Soil 251:279–289

Marschner P, Crowley DE, Lieberei R (2001a) Arbuscular mycorrhizal infection changes the bacterial 16S rDNA community composition in the rhizosphere of maize. Mycorrhiza 11:297–302

Marschner P, Yang CH, Lieberei R, Crowley DE (2001b) Soil and plant specific effects on bacterial community composition in the rhizosphere. Soil Biol Biochem 33:1437–1445

Marschner P, Kandeler E, Marschner B (2003) Structure and function of the soil microbial community in a long-term fertilizer experiment. Soil Biol Biochem 35:453–461

Marschner P, Solaiman MZ, Rengel Z (2005) Growth, phosphorus uptake, and rhizosphere microbial community composition of a phosphorus-efficient wheat cultivar in soils differing in pH. J Plant Nutr Soil Sci 168:343–351

Muyzer G, Smalla K (1998) Application of denaturing gradient gel electrophoresis (DGGE) and temperature gradient gel electrophoresis (TGGE) in microbial ecology. Antonie van Leeuwenhoek 73:127–141

Muyzer G, De Waal EC, Uitterlinden AG (1993) Profiling of complex microbial populations by denaturing gradient gel electrohphoresis analysis of polymerase chain reaction-amplified genes coding for 16S rRNA. Appl Environ Microbiol 59:695–700

Olsson PA, Bååth E, Jakobsen I (1997) Phosphorus effects on the mycelium and storage structures of an arbuscular mycorrhizal fungus as studied in the soil and roots by analysis of fatty acid signatures. Appl Environ Microbiol 63:3531–3538

Olsson PA, Francis R, Read DJ, Söderström B (1998) Growth of arbuscular mycorrhizal mycelium in calcareous dune sand and its interaction with other soil microorganisms as estimated by measurement of specific fatty acids. Plant Soil 201:9–16

Pankhurst CE, Yu S, Hawke BG, Harch BD (2001) Capacity of fatty acid profiles and substrate utilisation patterns to describe differences in soil microbial communities associated with increased salinity or alkalinity at three locations win South Australia. Biol Fertil Soils 33:204–217

Pennanen T, Fritze H, Vanhala P, Kiikkila O, Neuvonen S, Bååth E (1998) Structure of the microbial community in soil after prolonged addition of low levels of simulated acid rain. Appl Environ Microbiol 64:2173–2180

Pennanen T, Paavolainen L, Hantula J (2001) Rapid PCR-based method for the direct analysis of fungal communities in complex environmental samples. Soil Biol Biochem 33:697–699

Piceno YM, Lovell CR (2000) Stability in natural bacterial communities. I. Nutrient addition effects on rhizosphere diazotroph assemblage composition. Microb Ecol 39:32–40

Rosado AS, Duarte GF, Seldin L, Van Elsas JD (1998) Genetic diversity of nifH gene sequences in *Paenibacillus azotofixans* strains and soil samples analysed by denaturing gradient gel electrophoresis of PCR-amplified gene fragments. Appl Environ Microbiol 64:2770–2779

Sass H, Wieringa E, Cypionka H, Babenzien HD, Overmann J (1998) High genetic and physiological diversity of sulfate-reducing bacteria isolated from an oligotrophic lake sediment. Arch Microbiol 170:243–251

Sullivan TS, Stromberger ME, Paschke MW (2006) Parallel shifts in plant and soil microbial communities in response to biosolids in a semi-arid grassland. Soil Biol Biochem 38:449–459

Torsvik V, Salte K, Sorheim R, Goksoyr J (1990) Comparison of phenotypic diversity and DNA heterogeneity in a population of soil bacteria. Appl Environ Microbiol 56:776–781

Vainio EJ, Hantula J (2000) Direct analysis of wood-inhabiting fungi using denaturing gradient gel electrophoresis of amplified ribosmomal DNA. Mycol Res 104:927–936

Watanabe K, Kodama Y, Harayama S (2001) Design and evaluation of PCR primers to amplify bacterial 16S ribosomal DNA fragments used for community fingerprinting. J Microbiol Methods 44:253–562

Wawer C, Muyzer G (1997) Genetic diversity of *Desulfovibrio* spp in environmental samples analyzed by denaturing gradient gel electrophoresis of (NiFe) hydrogenase gene fragments. Appl Environ Microbiol 61:2203–2210

Wechter P, Williamson J, Robertson A, Kluepfel D (2003) A rapid, cost-effective procedure for the extraction of microbial DNA from soil. World J Microbiol Biotechnol 19:85–91

White DC (1993) In situ measurement of microbial biomass, community structure and nutritional status. Philos Trans R Soc Lond B 344:59–67

White DC, Davis WM, Nickels JS, King JC, Bobbie RJ (1979) Determination of the sedimentary microbial biomass by extractable lipid phosphate. Oecologia 40:51–62

White DC, Stair JO, Ringelberg DB (1996) Quantitative comparisons of in situ microbial biodiversity by signature biomarker analysis. J Ind Microbiol 17:185–196

Widmer F, Seidler RJ, Gillevet PM, Watrud LS, Di Giovanni GD (1998) A highly selective PCR protocol for detecting 16S rRNA genes of the genus *Pseudomonas* (*sensu stricto*) in environmental samples. Appl Environ Microbiol 64:2445–2453

Wintzigerode F, Göbel UB, Stackebrandt E (1997) Determination of microbial diversity in environmental samples: pitfalls of PCR-based rRNA analysis. FEMS Microbiol Rev 21:213–229

Yang CH, Crowley DE (2000) Rhizosphere miccrobial community structure in relation to root location and plant iron nutritional status. Appl Environ Microbiol 66:345–351

Zak DR, Pregnitzer KS, Curtis PS, Holmes WE (2000) Atmosperic CO_2 and the composition and function of soil microbial communities. Ecol Appl 10:47–59

Zak JC, Willig MR, Moorhead DL, Wildman HG (1994) Functional diversity of microbial communities: a quantitative aproach. Soil Biol Biochem 26:1101–1108

Zelles L, Rackwitz R, Bai QY, Beck T, Beese F (1995) Discrimination of microbial diversity by fatty acid profiles of phospholipids and lipopolysaccharides in differently cultivated soils. Plant Soil 170:115–122

13 Measurement of Microbial Biomass and Activity in Soil

Z. Solaiman

13.1 Introduction

Soil acts as a growth medium and can provide several biological functions such as transforming, storing and cycling energy-rich organic compounds. Soil microbial biomass is one of the most important soil biological properties. It regulates many critical processes in ecosystems, such as the biophysical integration of organic matter with soil solid, aqueous and gaseous phases. It also becomes vital in regulating the quantity and quality of components in the hydrologic cycle and in greenhouse gas emissions. The measurement of microbial biomass is useful for describing biomass turnover in different ecosystems. Several methods are used today to study soil microbial biomass. Among them the substrate-induced respiration and the fumigation–extraction methods are used frequently for biomass determination (Machulla 2003).

The best indicator of the whole metabolic activity of soil microbial populations is soil respiration, a robust parameter that can be rapidly and reproducibly determined. It allows gross comparisons of soils and reflects soil management changes, or the impact of elevated atmospheric CO_2 on soil microorganisms (Machulla 2003). In addition to microbial biomass and soil respiratory activity, soil enzymes can be determined. Enzymes in soil act as transformation agents for organic substances like cellulose, lignin, sugar and amino acids. Soil enzymes are mostly of microbial origin and are closely related to microbial biomass. Various enzymes have been measured for their suitability in soil investigations. This method was normally selected based on the specific element of interest such as

Zakaria Solaiman: School of Earth and Environmental Sciences, The University of Adelaide, SA 5005, Australia; Present address: School of Earth and Geographical Sciences, The University of Western Australia, Crawley, WA 6009, Australia,
email: solaiman@cyllene.uwa.edu.au

Soil Biology, Volume 11
Advanced Techniques in Soil Microbiology
A. Varma, R. Oelmüller (Eds.)

C, N or P. Dehydrogenase is one of the most frequently used enzyme tests for the measurement of total microbial activity in soil.

13.2 Protocols for Microbial Biomass Determination

The measurement of soil microbial biomass is one of the most accurate procedures used for a better understanding of the nutrient cycle in soil. Soil organic matter is the vital source of energy for microorganisms. It can be pooled into several fractions that vary in turnover time. The active fraction of organic matter consists of amino acids, proteins and carbohydrates, representing that small but dynamic portion of the huge and slowly changing background of stabile organic matter. This labile pool is readily available for microbial use and is mostly stored by soil microorganisms. A portion of such kind of organic substances can be quantified as an indicator of the actual amount of microbial populations. Thus, there has been increasing interest for definite measurements of the soil microbial biomass and several methods have been attempted to provide more accurate and useful procedures for microbial biomass measurements.

13.2.1 Chloroform Fumigation–Extraction Method for Microbial Biomass C and N

Here I outline microbial biomass C and N measurement by the fumigation–extraction procedure according to Vance et al. (1987) with some modifications. This chloroform fumigation–extraction method is theoretically based on the quantitative extraction of C and N held in the microbial biomass. The method outlined here is very commonly used due to its simplicity and applicability for a wide group of soils. In addition, various organic forms such as soluble free sugars, carbohydrates and proteins can also be measured in the same extracts (Joergensen et al. 1996). Ninhydrin, which is a reagent forming a purple complex with various molecules, such as amino acids, peptides and proteins (Moore and Stein 1948), has been lately used as a simple an reliable parameter in microbial biomass determinations (Joergensen and Brookes 1990).

Following soil fumigation with alcohol-free chloroform, a group of samples is extracted with 0.5 M K_2SO_4; and C, extractable ninhydrin-reactive N, total N and NH_4^+ N can be determined in fumigated and non-fumigated K_2SO_4 extracts. One of the advantages of this method is that it can determine microbial biomass C and N in the same soil extract. With this technique it is necessary to use either chloroform with low levels of amylene as the stabilising agent or chloroform with ethanol as the stabilising agent (in the latter case, you need to separate the ethanol out first to remove carbon contamination).

13.2.1.1 Equipment

Spectrophotometer, TOC analyser.

13.2.1.2 Reagents

Alcohol-free chloroform, 0.5 M K_2SO_4.

To produce alcohol-free chloroform: place 300 ml chloroform in a beaker, add a stirring bar, place on the stir-heater in a fume hood and heat carefully to 65 °C or until just below boiling, then boil for 5–10 min, allow the chloroform to cool, pour it into a dark glass bottle and store it in the refrigerator.

13.2.1.3 Protocol for Extraction

1. Prepare samples from freshly collected soil (0–10 cm) after removing large pieces of plant material before pre-treatment.
2. Pass all samples through a 2-mm sieve, adjust water-holding capacity to 45–50%, pre-incubate at 25 °C for 1 week and store at 4 °C before analysis.
3. Weigh 20 g (dry weight equivalent) soil into a glass beaker (50 ml) and label (using masking tape and a pencil). The chloroform may cause the marker pens to run out, so it is best to use a pencil.
4. Put into the dessicator, add some wet paper in the base to keep humidity up and to stop soil drying out, then add your glass beakers with soil and then add the chloroform in another beaker.
5. To ensure it boils quickly but not too vigorous, add approximately 30 ml chloroform to a 50 ml glass beaker containing some anti-bumping granules (at a big pinch, they can be re-used).
6. Place the 50-ml beaker containing chloroform and anti-bumping granules into a larger glass beaker (250 ml) that contains a little hot water. Do not put in too much as you do not want the water to flood into the 50-ml beaker. If the water and chloroform mix you will get very violent bubbling, which is likely to go over the soil. Then seal dessicator and apply a vacuum so that the chloroform boils for about 2 min.

7. Leave sealed dessiccator for 5–10 min, then re-apply the vacuum for a couple more minutes of boiling (just to ensure a good fumigation). Then seal dessicator and store at 22 °C or 25 °C (there are correction factors for these temperatures) for 24 h.
8. Extract the soil with 80 ml of 0.5 M K_2SO_4 for 20 g fumigated soil; and use deionised water-washed Whatman No. 42 filter papers (and also discard the first 5–10 ml of extract). Add one similar non-fumigated soil extraction on the first day. Keep some blanks of the 0.5 M K_2SO_4 that passed through the filter paper. If there is a high carbon background, then it may be necessary to acid-wash all new plastic ware, including storage vials. The extracts can be stored at –20 °C before analysis.

13.2.1.4
Measurement of Microbial Biomass C and N

Microbial biomass C can be analysed by TOC analyser and microbial biomass N either by persulfate digestion on an autoanalyser or by ninhydrin-positive compounds on a heating block using a spectrophotometer.

Extractable C in soil extracts could be measured with an automated carbon analyser. Organic C determination can be accelerated simply by using combustion, oxidation and infrared ray absorption processes (Shibara and Inubushi 1995; Wu et al. 1990).

Ninhydrin-reactive nitrogen can be determined in soil extracts, following Joergensen and Brookes (1990). Measure absorbance colorimetrically at 570 nm after the addition of ninhydrin solution (Turgay and Haraguchi 2000).

Measure extractable total N by the total persulfate oxidation procedure. This is based on the oxidation of total N to NO_3-N in alkali at elevated temperature by using persulfate as described by Cabrera and Beare (1993). The total N oxidised to NO_3-N is reduced to NO_2-N within a copperised cadmium reduction unit and can be measured according to the modified Gries–Ilosvay method (Keeney and Nelson 1982). The determination of extractable NH_4-N colorimetrically in soil extracts is based on the original indophenol blue procedure (Alef and Nannipieri 1995).

13.2.1.5
Calculation of Microbial Biomass C and N

The C and N flush due to fumigation can be calculated from the difference between the C and N content in fumigated and in non-fumigated samples. Factors to convert the flush into biomass (flush*factor) can be found in the literature (Sparling and Zhu 1993; Wu et al. 1990).

Biomass C (BC) can be calculated as indicated by Wu et al. (1990):

BC = EC: kEC,

where EC = [(extractable C in fumigated soil extracts) – (extractable C in non-fumigated soil extracts)] and kEC = 0.45 (extractable part of microbial C after fumigation).

Biomass N can be calculated according to Jenkinson (1988):

BN = EN: kEN,

where EN = [(total N, determined in fumigated extracts) – (total N, determined in non-fumigated soil extracts)] and kEN = 0.45 (extractable part of microbial N after fumigation).

Biomass ninhydrin-reactive N (BNRN) and extractable $NH4^+$-N (ENH4) can be calculated based on the same principle of the fumigation extraction method as (Joergensen and Brooke 1990; Turgay and Haraguchi 2000):

BNRN = [(ninhydrin-N in extracts of fumigated soils) – (ninhydrin-N in extracts of non-fumigated soils)]

and:

ENH4 = [($NH4^+$-N in extracts of fumigated soils) – ($NH4^+$-N in extracts of non-fumigated soils)].

13.2.2 Hexanol Extraction Method for Microbial P

The principle of the hexanol extraction method is essentially the same as the fumigation extraction method proposed by Brookes et al. (1982) and latter by Kouno et al. (1995), except that gas/liquid chloroform is used. The P in the microbial biomass P is solubilised by hexanol and then extraction is carried out by distilled water. Soil microbial biomass P is estimated from the amount of P absorbed by resin membranes after elution with NaCl/HCl solution.

13.2.2.1 Equipment and Materials

Balance, dispenser, horizontal shaker, photometer, polyethylene tubes (50 ml).

Fresh soil samples (or moist samples stored at 4 °C), from which large pieces of organic matter and roots have been removed.

13.2.2.2 Reagents

1. Anion-exchange resin membranes BDH 55164 2S (BDH Laboratory Supplies, Poole, UK). Cut each sheet (supplied as 12×12 cm sheets) into 12 strips, approximately 6×2 cm each.

2. 0.1 M NaCl/HCl (29.22 g NaCl, 49.1 ml HCl in 5 l H_2O), 0.5 M $NaHCO_3$, hexanol, P standard solution.
3. Colour reagent for P determination following Murphy and Riley (1962). This reagent must not show any trace of blue colour (Murphy and Riley 1962).

13.2.2.3
Protocol for Extraction

The method of measurement of microbial biomass P in soil outlined here is based on the fumigation–extraction method of Kouno et al. (1995) in which chloroform is used as a fumigant. Instead of liquid chloroform as suggested by Kouno et al. (1995), we use hexanol because chloroform may dissolve the anion-exchange membranes, which would change the composition of the membrane (Else Bünemann, personal communication). Hexanol has been shown to be as effective a fumigant as chloroform (McLaughlin et al. 1986). We also use 0.1 M NaCl/HCl to elute P from the resin strips as it is efficient and is less acidic than 0.5 M HCl, thus creating fewer problems with photometric P measurement.

Protocol to prepare resin strips to make bicarbonate form:

1. Shake resin strips for 1 h in 0.5 M HCl to remove any remaining P and wash with distilled water.
2. Shake for 1 h in 0.5 M $NaHCO_3$ (prepare freshly) and wash with water.
3. Shake for 1 h in 0.5 M $NaHCO_3$, leave in $NaHCO_3$.
4. Wash 3 times with water, put in water before use.
5. Weigh moist soil equalising 2 g dry soil into tubes.
6. Add 30 ml water to all samples.
7. Add one resin strip (6×2 cm) per sample.
8. Add 1 ml hexanol (1-hexanol) to the samples that are to be fumigated.
9. Add 1 ml of a P solution containing 20 μg/ml P to samples for sorption correction.
10. Include also blanks with H_2O only (no soil) or H_2O plus P spike.
11. Shake horizontally for 16 h.
12. Rinse resin strips with H_2O, remove adhering water by shaking, put strips into a clean vial.
13. Add 30 ml of 0.1 M NaCl/HCl; allow at least 30 min in the fumehood to debubble.
14. Shake for 2 h to elute P from resin membrane.
15. Take out resin strips and store resin in HCl.

13.2.2.4
Measurement of Microbial Biomass P

Measure P concentration in NaCl/HCl elute (e.g. 0.4 ml sample + 2 ml H_2O + 0.5 ml Murphy and Riley colour reagent) in spectrophotometer at 712 nm or 880 nm.

13.2.2.5 Calculation of Microbial Biomass P

The P flush due to fumigation can be calculated from the difference between available P content in fumigated and in non-fumigated samples. Factors to convert the flush into biomass (flush*factor) can be found in the literature (Kouno et al. 1995). The P content in fumigated samples is corrected by using a sorption curve between non-fumigated (no P added) and P-spiked samples (known amount of P added, chosen in the range of expected hexanol-labile P). Check linearity of sorption curve with a range of different P spikes, using single P spike if sorption curve is linear.

13.3 Protocol for Total Microbial Activity Determination

A sensitive and rapid method for the measurement of total microbial activity using fluorescein diacetate (FDA) is described here, after Adam and Duncan (2001), modified from Schnurer and Rosswall (1982). FDA hydrolysis is widely accepted as an accurate and simple method for measuring total microbial activity in a range of soils. Colourless FDA is hydrolysed by both free and membrane-bound enzymes, releasing a coloured end-product fluorescein which can be measured by spectrophotometry. The advantage of this method is that it is simple, rapid and sensitive. Note: some part of this protocol was copied from Adam and Duncan (2001), with permission from Elsevier.

13.3.1 Equipment

Spectrophotometer.

13.3.2 Reagents

1. Potassium sulfate buffer, 60 mM, pH 6.0: dissolve 8.7 g K_2HPO_4 and 1.3 g KH_2PO_4 in approximately 800 ml deionised water. Make volume up to 1 l with deionised water. Store in 4 °C fridge and check pH on day of use.
2. Chloroform/methanol (2:1): take 666 ml chloroform into a 1-l volumetric flask. Then make volume to 1 l with methanol and mix content thoroughly.

3. FDA stock solution (1000 μg/ml): dissolve 0.1 g fluorescein diacetate (3′6′-diacetyl-fluorescein; Sigma-Aldrich Co.) in approximately 80 ml of acetone and make volume to 100 ml with acetone. Store solution at –20 °C.
4. Fluorescein stock solution (2000 μg/ml): dissolve 0.2265 g fluorescein sodium salt in approximately 80 ml of 60 mM potassium phosphate buffer, pH 7.6, and then make up to 100 ml with buffer.
5. Fluorescein standard solution (20 μg/ml): add 1 ml of stock solution (2000 μg/ml) to a 100-ml volumetric flask and then make up to 100 ml with 60 mM potassium phosphate buffer, pH 7.6.
6. Standard curve: prepare standard curve 1–5 μg/ml from 20 μg/ml standard solution by appropriate dilution in 60 mM potassium phosphate buffer, pH 7.6.

13.3.3 Protocol for Extraction

1. Weigh 2 g of moist soil (sieved <2 mm) into a 50-ml conical flask.
2. Add 15 ml of 60 mM potassium phosphate buffer, pH 7.6.
3. Add 0.2 ml of 1000 μg/ml FDA stock solution to start the reaction; prepare blanks without the addition of FDA substrate.
4. Put stoppers on the flasks and shake contents by hand. Then place flasks in an orbital incubator (100 rpm) at 30 °C for 20 min.
5. Add 15 ml of chloroform/methanol (2:1, v/v) immediately to terminate the reaction (should be done in a fume cupboard).
6. Transfer contents to 50-ml centrifuge tubes and centrifuge at 2000 rpm for approximately 3 min.
7. Filter supernatant (Whatman No. 42) into 50-ml conical flasks, then measure at 490 nm on a spectrophotometer.
8. Calculate the concentration of fluorescein using standard curve and express as μg/g fluorescein oven-dry soil.

13.4 Protocol for Soil Dehydrogenase Enzyme Analysis

Soil enzymes play an important role by catalysing many reactions and have potential as an indicator of microbial activity in soils. Dehydrogenases are oxidoreductase enzymes that take part in respiration in microbial cells (Mosher et al. 2003). Dehydrogenase activity can be measured by reduction of 2,3,5-triphenyltetrazolium chloride (INT) to a red-coloured iodonitrotetrazolium formazan (INTF), which can be measured by a colorimetric method (Friedel et al. 1994). A simple assay is outlined here, after von Mersi and Schinner (1991) and Mathew and Obbard (2001) with some modifications.

13.4.1 Equipment

Spectrophotometer.

13.4.2 Reagents

1. 2,3,5-Triphenyl tetrazolium chloride (INT): dissolve 500 mg of INT in 2 ml of dimethylformamide, make-up the volume to 100 ml with 0.1 M Tris/HCl buffer (pH 7.9) and dissolve in an ultrasonic bath (von Mersi and Schinner 1991).
2. Iodonitrotetrazolium formazan (INTF): prepare standard curve using 0–24 μg/ml INTF.

13.4.3 Protocol for Extraction

1. Collect soil cores of 0–15 cm, air-dry, sieve through 2 mm, homogenise, analyse on same day or keep at 4 °C for a short period of time before analysis.
2. Weigh 5 g of soil into a 50-ml screw-cap centrifuge tube and add 2.5 ml of deionised water.
3. Add 1 ml of Tris/INT to start the reaction; prepare blanks without the addition of INT substrate.
4. Cap the tubes and place them in an orbital incubator (100 rpm) at 40 °C for 2 h in the dark.
5. Add 10 ml of extracting solution (dimethylformamide:ethanol in a 1:1 ratio).
6. To extract the developed INTF, keep samples in the dark for 1 h, shake vigorously every 20 min and filter solution with Whatman No. 42.
7. Measure the INTF in filtrate on a spectrophotometer at 464 nm in the dark as tetrazolium compound is light-sensitive.
8. Calculate the concentration of INTF using standard curve and express as μg INTF/g oven dry soil.

13.5 Conclusions

Measurement of microbial biomass and activity is essential to investigate the functions of a microbial community. These approaches have the resolution to get a comprehensive view of various stages of microbial community changes due to anthropogenic disturbances or sustainable farming systems. While protocols are briefly outlined here, the reader should consult the references listed for greater exploration of the techniques.

References

Adam G, Duncan H (2001) Development of a sensitive and rapid method for the measurement of total microbial activity using fluorescein diacetate (FDA) in a range of soils. Soil Biol Biochem 33:943–951

Alef K, Nannipieri P (1995) Soil nitrogen. In: Alef K, Nannipieri P (eds) Methods in applied soil microbiology and biochemistry. Academic, London, pp 79–87

Brookes PC, Powlson DS, Jenkinson DS (1982) Measurement of microbial biomass phosphorus in soil. Soil Biol Biochem 14:319–329

Cabrera MR, Beare MH (1993) Alkaline persulphate oxidation for determining total nitrogen in microbial biomass extracts. Soil Sci Soc Am J 57:1007–1012

Friedel JK, Molter K, Fischer WR (1994) Comparison and improvement of methods for determining soil dehydrogenase activity by using triphenyltetrazolium chlorite. Biol Fertil Soils 18:291–296

Jenkinson DS (1988) Determination of microbial biomass carbon and nitrogen in soil. In: Wilson JR (ed) Advances in nitrogen cycling in agricultural ecosystems. CAB International, Wallingford, pp 368–386

Joergensen RG, Brookes PC (1990) Ninhydrin reactive nitrogen measurements of microbial biomass in 0.5 M K_2SO_4 soil extracts. Soil Biol Biochem 22:1023–1027

Joergensen RG, Mueller T, Volkmar W (1996) Total carbohydrates of the soil microbial biomass in 0.5 M K2SO4 soil extracts. Soil Biol Biochem 28:1147–1153

Keeney DR, Nelson MH (1982) Nitrogen – inorganic forms In: Page AL, Miller DR, Keeney DR (eds) Methods of soil analysis, part 2, chemical and microbiological methods. ASA/SSSA, Madison, pp 643–698

Kouno K, Tuchiya Y, Ando T (1995) Measurement of soil microbial biomass phosphorus by an anion exchange membrane method. Soil Biol Biochem 27:1353–1357

Machulla G (2003) Soil microbial indicators and their environmental significance. J Soil Sediment 3:229

Mathew M, Obbard JP (2001) Optimisation of the dehydrogenase assay for measurement of indigenous microbial activity in beach sediments contaminated with petroleum. Biotechnol Lett 23:227–230

McLaughlin MJ, Alston AM, Martin JK (1986) Measurement of phosphorus in the soil microbial biomass: a modified procedure for field soils. Soil Biol Biochem 18:437–443

Mersi W von, Schinner F (1991) An improved and accurate method for determining the dehydrogenase activity in soils with iodonitrotetrazolium chloride. Biol Fertil Soils 11:216–220

Moore S, Stein WH (1948) Photometric ninhydrin method for use in chromatography of amino acids. J Biol Chem 243:6281–6283

Mosher JJ, Levison BS, Johnston CG (2003) A simplified dehydrogenase enzyme assay in contaminated sediment using 2-(p-iodophenyl)-3(p-nitrophenyl)-5-phenyl tetrazolium chloride. J Microbiol Methods 53:411–415

Murphy J, Riley JP (1962) A modified single solution method for the determination of phosphate in natural waters. Anal Chem Acta 27:31–36

Schnurer J, Rosswall T (1982) Fluorescein diacetate hydrolysis as a measure of total microbial activity in soil and litter. Appl Environ Microbiol 43:1256–1261

Shibara F, Inubushi K (1995) Measurement of microbial biomass C and N in paddy soils by the fumigation–extraction method. Jpn J Soil Sci Plant Nutr 41:681–689

Sparling GP, Zhu C (1993) Evaluation and calibration of methods to measure microbial biomass C and N in soils from Western Australia. Soil Biol Biochem 25:1793–1801

Turgay OC, Haraguchi A (2000) Different indices in soil microbiological activities: measurements of soil microbial biomass carbon, nitrogen and NRN. Proc Int Symp Desertification 2000

Vance ED, Brookes PC, Jenkinson DS (1987) An extraction method for measuring soil microbial biomass. Soil Biol Biochem 19:703–707

Wu J, Joergensen RG, Pommerening B, Chaussod R, Brookes PC (1990) Measurement of soil microbial biomass C by fumigation–extraction – an automated procedure. Soil Biol Biochem 22:1167–1169

14 Immuno-Technology for the Localization of Acid Phosphatase Using Native Gel Bands in *Piriformospora indica* and Other Soil Microorganisms

R. Malla, U. Pokharel, R. Prasad, R. Oelmüller, and A. Varma

14.1 Introduction

Piriformospora indica a newly discovered model endophyte, has been shown to transfer phosphate (P) from the external medium into the roots of the plants (Varma et al. 2000, 2004; Malla et al. 2002). The axenic cultivability of this fungus provides an opportunity to study the enzyme involved in phosphate metabolism, purification and the biochemical and immunological characterization of this enzyme, including a comparative study in isozyme polymorphism, the use of tools like two-dimensional PAGE and molecular markers like random amplified polymorphic DNA (RAPD) to establish the variability between *P. indica* and the closely related organism *Sebacina vermifera*.

14.1.1 Taxonomic Status

Proteomics and genomics data about this fungus were recently presented (Peskan et al. 2004; Kaldorf et al. 2005; Shahollari et al. 2005). Extrapolating from known rDNA sequences in the Sebacinaceae, it is evident that there is a cosm

Utprekshya Pokharel: Department of Microbiology, Punjab University, Chandigarh, India

Ralf Oelmüeller: Institute of General Botany, Department of Environmental Sciences, University of Jena, Dornburger Street 159, D07743, Jena, Germany

Ram Prasad; Ajit Varma: Amity Institute of Microbial Science, Amity University, Uttar Pradesh, Sector-125, Noida, Uttar Pradesh, India

Rajani Malla: Department of Microbiology, Tri-Chandra Campus, Tribhuvan University, Kathmandu, Nepal, Tel: 009771-4670857, 00977-9851013734,
email: rajanimalla2000@yahoo.com

Soil Biology, Volume 11
Advanced Techniques in Soil Microbiology
A. Varma, R. Oelmüller (Eds.)

of mycorrhizal biodiversity yet to be discovered in this group. Taxonomically, the Sebacinaceae is recognized within a new order, the Sebacinales (Weiß et al. 2004). The order primarily contains the genera *Sebacina*, *Tremelloscypha*, *Efibulobasidium*, *Craterocolla* and *Piriformospora*.

14.1.2 Phosphatases

Phosphatases play an important role in the P metabolism of the organism by hydrolysis of polyP and organic phosphates (Pasqualini et al. 1992). Phosphatases are enzymes of wide specificity, which cleave phosphate ester bonds, and this plays an important role in the hydrolysis of polyP and organic phosphates. In fungi, these phosphatases may be located in the periplasmic space, cell wall, vacuoles and culture medium. Acid and alkaline phosphatases are the two forms of active phosphatase. Alkaline phosphatase (ALPase) occurs in roots mainly after mycorrhizal colonization and has been proposed as a marker for analyzing the symbiotic efficiency of root colonization (Tisserant et al. 1993). However, the purification of enzyme has so far been unsuccessful (Kojima et al 2001). Extra- and intracellular phosphatases are responsible for the hydrolysis of various forms of phosphates, including complex and insoluble phosphates. Acid phosphatase (ACPase) has been found to be mainly involved in the uptake of P by fungal mycelia and ALPase is linked with its assimilation (Fries et al. 1998). ACPase in soil originates from both plants and fungi, while ALPase is believed to be of purely microbial origin (Gianinazzi-Pearson and Gianinazzi 1978; Tarafdar and Rao 1996). Studying ACPases is difficult due to their multiform occurrence in organisms, their relative nonspecificity, small quantity and their instability in dilute solution. Their study is also complicated by wide variations in the activity and property of isozyme between species and between different developmental stages (Alves et al. 1994).

14.1.2.1 Acid Phosphatase

Extracellular ACPase is usually localized in the cell wall, outer surface of root epidermis cells and the root apical meristem. Intracellular ACPase appears to be much less stable than extracellular forms, which remain stable for hours to days (Miller et al. 2001). In plants ACPases activity is increased by salt and osmotic stress (Ehsanpour and Amini 2003).

The first truly secreted *ACPase* gene to be characterized was from *Arabidopsis* (*AtsAPase*; Haran et al. 2000). sAPase, the white lupin ACPase, is a glycoprotein.

Protein blots probed with antibodies for sAPase showed rapid accumulation of the protein in P-deficient roots accompanied by secretion into the rhizosphere (Miller et al. 2001).

14.1.2.2 Alkaline Phosphatase

Alkaline phosphatase occurs in roots mainly after mycorrhizal colonization and has been proposed as a marker for analyzing the symbiotic efficiency of mycorrhizal colonization (Tisserant et al. 1993). ALPase is active in alkaline conditions; and ALPase activity is shown to increase sharply prior to mycorrhizal stimulation of plant growth and then to decline as the mycorrhizal colonization ages and P accumulates within the host. Saito (1997) found that ALPase activity is localized in the arbuscular hyphae of *Gigaspora margarita* and that glucose is one of the carbon sources from host plant to arbuscular mycorrhizal (AM) fungi. Phosphate efflux from the fungi to the host plant at arbuscules is supported by the recent discovery of novel plant P_i transporters that are localized around arbuscules and acquire P_i from the fungi (Rausch et al. 2001; Harrison et al. 2002; Paszkowski et al. 2002).

Purification of this enzyme has so far been unsuccessful (Kojima et al. 2001); and little is known about the enzymatic characteristics of the ALPase in AM fungi. The AM ALPase is expressed under symbiotic conditions and it may have a role in nutrient exchange with host plants rather than in nutrient uptake from the rhizosphere (Aono et al. 2004). However, the function of arbuscular ALPase in symbiosis is still little known, and cloning of the enzyme may shed light on its unknown function. A cDNA clone showing similarity to the yeast ALPase gene, *PHO8* (Keneko et al. 1987), was found in an expressed sequence-tagged (EST) library constructed from the extraradical hyphae of *Glomus intraradices*. Using this clone, Aono et al. (2004) cloned the ALPase gene from the AM fungi *Gl. intraradices* and *Gi. margarita* for the first time.

14.1.2.3 Acid Phosphatases in *P. indica*

The fungus *P. indica* produces only one form of intracellular ACPase irrespective of the phosphate concentration. The enzyme is possibly a constitutive enzyme showing a molecular mass of 66 kDa, as separated by SDS-PAGE (Malla et al. 2004). The enzyme shows the pH and temperature optima of 5.3 and 40 °C, respectively. The K_m for *p*-NPP (monoester) is 0.35 mM. Antibodies were raised against cytosolic ACPase, and using a gel band in native PAGE after selective precipitation with ammonium sulfate, followed by gel filtration and ion exchange chromatography, gave sufficient quantities of antibodies based on immunoblot analysis. Its reaction with native protein as well as denatured protein is significant. The antibody immuno-precipitates a single band of approximately 66 kDa protein in SDS gel. The antiserum localized the enzyme on the vacuoles, cell wall and cytoplasm of the mycelium, indicating the possible sites of phosphate metabolism (Malla and Varma 2004).

14.2 Immunotechnology for the Detection and Localization of Acid Phosphatase in *P. indica*

14.2.1 Extraction of Protein and Enzyme Assay

1. Materials and equipment
   Hill and Kafer medium, centrifuge, spectrophotometer.
2. Reagents and solutions
   Disodium *P*-nitro phenyl phosphate (Sigma N-2640), sodium acetate buffer (0.05 M, adjusted to pH 5.3), 0.05 M NaOH, phosphate buffer saline (PBS; pH 7.4), protein extraction buffer (Rosendahl 1994), Tris-HCl (10 mM), $NaHCO_3$ (10 mM), $MgCl_2$ (10 mM), Na_2EDTA (0.1 mM), β-mercaptoethanol (10 mM), sucrose (150 g/l), Triton X-100 (1 ml/l), protease inhibitors from stock (–80 °C), dissolved in distilled water and pH adjusted to 8.0.
3. Experimental procedures
   a. Inoculate usually 6–8 actively growing agar discs of fungus into 500-ml Erlenmeyer flasks containing 250 ml of Hill and Kafer broth. Incubate the flasks at 28±2 °C, with constant shaking at 120 rpm on a rotary shaker (GFL 3019; Germany).
   b. Harvest biomass and homogenize using extraction buffer, pH 7.4, with liquid nitrogen.
   c. Centrifuge at 12 000 g for 20 min. The crude enzyme extract can be stored at –80 °C in aliquots. Using those aliquots, acid phosphatase activities are determined spectrophotometrically using *P*-nitro phenyl phosphate as substrate.
4. Calculation
   Enzyme activity unit/mg = (EΔ410 nm/min × Total volume × D.F.)/18.5 (Enz. Vol) mg/ml
   Specific Activity = [Protein of interest (units)]/Total Protein conc. Mg

14.2.2 Purification of Protein by Column Chromatography

This method follows Cutler (2001). Materials: Sephadex G-100 (Sigma; fractionation range: 5000–100 000), chromatography column (22.5×38.0 cm) plugged with a small amount of glass wool

14.2.2.1 Experimental Procedure

1. Packing of the column:
   a. Allow the dry gel to swell in excess water and leave to stand for 3 days with intermittent stirring and decantation. Wash with buffer (PBS) and pack into the column. Prior to packing the column the matrix should made in the form of thick slurry.
   b. Pour the slurry into the column slowly with the help of glass rod and allow to settle. Repeat this until the required volume is reached. Note: special care should be taken to avoid introduction of air bubbles while the column is packing.
2. Application of protein sample:
   a. Once the column is packed, allow the buffer to run through the column overnight for equilibration and stabilization. Disconnect the column from the buffer reservoir and allow the gel to just run.
   b. Carefully apply the protein sample (partially purified by ammonium sulfate precipitation, then dialyzed; Fig. 14.1) before drying the gel and running it down the inside wall of the column so that the gel bed is not disturbed. When the samples enter the bed, gently overlay the gel with buffer and reconnect the column to the buffer reservoir.
   c. Record the flow rate approximately 0.5 ml/min. Collect fractions of 4 ml each and test for enzyme activity and protein concentration.
   d. Store the column in a buffer containing 0.02% sodium azide to prevent microbial growth, while not in use.

14.2.3 Purification of Protein by Ion Exchange Chromatography

This method follows Bollag et al. (1996). Materials: DEAE-Sephadex Tris buffer, pH 7.4.

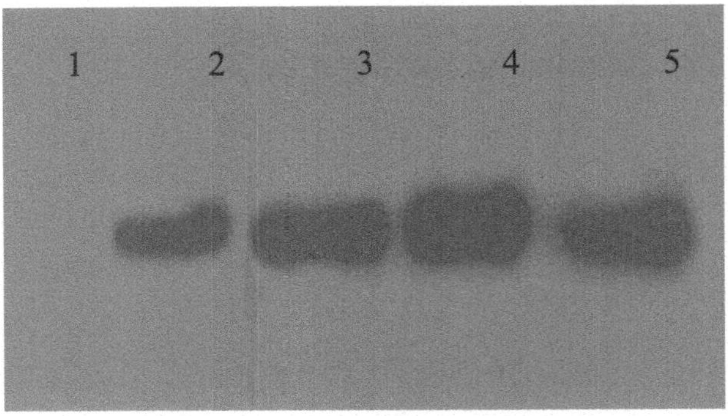

Fig. 14.1 Fractionation of ACPase by ammonium sulfate precipitation. Saturation in each lane: *lane 1* 50%, *lane 2* 60%, *lane 3* 70%, *Lane 4* 80%, *lane 5* 90%. The crude extract fractionated by each different % saturation of ammonium sulphate was centrifuged at 12 000 rpm for 30 min. The pellet obtained after centrifugation was dissolved in 80 mM phosphate buffer saline (PBS). Total protein was estimated by the method of Bradford (1976). 50 μg of total protein was loaded in each well of 10% native PAGE. After separation for 6 h the gel was neutralized with 50 mM sodium acetate buffer and assayed with 2 mg/ml *p*-nitrophenyl phosphate (*p*-NPP). The optimum % saturation of ammonium sulfate for ACPase is 80%

14.2.3.1 Experimental Procedure

The appropriate pH for ion exchange chromatography can be determined by test tubes experiments. Take nine test tubes, each containing 1 ml of ion exchanger DEAE Sephadex already soaked and washed with appropriate buffer, pH ranging from 5.5 to 9.0. Remove the excess buffer and add 100 μl of protein solution to each tube. Mix the tube and allow the matrix to settle for a few minutes. Test the supernatant for protein of interest. The best pH which bound the protein can selected for ion exchange.

1. Preparation of matrix and packing of column:
   As above.
2. Sample application:
   a. Drain the column until the buffer reach the surface of the matrix bed and close the column outlet.
   b. Gently apply the sample pooled from gel filtration chromatography onto the bed surface with the help of a pipette.
   c. Open the column outlet until the sample enters the matrix. Add buffer gently to the bed surface and then hook up buffer reservoir.
   d. Remove the unbounded protein by washing. After washing the protein elute with 0.1 M NaCl.

14.2.4 Native Polyacrylamide Gel Electrophoresis

This native PAGE method follows Walker (1994). Reagents and materials: 10% acrylamide solution from stock, separating gel buffer (1.5 M Tris-HCl, pH 8.8), stacking gel buffer (0.5 M Tris-HCl, pH 6.8), 10% ammonium persulfate in water, 0.04% TEMED sample buffer (5×), electrophoresis buffer, staining solution, de-staining solution, micro-syringe for loading samples.

14.2.4.1 Experimental Procedure

1. Transfer 10% polyacrylamide gel solution to the gel cassette by running the solution carefully down one edge between the glass plates until it reaches a position 1 cm from the bottom of the sample loading comb.
2. To obtain a smooth surface carefully run a distilled water and butanol mixture down one edge into the cassette, using a Pasteur pipette.
3. Allow the gel to polymerize. After polymerization of separating gel, pour-off the overlaying water mixture and dry the surface with the application of Whatman filter paper.
4. Apply 4% stacking gel to the gel cassette until the solution reaches the cut away edge of the gel plate.
5. Place well forming comb into this solution and allow to polymerize. This preparation takes about 30 min.
6. Carefully remove the comb and spacer after the gel sets and assemble the cassette in the electrophoresis tank filled with electrophoresis buffer.
7. Mix 50 µg each samples (1 µg/ml) with 5× sample loading buffer.
8. Centrifuge for 5 min at 5000 rpm in micro-centrifuge.
9. Load the samples onto the gel wells with the help of a Hamilton micro-syringe or gel loading tips. The dense sample settles to the bottom of the loading well.
10. Connect the power pack to the apparatus and run the proteins in stacking gel at a constant voltage of 70 V; and run in the separating gel at 120 V until the dye front reaches the bottom of the plate, 1 cm above the edge. Note: The native PAGE is run in a cold room maintained at 4 °C.

After completion of electrophoresis, the gel is subjected to:

a. ACPase enzyme assay.
b. Gel staining for 60 m in Coomassie blue.

Observe the resulting bands and compare with bands in gel enzyme assay. Note: One can identify the desired protein in Coomassie-stained gel. Caution: Acrylamide is neurotoxic even at minimal doses. Normally a small portion of gels remains unpolymerized even after electrophoresis, Gloves must be worn at all times when making or handling gels.

14.2.5 Detection of Enzyme in Native PAGE

This protocol follows Walker (1996). Reagents: sodium acetate buffer (50 mM, pH 5.3), *p*-nitrophenyl phosphate di-sodium salt (2 mg/ml; Sigma Chemical Co.).

14.2.5.1 Experimental Procedure

1. For the detection of protein for their biological activity, duplicate samples can be run in native gel. One set of samples can bestained by Coomassie for all protein bands and another set for phosphatase activity.
2. Equilibrate the gel in 50 mM sodium acetate buffer, pH 5.3, for 30 min at 4 °C in cold room.

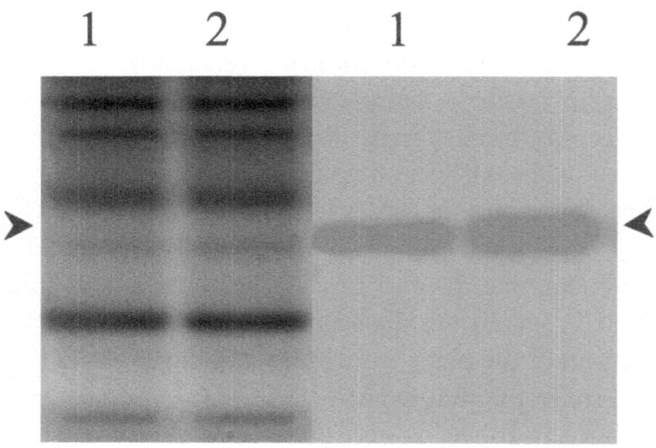

Fig. 14.2 Location of acid phosphatase in native gel. Native PAGE of *Piriformospora indica* was stained by Coomassie blue (*left*) and gel assayed (*right*) using *p*-NPP as substrate. *Lane 1* CF, *lane 2* W/MF. Separation was done in 8% gel at 4 °C for 6 h. Duplicate samples were run. One set of samples was stained for protein profile with Coomassie blue (*left*) and the other set for acid phosphatase activity, washing the gel in 2 mg/ml substrate solution that gave a yellow colored *p*-nitrophenol product at the site of enzyme. The *arrows* represent the bands of acid phosphatase

3. Immerse the gel in solution containing 2 mg/ml of the enzyme substrate (*p*-NPP) in a shaking water bath until a yellow color develops (Fig. 14.2).

14.2.5.2
Elution of Enzyme

This process is modified from that of Summers and Szewczyk (1996). Materials and reagents: transilluminator, sharp razor blade, dissecting scissors, electro-blotting apparatus, nitrocellulose membrane (NC), Ponceau S stain, Ponceau S (0.5%), acetic acid (10%), elution buffer 1 and elution buffer 2.

1. Elution Buffer 1: NH_4HCO_3 (50 mM), SDS (4%), PMSF (2 mM), DTT (2 mM), TPCK (50 μM), benzedene (50 μM), DTT (2 mM).
2. Elution Buffer 2: SDS (2%), Tritan X-100 (1%), Tris-HCl (50 mM), pH 9.5.

14.2.5.3
Experimental Procedure

Run 6% native polyacrylamide gel and assay for ACPase in presence of 2 mM *p*-NPP. The yellow band formed can be subjected to either of the following processes.

1. Manual cutting of bands:
   a. Cut the gel into small pieces and pass through different pore sized needles with the help of plastic disposable syringe along with elution buffer 1.
   b. Transfer the gel to a Falcon tube. Boil the mixture over a boiling water bath for 6 min then keep at 60 °C overnight in a water bath.
   c. Centrifuge at 13 000 rpm in a spin filter (Fig. 14.3).
   d. Collect the supernatant and separate in 12% SDS-PAGE with molecular marker.

2. Electro-blotting:
   a. Electro-blot the gel onto NC membrane (see the methodology for Western blot, Section 14.2.10).
   b. After transfer, stain the NC membrane with Ponceau S for 5 min. Excise the bands of interest with scissors.
   c. Destain the membrane with distilled water. Place the membrane in an Eppendorf tube and add 0.2 ml of elution buffer 2 per cm^2 of membrane.
   d. Mix well by vortexing the immobilon in eluant for 10 min.
   e. Spin down the immobilon for 5 min. Collect the supernatant.

Note: Elution is necessary for the determination of molecular size of the protein, which can be achieved by SDS-PAGE.

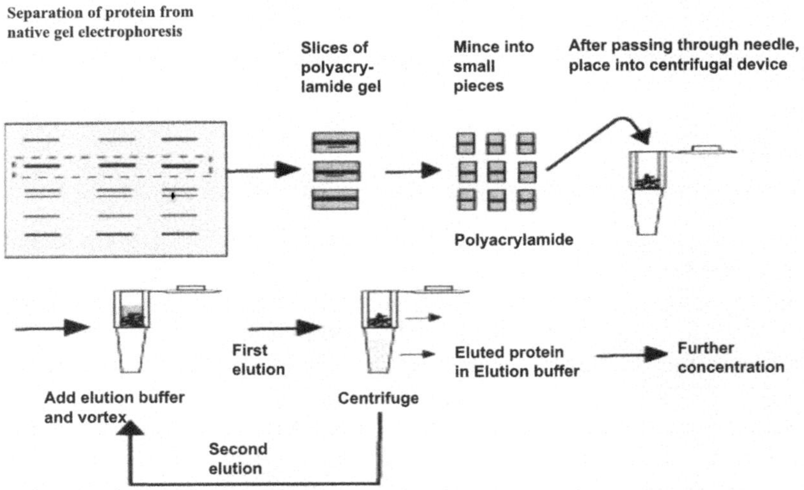

Fig. 14.3 Diagrammatic protocol for the elution of protein from polyacrylamide gel electrophoresis

14.2.6 Isolation of Acid Phosphatase for Raising Antibody

Reagents: same as above (14.2.4 Native PAGE).

14.2.6.1 Experimental Procedure

The overall process of separation of protein in native gel is given in Section 14.2.4. The only difference is that the comb in stacking gel is inserted in an inverted position to obtain a big well which can hold about 2 ml of the partially purified fraction of ACPase from ion exchange chromatography.

14.2.7 Production of Antibodies using Acid Phosphatase in Native Gel

The method used here (Malla and Varma 2004a) is essentially a modification of Amero et al. (1996). The difference is that, instead of SDS-PAGE, the protein is separated in native PAGE or in a nondenatured form.

Reagents: glutaraldehyde, Freund's complete and incomplete adjuvant, PBS, pH 7.4, *p*-nitrophenyl phosphate di-sodium salt (Sigma; 2 mg/ml).

14.2.7.1
Experimental Procedure

1. After completion of electrophoresis assay, view the gel for ACPase on a trans-illuminator and cut out the yellow bands of interest manually with a razor blade.
2. Then crosslink proteins in the gel by immersing the gel by gentle shaking in 2% glutaraldehyde for 40 min (Reichli 1980). This step minimizes loss of proteins during subsequent washing of the gel and enhances the immunological response by polymerizing the proteins.
3. Glutaraldehyde can be removed by washing with PBS, several times at a regular interval of 20 min. Caution: Any residual glutaraldehyde is toxic to animals. Residual glutaraldehyde can easily be detected by smell. Subsequent washing also removes nitrophenol (yellow color) produced during the enzyme assay and the process may also remove lots of undesirable elements from the gels, including unpolymerized acrylamide, which is very harmful to animals.
4. Preparation of antigen:
   a. Pass the polyacrylamide and PBS mixture through different pore sized needles ranging from 18G to 24G.
   b. Mix approximately 200 μg of protein from one syringe with 500 μl of Freund's complete or incomplete adjuvant into another 3-ml syringe with the help of a disposable emulsifying adapter until a uniform white and viscous emulsion is formed (Fig. 14.4).

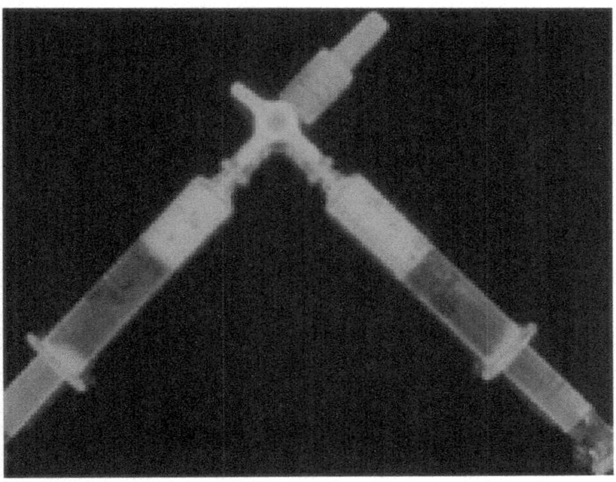

Fig. 14.4 Emulsification of antigen. The protein is mixed with the help of a disposable emulsifying adapter until a uniform white and viscous emulsion is formed

Table 14.1 Immunization schedule for the production of antibodies against purified acid phosphatase. The adjuvants used were from Sigma (Hahn et al. 1998). *s.c.* Subcutaneous

Day	Target	Amount of antigen	Adjuvant
0	s.c.	Approx. 200 μg of antigen in polyacrylamide gel	Freund's complete, 500 μl
14	s.c.	Approx. 200 μg of antigen in polyacrylamide gel	Freund's incomplete, 500 μl
28	Bleeding for antiserum		
29	s.c.	Approx. 200 μg of antigen in polyacrylamide gel	Freund's incomplete, 500 μl
42	Bleeding for antiserum		
43	s.c.	Approx. 200 μg of antigen in polyacrylamide gel	Freund's incomplete, 500 μl
56	Bleeding for antiserum		
57	s.c.	Approx. 200 μg of antigen in polyacrylamide gel	Freund's incomplete, 500 μl
71	Bleeding for antiserum		

c. Inject the prepared antigen sub-cutaneously in rabbits, according to the schedule given in Table 14.1.

14.2.8 Antiserum Preparation

1. Apparatus: centrifuge
2. Experimental procedure
   a. Allow the collected blood to clot normally for 2 h at room temperature followed by overnight at 4 °C to allow clot to retract.
   b. Loosen the clot from the side of the tube walls gently with a wooden applicator stick.
   c. Separate the upper straw-colored liquid, centrifuge at 8000 rpm for 30 min at 4 °C in a micro-centrifuge to remove remaining blood cells and debris.

d. The supernatant thus obtained can be used as raw serum, which may be stored frozen for long period of time in screw-top-tubes, at least 6 months at –20 °C and for several years at –70 °C in aliquots.

Note: assay of antibody titer (detection of antibodies) can be done by double and single immunodiffusion, enzyme linked immunosorbent assay (ELISA).

14.2.9 Purification of Immunoglobulin from Serum

14.2.9.1 Fractionation by Ammonium Sulphate

This method follows Heide and Schwick (1978).

1. Reagents: saturated ammonium sulfate solution, PBS, pH 7.4.
2. Experimental procedure
   a. Precipitate the immunoglobulin (IgG) fraction in the antiserum up to 50% by slowly adding an equal amount of saturated ammonium sulfate solution drop-wise while gently stirring the sample at 4 °C for 2 h.
   b. Centrifuge at 8000 rpm for 20 min at 4 °C. Discard the supernatant and drain the pellet (carefully invert the tube over a paper tissue).
   c. Dissolve the precipitate in 10–20% of the original volume in PBS or other buffer by careful mixing with a wide-gage Pasteur pipette.
   d. When fully dispersed, add more buffer to give 20–50% of the original volume and dialyze against the required buffer (e.g., PBS) at 4 °C overnight with three changes of buffer.

Caution: Immunoglobulin can be stored at –20 °C in aliquots (Harris 2001), for later use or further purification by DEAE-Sepharose CL-6B.

14.2.9.2 Purification by DEAE-Sephadex CL-6B Chromatography

DEAE-Sephadex ion exchange chromatography yields IgG purified from other immunoglobulin subclasses and most serum proteins (Johnstone and Thorpe 1996).

1. Reagents and materials: 0.02 M sodium phosphate equilibration buffer, pH 7.4 ($NaH_2PO_4.H_2O$ and Na_2HPO_4), 0.1 M NaCl, sodium azide, chromatography column, DEAE-Sephadex CL-6B.
2. Experimental procedure, sample application and elution

a. After equilibration of the column with equilibration buffer, disconnect the column from buffer and apply about 5 ml of the dialyzed serum sample to the column without disturbing the gel bed.
b. After applying the sample to the gel base, gently overlay the gel with buffer and reconnect the buffer to the reservoir.
c. Elute the fractions with 0.1 M NaCl. Collect the fractions (4 ml each) which contain IgG and monitor the absorbance of the elute at 280 nm.
d. After use, store the column with sodium azide solution to prevent bacterial contamination.

14.2.10 Western Blot

This Western blot protocol is after Towbin et al. (1979).

1. Reagents: transfer buffer (containing Tris-base, glycine, methanol, SDS), washing buffer [Tris buffer saline with Tween-20 (TBST)], blocking buffer [5% BSA (A4503; Sigma, St. Louis,USA) in 25 mM TBS], dilution buffer (1% BSA in 25 mM TBS), Ponceau S stain, substrate solution [3,3'-diamino-benzedine tetra hydrochloride (DAB; Sigma), in combination with urea peroxide], antiserum (diluted in 1% BSA in TBS), enzyme conjugated secondary antibody (HRPO, anti-rabbit IgG; A-9169, Sigma,).
2. Materials and apparatus: Bio-Rad trans-blot apparatus, orbital shaker, nitrocellulose sheet (0.45 μm pore size; Schnieder & Schuell, Germany), filter paper (Whatman 3MM, Maidstone, UK), piece of polyacrylamide gel with protein of interest or acid phosphatase, SDS-PAGE, native PAGE.

14.2.10.1 Experimental Procedure

a. Place four pieces of wetted 3MM filter paper on the cathodal side of the cassette on top of the wetted sponge pad.
b. Place the gel of protein separated in SDS- or native PAGE onto the filter papers (keep them wet), then place the nitrocellulose sheet on the gel, i.e., on the anodal side, carefully avoiding air bubbles throughout the process.
c. Place the remaining four filter papers over nitrocellulose membrane and expel all air bubbles between the nitrocellulose membrane and gel. This can be achieved by soaking the gel/nitrocellulose/filter paper assembly liberally with transfer buffer and then pressing with the help of a Teflon rod.
d. Finally place a wet sponge pad on top of the filter paper and clamp securely and tightly in the perforated cassette.
e. Then submerge the sandwich assembly in transfer tank filled with transfer buffer in a cold room at 4 °C.

f. Connect the transfer assembly to the power supply. Run electrophoresis at 40 V for 3 h.

1. Ponceau S stain
   a. Stain the stripe of the blot with 100 ml of Ponceau S for 15 min immediately after electroblotting to confirm that the polypeptides have been transferred successfully onto the filter and mark the position of markers.
   b. De-stain with de-ionized water and TBS.

2. Immuno-detection

Incubate the blot at 4 °C with the following solutions, with intervals of five washings with TBST (15 min each with gentle shaking):
   a. 50 ml of blocking buffer (5% BSA in TBS) for 2 h to block the remaining protein binding sites on the nitrocellulose.
   b. 50 ml of 1:10 dilution of antiserum purified by DEAE-Sephadex CL-6B diluted in 1% BSA in TBS overnight.
   c. 50 ml of 1:100 dilution of enzyme conjugated secondary antibody diluted in 1% BSA in TBS for 2 h.

3. Visualization of antigen–antibody complex
   a. Briefly rinse the blots twice with 50 ml sodium acetate buffer.
   b. Incubate the blot with 50 ml of diaminobenzidine (DAB) combined with urea peroxide (Sigma) until red-brown bands appear.
   c. Stop the reaction by rinsing the blot repeatedly with distilled water (Fig. 14.5).

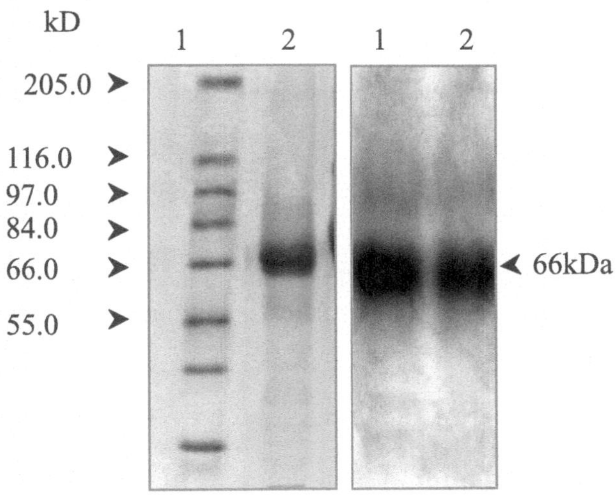

Fig. 14.5 *Left* IgG fractionation and purification steps. The antiserum raised was collected by retro-orbital bleeding, kept at room temperature for 2 h, then overnight at 4 °C. The clear serum was separated by centrifugation and purified by ion exchange chromatography using DEAE Sephadex CL-6B (*lane 2*). *Lane 1* Molecular marker (Sigma). *Right* Immunoblotting analysis of acid phosphatase with protein separated in native PAGE. *Lane 1* Cytoplasmic fraction, *lane 2* wall membrane fraction

14.2.11 Immuno-Fluorescence

This immuno-fluorescence method follows Meyberg (1998). Reagents: 3.7% paraformaldehyde, filtered through an 0.4 µm Millipore filter and mixed with an equal amount of double strength buffer, washing buffer 1 (PBS containing 100 mM glycine), permeabilizing buffer (0.1% Triton X-100 in PBS), washing buffer 2 (PBST), blocking buffer (1% BSA), PBS, pH 7.4.

14.2.11.1 Experimental Procedure

1. Cell culture
   a. Immerse the cover glasses in 50% H_2SO_4. Wash in running tap water, Sterilize under UV light for 4 h.
   b. Grow the culture in culture dishes with the cover glasses for 48 h. Drain-off culture medium and rinse cover slips with PBS.
2. Fixation
   a. Fix cells in 3.7% para-formaldehyde in PBS for 15 min at room temperature. Wash three times for 5 min each with PBS containing 100 mM glycine.
   b. Permeabilize the cells with 0.1% Triton X-100 in PBS for 4 min. Rinse with PBS.
   c. Incubate in 1% BSA in PBS, pH 7.4, for 30 min to block unspecific binding. Wash with PBST, 3×10 min. Incubate with primary antibody diluted 1:100 in 1% BSA, PBS, pH 7.4 for 60 min.
   d. Wash with PBST, 3×10 min.
   e. Incubate with FITC (F-0382; Sigma Aldrich), conjugated secondary antibody developed in goat diluted 1:100 in 1% BSA, PBS, pH 7.4, for 60 min at 37 °C.
   f. Wash with PBST, 3×10 min. Mount in PPD-mounting medium (or 90% glycerol). Observe under fluorescence microscope (model, FV-300; Olympus; Fig. 14.6).

14.2.12 Localization of ACPase by Immunogold Technique

This technique follows Botton and Chalot (1991). Reagents: phosphate buffer saline (PBS, 0.1 M), fixative (1% glutaraldehyde, 2% paraformaldehyde) filtered through a 0.45 µm pore-sized filter paper, stock solution of 4% osmium

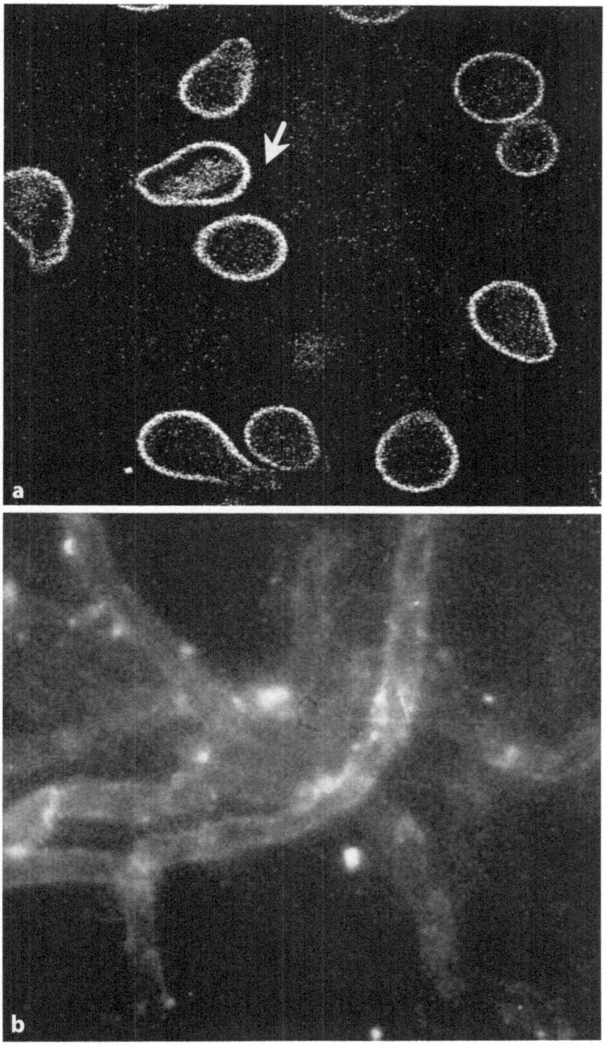

Fig. 14.6 Immunofluorescence of *Piriformospora indica* using FITC conjugated antibody. Immunofluorescence of *P. indica*. Chlamydospore. **a/b** Hyphae seen under blue filter in confocal microscopy (Olympus), using FITC (F-0382; Sigma Aldrich) conjugated second antibody developed in goat. The characteristic fluorescence pigments restricted at the chlamydospore cell wall may be due to low penetration power. The characteristic fluorescence is distributed uniformly in the mycelium

tetroxide, primary antibody (raised against acid phosphatase in rabbit), secondary antibody (anti-rabbit Goat IgG conjugated with gold particles), stain (uranyl acetate, lead citrate).

14.2.12.1
Experimental Procedure

1. Fixation:
   a. Fix the 4-day-old samples at 4 °C for 18 h.
   b. Wash the tissues in fresh buffer and post-fix for 2 h in 1% osmium tetroxide (Palade 1952) in the same buffer at 4 °C.
2. Embedding procedure:
   After fixation, dehydrate the specimens in graded alcohol/acetone solutions and embed in LR white resin.
3. Ultra-thin sectioning and staining:
   a. Cut ultra-thin sections (60–90 nm thick) on an ultra-microtome diamond knife. From the good portion of the knife-edge, cut silver sections.
   b. Spread the ribbons containing silver sections by toluene.
   c. Pick up the ribbons on the shining surface of the nickel grids.
   d. Carefully rinse the grid in perfectly clean, de-ionized water several times to remove all dirt from the ribbon.
4. Labeling the grid with primary antibody:
   a. Add primary antibody IgG (1:100) raised in rabbit against acid phosphatase to the nickel grid containing ultra-thin sections and keep overnight at 4 °C.
   b. Wash with 0.1 M PBS four times. Dilute IgG with 0.1% BSA in 0.1 M PBS.
5. Labeling with secondary antibody:
   a. Add anti-rabbit-goat IgG conjugated with 15 nm gold particles (1:100).
   b. Wash with 0.1M PBS and diluted with 0.1% BSA in 0.1M PBS.
   Note: To keep background labeling low short incubation time was preferred.
6. Staining the grid:
   Stain the sections then stain in 0.5% aqueous uranyl acetate (Watson 1958), for 10 min and in lead citrate (Reynolds 1963) for 5 min. The staining can be carried out in the following way:
   a. Keep one or two drops of stain on a parafilm M sheet. As this material is hydrophobic, the stain remains as a drop. The grid with a ribbon is then held over the drop of stain, keeping the shining surface of the grid downward so that the ribbon is immersed in the stain.
   b. After staining, the grid is taken out of the stain with the help of fine tweezers. Then hold the grid over a beaker vertically and run de-ionized water carefully over it.

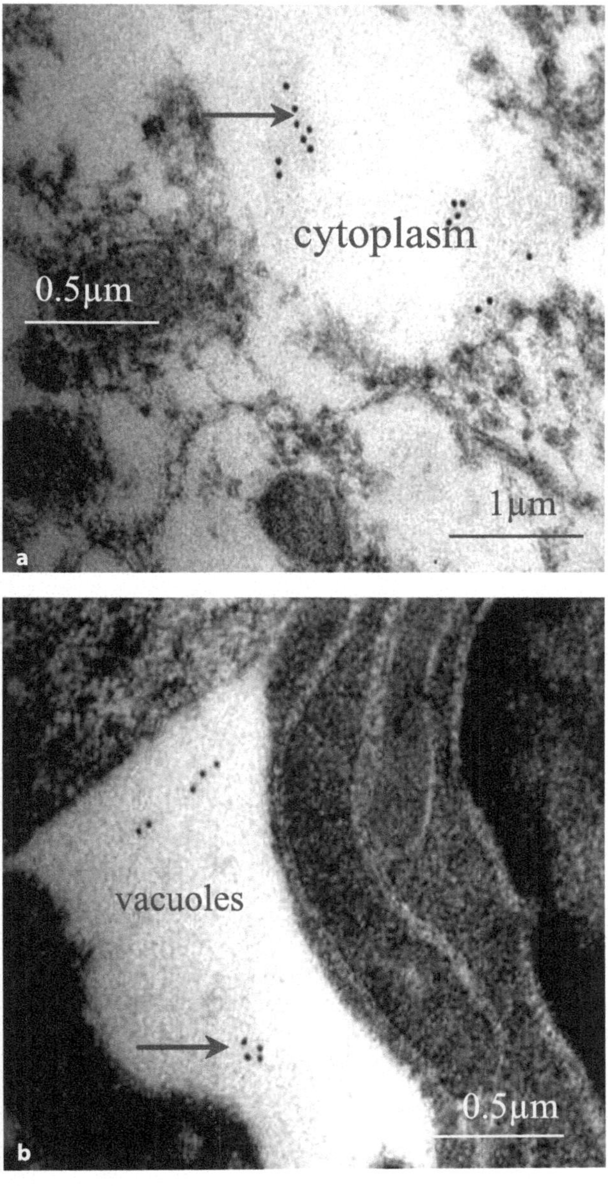

Fig. 14.7 Immunolocalization of acid phosphatase in *P. indica*, shown by electron micrographs (**a**, **bM**) of an ultrathin section of *P. indica* mycelium treated with secondary antibody (goat anti-rabbit) coupled to colloidal gold (15 nm size). *Dark dots* are gold particles indicating localization of the enzyme acid phosphatase. Localization is prominent in vacuoles, cell wall and cytoplasm. The cells were fixed with 1% glutaraldehyde and post-fixed with 1% osmium tetroxide. The primary antibody was raised against acid phosphatase

c. Continue the process for 5–6 min for complete removal of the excess stain.
d. Drain off the excess water with the help of a filter paper. Place the grid in a grid box. Observe the stained sections with a Philips CM-10 electron microscope. Operate the microscope at 60–80 kV (Fig. 14.7).

14.3 Troubleshooting

An inability to attain a high titer of antiserum after several booster injections may be due to the use of inappropriate adjuvant. Some experimentation may be necessary to optimize the antigen/adjuvant ratio for different antigens. Inadequate antigen emulsification may also result in a poor antiserum titer. Repeat the emulsification process. Be sure to use phosphate buffer saline. Avoid plastic syringes. The antigen injected may be of a poor immunogen. This method is very useful when purified protein is not available or difficult to obtain. Since nondenatured protein in native gel is used as immunogen, its reaction with native protein is very strong or it easily immunoprecipitates native protein (Malla and Varma 2004). The advantage of using native protein in gel over denatured protein in SDS gel is that the antibody generated by this method is of a high titer. When antibodies are raised against denatured proteins, they may only react with the denatured protein and may not immunoprecipitate native proteins.

The idea is to inject as much antigen as possible with a minimum amount of gel. Injection of too much gel may harm animals and also creates persistent wounds at the injection site. The animal may die during the project. Acrylamide is neurotoxic even at minimal doses. Normally a small portion of gel remains unpolymerized even after electrophoresis.

14.4 Conclusions

The immune system and the products resulting from an immune response as well as their interactions with other cellular molecules can provide powerful tools if one's conceptual approach is sound. Antibodies raised against cytosolic acid phosphatase of *P. indica* using gel bands in native PAGE, after selective precipitation of ammonium sulfate followed by gel filtration and ion exchange chromatography, gave productive antibody and immunoblotting analysis. The antibody localized the enzyme at different locations within the cell structure. Its reaction with native protein as well as denatured protein was significant.

References

Alves JM, Siachakr CD, Allot M, Tizroutine S, Mussio I, Servaes A (1994) Isozyme modifications and plant regeneration through somatic embryogenesis in sweet potato. Plant Cell Rep 13:437–441

Amero SA, James TC, Elgin SCR (1996) Production of antibody using proteins in gel bands. In: Walker JM (ed) The protein protocols handbook. Humana, Totowa, pp 717–720

Aono TIE, Dewbre GR, Maldonado-Mendoza, Harrison MJ, Saito M (2004) Expression of alkaline phosphatase genes in arbuscular mycorrhizas. New Phytol 162:525–534

Bollag DM, Rozycki MD, Edelstein SJ (1996) Protein methods, 2nd edn. Wiley–Liss, New York, pp 231–269

Botton B, Chalot M (1991) Techniques for the study of nitrogen metabolism in ectomycorrhiza. Methods Microbiol 23:203–252

Bradford MM (1976) A rapid and sensitive method for the quantitation of microgram quantities of protein utilizing the principle of protein–dye binding. Ann Biochem 72:248–252

Cutler P (2001) Chromatography on the basis of size. In: Simon RA (ed) Protein purification techniques. A practical approach, 2nd edn. Oxford University Press, Oxford, pp 187–211

Ehsanpour AA, Amini F (2003) Effect of salt and drought stress on acid phosphatase activities in alfalfa (*Medicago sativa* L.) explants under in vitro culture. Afr J Biotechnol 2:133–135

Fries LL, Pacovsky MRS, Safir GR, Kaminski J (1998) Phosphorus effect on phosphatase activity in endomycorrhizal maize. Physiol Plant 103:162–171

Gianinazzi-Pearson V, Gianinazzi S (1978) Enzymatic studies on the metabolism of vesiculararbuscular mycorrhiza. II. Soluble alkaline phosphatase specific to mycorrhizal infection in onion roots. Physiol Plant Pathol 12:45–53

Hahn AC, Goebel, Hock B (1998) Polyclonal antibodies for the detection of arbuscular mycorrhizal fungi. In: Varma A (ed) Mycorrhiza manual. Springer, Berlin Heidelberg New York, pp 255–270

Haran S, Logendra S, Saskar M, Bratanova M, Raskin I (2000) Characterization of *Arabidopsis* acid phosphatase promoter and regulation of acid phosphatase expression. Plant Physiol 124:615–626

Harris ELV (2001) Concentration of the extract. In: Simon R (ed) Protein purification techniques, a practical approach, 2nd edn. Oxford, University Press, Oxford, pp 111–154

Harrison MJ, Dewbre GR, Liu JY (2002) A phosphate transporter from *Medicago truncatula* involved in the acquisition of phosphate released by arbuscular mycorrhizal fungi. Plant Cell. 14:2413–2429

Heide K, Schwick HG (1978) Salt fractionation of immunoglobulins. In: Weir DM (ed) Handbook of experimental immunology, vol 1: immunochemistry. Blackwell Scientific, Oxford, pp 7.1–7.11

Johnstone A, Thorpe R (1996) Immunochemistry in practice. Blackwell Science, Oxford, pp 35–40

Kaldorf M, Koch B, Rexer K-H, Kost G, Varma A (2005) Patterns of interaction between *Populus* Esch5 and *Piriformospora indica*: a transition from mutualism to antagonism. Plant Biol 7:210–218

Keneko Y, Hayashi N, Toh-e A, Banno I, Oshima Y (1987) Structural characteristics of the Pho8 gene encoding repressible alkaline phosphatase in *Saccharomyces cerevisiae*. Gene 58:137–148

Kojima T, Hayatsu M, Saito M (2001) Electrophoretic detection and partial purification of the phosphatase specific for arbuscular mycorrhizal symbiosis. Bull Natl Grassl Res Inst 60:9–18

Malla R, Varma A (2004) Phosphatase(s) from microorganisms. In: Varma A, Podila GK (eds) Biotechnological applications of microbes. IK International, New Delhi, pp 125–150

Malla R, Singh A, Md Z, Yadav V, Verma SA, Rai M, Varma A (2002) *Piriformospora indica* and plant growth promoting Rhizobacteria: an appraisal. In: Rao GP, Bhat DJ, Lakhanpal TN, Manoharichari C (eds) Frontiers of fungal diversity in India. International Book Distributing, Lucknow, pp 401–419

Malla R, Prasad R, Giang PH, Pokharel U, Oelmueller R, Varma A (2004) Characteristic features of symbiotic fungus *Piriformospora indica*. Endocytobios Cell Res 15:579–600

Meyberg M (1998) Selective staining of fungal hyphae in parasitic and symbiotic plant fungus associations. Histochemistry 88:197–199

Miller SS, Liu J, Allan DL, Menzhuber CJ, Fedorova M, Vance CP (2001) Molecular control of acid phosphatase secretion in to the rhizosphere of proteoid roots from phosphorus-stressed white lupin. Plant Physiol 127:594–606

Palade GE (1952) A study of fixation for electron microscopy. J Exp Med 95:285–297

Pasqualini, Panara SF, Antoneilli M (1992) Acid phosphatase activity in *Pinus pinea*, *Tuber albidum* ectomycorrhizal association. Can J Bot 70:1377–1383

Paszkowski U, Kroken S, Roux C, Briggs SP (2002) Rice phosphate transporters includes an evolutionarily divergent gene specifically activated in arbuscular mycorrhizal symbiosis. Proc Natl Acad Sci USA 99:13324–13329

Peskan T, Shahollari B, Pham HG, Hehl S, Markent C, Blank V, Kost G, Varma A, Oelmueller R (2004) Association of *Piriformospora indica* with *Arabidopsis thaliana* roots represent a novel system to study beneficial plant–microbe interactions and involve in early plant protein modifications in the endocytoplasmic reticulum and in the plasma membrane. Physiol Plant 122:465–471

Rausch C, Daram P, Brunner S, Jansa J, Laloi M, Leggewie G, Amrhein N, Bucher M (2001) A phosphate transporter expressed in arbuscule containing cells in potato. Nature 414:462–466

Reichli M (1980) Use of glutaraldehyde as a coupling agent for proteins and peptides. Methods Enzymol 70:159–165

Reynolds ES (1963) The use of lead citrate at high pH on an electron opaque stain in electron microscopy. J Cell Biol 17:208–212

Rosendahl S (1994) Isozyme analysis of mycorrhizal fungi and their mycorrhiza. In: Norris JR, Read D, Varma A (eds) Techniques for mycorrhizal research. Academic, London, pp 629–654

Saito M (1997) Regulation of arbuscular mycorrhizal symbiosis: hyphal growth in host roots and nutrient exchange. JARQ 31:179–183

Shahollari B, Varma A, Oelmueller R (2005) Expression of receptor kinase in roots is stimulated by the basidiomycete *Piriformospora indica* and the protein accumulates in Triton-X-100 insoluble plasma membrane microdomains. J Plant Physiol 162:945–958

Summers DF, Szewczyk B (1996) Elution of SDS-PAGE separated proteins from immobilon membranes for use as antigens. In: Walker JM (ed) The protein protocols handbook. Humana, Totowa, pp 699–702

Tarafdar JC, Rao AV (1996) Contribution of *Aspergillus* strains to acquisition of phosphorus by wheat (*Triticum aestivum* L.) and chick pea (*Cicer arietinum* L.) grown in a loamy soil. Appl Soil Ecol 3:190–194

Tisserant B, Gianinazzi-Pearson V, Gianinazzi S, Gollotte A (1993) Inplant histochemical staining of fungal alkaline phosphatase activity for analysis of efficient arbuscular mycorrhizal infections. Mycol Res 97:245–250

Towbin H, Staehlin T, Gordon J (1979) Electrophoretic transfer of proteins from polyacrylamide gels to nitrocellulose sheets: procedure and some applications. Proc Natl Acad Sci USA 76:4350–4354

Varma A, Rai MK, Sudha, Sahay N (2000) Microbial biotechnology: new-paradigms and role in sustainable agriculture. In: Rajak RC (ed) Microbial biotechnology for sustainable development and productivity, Scientific Publishers, Lahore, pp 22–37

Varma A, Abbott LK, Werner D, Hampp R (2004) The state of art. Plant surface microbiology. In: Varma A, Abbott L, Werner D, Hampp R (eds) Plant surface microbiology. Springer, Berlin Heidelberg New York, pp 1–11

Walker JM (1994) Basic protein and peptide protocols. In: Walker JM (ed) The protein protocols handbook. Humana. Totowa, pp 186–187

Walker JM (1996) Non-denaturing polyacrylamide gel electrophoresis of protein. In: Walker JM (ed) The protein protocols handbook. Humana, Totowa, pp 51–54

Watson ML (1958) Staining times section for electron microscopy with heavy metals. J Biophys Biochem Cytol 4:475–478

Weiß M, Selosse MA, Rexer KH, Urban A, Oberwinkler F (2004) Sebacinales: a hitherto overlooked cosm of heterobasidiomycetes with abroad mycorrhizal potential. Mycol Res 108:1003–1010

15 Use of Short Oligonucleotide Primers in Random Amplified Polymorphic DNA Techniques for Species Identification

R. Malla and A. Varma

15.1 Introduction

The introduction of molecular techniques in biology has been a major force in the areas of systematics and population biology of the fungi. The introduction of PCR-based methods has significantly increased the level of activity. The simplicity of the techniques, coupled with the general use of particular regions of the genomes, has resulted in many important advances in our understanding of taxonomic grouping as well as the evolutionary histories and functional properties associated with them. The nuclear genomes of fungi are small, intermediate between that of prokaryotes and the higher eukaryotic plants and animals. Compared with higher plants and animals, fungi have a much lower percentage of redundant DNA. Typically 10–20% of the DNA in fungi is redundant, while as much as 80% of the DNA may be redundant in other eukaryotes (Dutta 1974). The baker's yeast *Saccharomyces cerevisiae* contains a genome of 16 chromosomes, including 13.4 million bases. Its genome displays significant redundancy, with 53 duplicated gene clusters among the 16 chromosomes. These duplicated regions represent more than 30% of the entire genome (Mewes et al. 1997). Fungi have extrachromosomal genetic elements, the most important of which are found in mitochondria. Mitochondrial (mt) genomes provide another source of genetic variability that is independent of sexual reproduction. The mitochondrial genome in fungi is usually uniparentally (maternally) inherited. The mtDNA is the useful tool for the taxonomic studies because it is relatively small, making it possible to analyze the entire genome, and its composition is not complicated by the recombination that occurs (Taylor 1986).

Variation within species can be assayed using the random amplified polymorphic DNA (RAPD) method (Welsh and McClelland 1990; Williams et al.

Rajani Malla: Department of Microbiology, Tri-Chandra Campus, Tribhuvan University, Tel: 009771-4670857, 00977-9851013734, email-rajanimalla2000@yahoo.com

Soil Biology, Volume 11
Advanced Techniques in Soil Microbiology
A. Varma, R. Oelmüller (Eds.)

1990), in which arbitrary short oligonucleotide primers, targeting unknown sequences in the genome, are used to generate amplification products that often show size polymorphism within species. RAPD analysis offers the possibility of creating polymorphism without any prior knowledge of the DNA sequences of the organism investigated. The method is fast and economic for screening large number of samples. The RAPD band pattern has been used to define some fungal species in which species-specific bands or combinations of bands have been considered. In these techniques there is the assumption that bands with identical mobility and staining intensity are of the same or very similar sequences. Characterization of species at morphological and protein level is not fully reliable since environmental conditions influence the nature of the organism to a great extent. The use of molecular markers such as RAPDs along with morphological and protein traits may provide a more clear concept of the species. RAPD markers are randomly distributed through out the genome and can be efficiently and randomly sampled using established procedures. The RAPD procedure developed by Welsh and McClelland (1990) and Williams et al. (1990) involves simultaneous amplification of several anonymous loci in the genome and has been used for genetic, taxonomic and ecological studies of several fungi (Zinno et al. 1998).

PCR-based techniques have already been applied to endo-and ectomycorrhizal fungi where morphological characters are in conflict, ambiguous and missing (Podila and Lanfranco 2004). This approach has allowed the development of molecular tools for their identification and increased the level of understanding in the molecular taxonomy of microorganisms (Solaiman and Abbott 2004; Varma et al. 2004). The most commonly used PCR-based techniques include amplification of variable regions in the ribosomal genes, restriction fragment length polymorphism (RFLP), amplification of short repeated sequences (microsatellites) and random amplification of polymorphic DNA (RAPD; Erlich et al. 1991). These techniques provide a different degree of resolution in the study of genetic polymorphisms. RAPD reveals intraspecific differences by originating DNA fingerprints, which may be unique for a single isolate (Perotto et al. 1996). Identification of individual clones is essential for the better understanding of the diversity, structure and dynamic of populations of ectomycorrhizal fungi. Unfortunately, this approach is time-consuming. RAPD (Welsh and McClelland 1990; Williams et al. 1990) has therefore been used for the analysis of populations of *Suillus granulatus* (Jacobson et al. 1993) and *Laccavia bicolor* (Buschena et al. 1992). However, this technique has been reported to be very sensitive to experimental variables and the RAPD assay conditions described for one species may not be suitable for another. Huai et al. (2003) studied the genetic variation and spatial distribution of the ectomycorrhizal fungus *Tricholoma terreum*. The 33 sporophores studied belonged to distinct genotypes, based on the analysis of RAPD markers. The genets of *T. terreum* were small and not larger than 0.5 m. Two major phenetic groups, i.e., eight individuals in group 1 and 25 in group 2, were identified by principal component analysis and by the unweighed pair group method with arithmetic means of simple matching coefficients, respectively. The application of RAPD analysis was investigated for the identification of ectomycorrhizal symbionts of spruce (*Picea abies*) belonging to the genera *Boletus*, *Amanita* and *Lactarius* at and below the species level. Using both fingerprinting [M13, $(GTG)_5$, $(GACA)_4$] and random oligonucleotide primers (V1, V5), a high degree of variability of amplified DNA fragments (band-sharing index 65–80%) was detected between different strains of the same species, hence enabling the identification of individual strains within the same species. The band-sharing index between different species of the same genus (*Boletus*, *Russula*, *Amanita*) was in the range 20–30% and similar values were obtained when strains from different taxa were compared. Thus RAPD is too sensitive at this level of relationship and cannot be used to align unknown symbionts to a given taxon. They therefore conclude that RAPD is a promising tool for the identification of individual strains and could thus be used to distinguish indigenous and introduced mycorrhizal strains from the same species in natural ecosystem. The genetic variability of *Trichoderma* isolates using the RAPD were analyzed by Góes et al. (2002), who found high intra-specific genetic variation among those fungi.

15.2 Polymorphism between *Piriformospora indica* and *Sebacina vermifera*, Members of the Order Sebacinales

Piriformospora indica, a new endophyte (Verma et al. 1998) has the ability to grow axenically. The cultivability of this fungus in different synthetic media, like *Aspergillus* medium (Malla et al. 2002; Pham et al. 2004), provided an opportunity to study the comparative isozyme polymorphism and a molecular marker like RAPD to establish variability in between *P. indica* and the closely related organism *Sebacina vermifera* (Malla et al. 2004b). The analysis of enzymes, isozymes like laccase, malate dehydrogenase, esterase, peptidase, peroxidase, acid and alkaline phosphatase and non-enzymic proteins and their mobility displayed clear variations among different species (Malla et al. 2004a).

Proteomics and genomics data about this fungus were recently described (Peskan et al. 2004; Kaldorf et al. 2005; Shahollari et al. 2005). *S. vermifera* sensu stricto consists of a broad complex of species possibly including mycobionts of jungermannioid and ericoid mycorrhizas. Extrapolating from the known rDNA sequences in the Sebacinaceae, it is evident that there is a cosm of mycorrhizal biodiversity yet to be discovered in this group (Weiß et al. 2004).

The acid phosphatases (ACPases; Fig. 15.1) in *P. indica* and *S. vermifera* sensu stricto are similar in their molecular mass. The antibody raised against the ACPase of *P. indica* showed a maximum ELISA reading with *S. vermifera* sensu stricto, supporting a strong relationship between these two fungi. The immunoblot analysis showed a strong reactivity of *P. indica* antiserum with *S. vermifera*

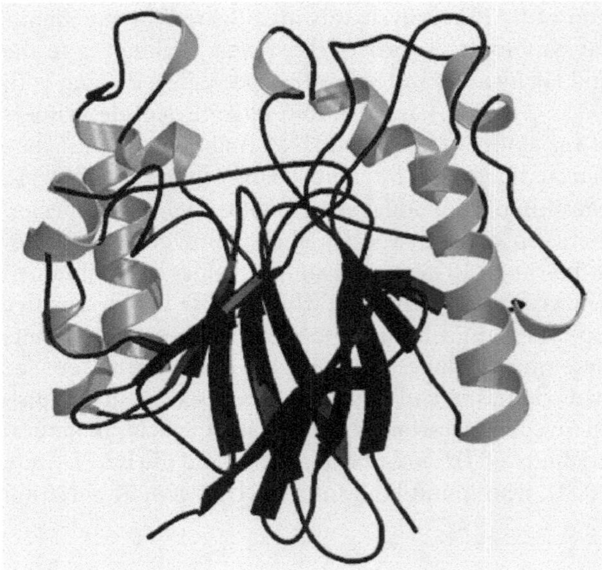

Fig. 15.1 3D structure of acid phosphatase

sensu stricto. The antiserum blotted bands at 66 kDa with *S. vermifera* separated in denatured PAGE and at a similar location with *P. indica* in non-denatured PAGE. The antiserum also localized the enzyme in *S. vermifera* by an immunofluorescence technique, showing a strong relationship of this fungus with *P. indica* (Fig. 15.2). The immunogold labeling of antiserum from *P. indica* precisely localized the enzyme in the cytoplasm and vacuoles of *S. vermifera*, supporting the strong immunological link between these two fungi. Two-dimensional maps of crude protein of these two fungi showed some differences in minor proteins. *P. indica* and *S. vermifera* sensu stricto belonging to same taxonomic group show similar morphology, functions and isozymes. However, they show distinct genetic variation based on the RAPD analysis and can be considered to be placed within species from the same ancestral root.

The study aimed to establish genetic diversity between the two species *P. indica* and *S. vermifera* sensu stricto belonging to the same order, Sebacinales. Seven random 10-bp oligonucleotide primers of different origin were used. Clustering of similarity matrices was done by the un-weighted pair group method with arithmetic mean (UPGMA) and projection by the TREE program of NTSYS-pc (Numerical Taxonomy System, Applied Biostatistic). Out of seven primers, six gave scorable, reproducible DNA products (bands) suitable for establishing a genetic diversity. UPGMA cluster analysis clustered the isolates into two distinct groups. The average genetic similarity between both fungi was 0.58 (i.e., 58%) and can be considered to place them within species from the same ancestral root. These results illustrated the potential value of RAPD techniques for detecting polymorphism among fungal isolates.

66kDa ➤
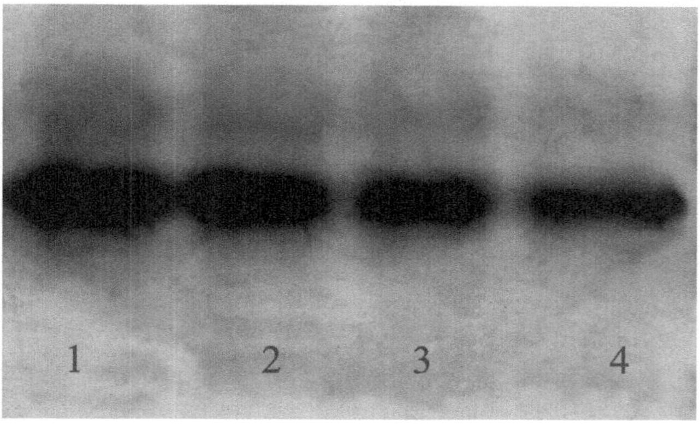

Fig. 15.2 Western blot analysis of *P. indica* and *S. vermifera* sensu stricto. Protein separated by 10% SDS PAGE transferred to nitrocellulose membrane by electroblotting. The blot were blocked using 5% bovine serum albumin and reacted with acid phosphatase antiserum and peroxidase conjugated secondary antibody, visualized using DAB. *Lanes 1, 2* Cytoplasmic fraction (CF; *lane 1*) and wall membrane fraction (W/MF; *lane 2*) of *P. indica* reacted with homogenous antiserum. *Lanes 3, 4* CF (*lane 3*) and W/MF (*lane 4*) of *S. vermifera* sensu stricto cross-reacted with *P. indica* antiserum. The result shows precisely defined bands in all samples. All blotted bands represent similarities in their molecular mass, supporting immunologically highly related species

15.3 General Protocol for RAPD Technique to Show Polymorphism

1. Equipment: thermal cycler, gel electrophoresis apparatus, band analysis software, UV transilluminator and gel documentation system. Caution: UV rays are dangerous. Protect eyes with a plastic shield.
2. Reagents (all the chemicals, primers and enzymes were obtained from Operon Technology): DNA isolation buffer (Moller et al. 1992), 2% hexadecyltrimethyl ammonium bromide (CTAB), NaCl (1.4 M), EDTA (20 mM), Tris HCl (100 mM), chloroform:isoamylalcohol (20:1), isopropanol, sodium acetate, ethanol (70%), Tris EDTA (TE, pH 8.0), Tris-HCl (pH 8.0, 10 mM), EDTA (pH 8.0, 1 mM), DNA amplification mixture for PCR, RNase A, loading buffer, bromophenol blue (0.25%), sucrose in water (40%, w/v; store in small aliquots at 4 °C), primers (short oligonucleotide), ethidium bromide (caution: ethidium bromide is a powerful mutagen; wear gloves and masks when handling and weighing). Note: all buffers, pipette tips and Eppendorfs should be sterilized at 121 °C for 15 min. Sterilize by autoclaving at 15 psi (ca. 103 kPa) for 15 min.
3. DNA amplification mixture for PCR (25 µl; Operon Technologies, Alameda, Calif.): 10× buffer (2.5 µl), $MgCl_2$ (2.5 µl), dNTPs (10 mM; 0.8 µl), primer (30 ng/µl; 1.0 µl), Taq polymerase (3 units/µl; 0.5 µl), template DNA (1 µl), Milli Q water (ultrapure, 16.7 µl).

15.3.1 Experimental Procedures

15.3.1.1 DNA Extraction

1. Carry out isolation and purification of fungal DNA following the modified CTAB protocol of Moller et al. (1992).
2. Inoculate flasks containing 100 ml Hill and Kafer medium with axenic culture of *P. indica* and place in a growth chamber at 28 °C for 6–8 days. Collect the hyphae by filtration. Keep the mycelial network at –20 °C.
3. Grind the freeze-dried mycelia (5 g) using liquid nitrogen and transfer the powdered mycelium into Eppendorf tubes (2 ml). Add equal amounts of pre-warmed isolation buffer (2% CTAB, 1.4 M NaCl, 20 mM EDTA, 100 mM Tris-HCl) as fast as possible and incubate for 30 min at 60 °C in a water bath. Gently mix after every 10 min. Add one volume of chloroform: isoamyl-alcohol (24:1).
4. Cap the tubes and shake for 10 min by hand. Mix gently but thoroughly to ensure emulsification of the phase.
5. Centrifuge the emulsion for 10 min (15 000 rpm at room temperature). Extract the upper aqueous phase with fresh chloroform isoamyl alcohol and transfer the final aqueous phase to a new Eppendorf tube using a large bore pipette.
6. After adding 0.6 vol. of ice-cold isopropanol and 0.1 vol. of sodium acetate, cap the tubes and place at –20 °C overnight and then centrifuge again at 15 000 rpm for 10 min.
7. Transfer the precipitated whitish network of DNA-CTAB complex to a new Eppendorff tube. Add washing solution (70% ethanol) and incubate for 30 min.
8. Mix gently but thoroughly by hand and centrifuge for 5 min at 8000 rpm at 4 °C. Remove residual CTAB at this step.
9. Decant the washing solution and dry the pellet at 37 °C for 3 h to ensure the removal of all parts of ethanol.
10. Add appropriate volume of 1× TE buffer and allow the pellet to dissolve at 4 °C without agitation.
11. After extraction, purify the DNA by using RNase A. Dilute the DNA in TE buffer (1×) for RAPD and store at –20 °C for further use.
12. DNA concentration can be quantified by UV spectrophotometer at 260 nm and by comparison to DNA standards by agarose gel electrophoresis.

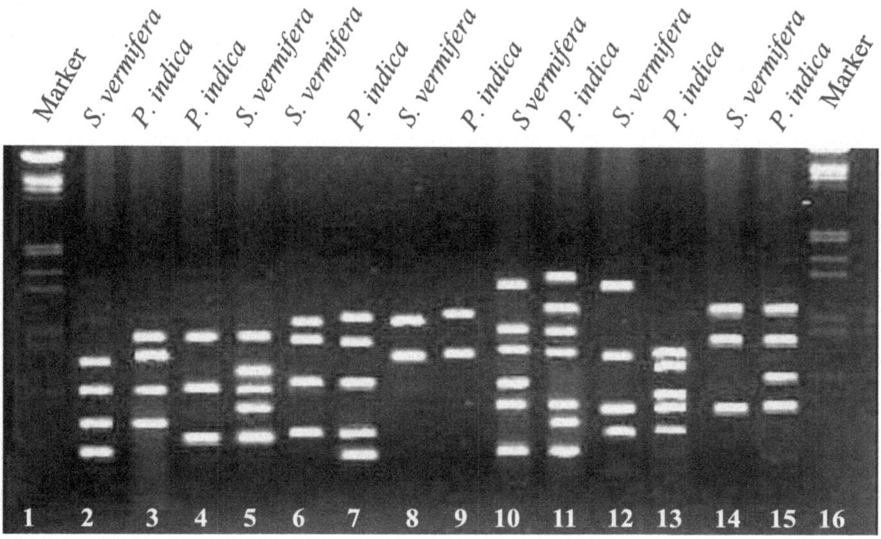

Fig. 15.3 The RAPD analysis of *P. indica* and *S. vermifera* sensu stricto to show genetic variation between these two fungi. Out of seven primers used for amplification, six have given a productive polymorphism. *Lanes 1, 16* Marker (λDNA *Eco*R1, *Hind*III). *Lanes 2, 3* Primer OPA10, *lanes 4, 5* OPD01, *lanes 6, 7* OPC06, *lanes 8, 9* OPC10, *lane 10, 11* OPC01, *lanes 12, 13* OPI04, *lanes 14, 15* OPI10. No polymorphism was observed when the genomic DNA was amplified with OPC10 (*lanes 8, 9*)

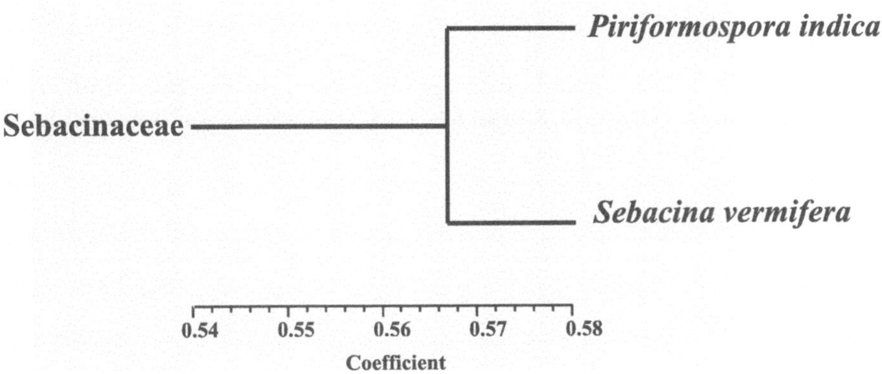

Fig. 15.4 Dendrogram showing phylogenetic relationship between *P. indica* and *S. vermifera* sensu stricto. The NTSYS-pc computer program (Numerical Taxonomy System, Applied Biostatistics) was used for data analysis

15.3.1.2
RAPD Analysis

RAPD analysis is done following Zinno et al. (1998).

1. DNA amplification is done in a total volume of 25 µl, containing 2.5 µl buffer (10× without $MgCl_2$), 2.5 µl $MgCl_2$, 0.8 µl dNTPs (10 mM), 1.0 µl primer (30 ng/µl), 0.5 µl *Taq* polymerase (3 units/µl) and DNA according to concentration use. Random 10-bp oligonucleotide primers (Operon Technologies Alameda, Calif.) are used to produce amplification: OPA10 (GTGATCGCAG), OPD01 (ACCGCGAAGG), OPC06 (GAACGGACTC), OPC10 (TGTCTGGGTG), OPC01 (TTCGAGCCAG), OPI04 (CCGCCTAGTC), OPI10 (ACAACGCGAG).
2. Each isolate is tested at a range of DNA concentrations from 0.5 µl to 2.5 µl and the clearest amplification of RAPD bands is used.
3. DNA is amplified in a PTC-200 thermal cycler (Techne, UK) with the following thermal profile: 95 °C for 5 min (initial denaturation cycle), then 36 cycles of 94 °C for 30 s (denaturation cycle), 36 °C for 2 min (annealing) and 72 °C for 2 min (extension); and a final extension at 72 °C for 5 min.
4. For separation, the amplified DNA samples are mixed with 6× loading dye and electrophoresed on 1.5% agarose gel in 1% TBE at 3.5 V/cm for 2 h, then stained with ethidium bromide and photographed under a transilluminator (Figs. 15.3, 15.4).

15.4
Troubleshooting

Only amplification products that are reproducible over two amplifications should be included. The variation in the intensity of fluorescence of different ethidium bromide-stained PCR products across the isolates was not considered for the purpose of data analysis.

15.5
Conclusions

The RAPD data confirmed that, even between these two species of Sebacinales belonging to same morpho-zymographical groups and with minor protein differences shown by 2-D PAGE, the level of variation was substantially high according to RAPD. Thus, it is suggested that such isolates should be considered as separate species. Molecular characterization offers an alternative approach for more reliable and reproductive identification at species level. By using molecular markers like RAPDs, genetic polymorphism within species can be assayed. The use of immunological, molecular and enzymological techniques has opened an important area of research in *P. indica*. This study has opened up several novel pathways which can be explored to fill some lacunae in the molecular aspects of arbuscular mycorrhizal research, since *P. indica* is an axenically cultivable fungus mimicking various AM characters.

References

Buschena CA, Doudrick RL, Anderson NA (1992) Persistence of *Laccaria* spp as mycorrhizal symbionts of container-grown black spruce. Can J For Res 22:1883–1887

Dutta SK (1974) Repeated DNA sequences in fungi. Nucleic Acids Res 1:1411–1419

Erlich HA, Gelfand DH, Sninsky JJ (1991) Recent advances in the polymerase chain reaction. Science 252:1643–1651

Góes LB, Costa ABL, Freire LLC, Oliveira NT (2002) Randomly amplified polymorphic DNA of *Trichoderma* isolates and antagonism against *Rhizoctonia solani*. Braz Arch Biol Technol 45

Huai WX, Guo LD, He W (2003) Genetic diversity of an ectomycorrhizal fungus *Tricholoma terreum* in a *Larix principis-rupprechtii* stand assessed using random amplified polymorphic DNA. Mycorrhiza 13:265–270

Jacobson KM, Miller OK, Turner BJ (1993) Randomly amplified polymorphic DNA markers are superior to somatic incompatibility tests for discriminating genotype in natural populations of the ectomycorrhizal fungus *Swillus granulatus*. Proc Natl Acad Sci USA 90:9159–9163

Kaldorf M, Koch B, Rexer KH, Kost G, Varma A (2005) Patterns of interaction between *Populus* Esch5 and *Piriformospora indica*: a transition from mutualism to antagonism. Plant Biol 7:210–218

Malla R, Varma A (2004a) Phosphatase(s) from microorganisms. In: Varma A, Podila GK (eds) Biotechnological applications of microbes. IK International, New York, pp 125–150

Malla R, Prasad R, Giang PH, Pokharel U, Oelmueller R, Varma A (2004b) Characteristic features of symbiotic fungus *Piriformospora indica*. Endocytobiosis Cell Res 15:579–600

Malla R, Singh A, Md Z, Yadav V, Suniti, Verma A, Rai M, Varma A (2002) *Piriformospora indica* and plant growth promoting Rhizobacteria: an appraisal. In: Rao GP, Bhat DJ, Lakhanpal TN, Manoharichari C (eds) Frontiers of fungal diversity in India. pp 401–419

Mewes HW, Abbermann K, Bahr M, Frishman D, Gleissner A, Hani J, Heumann K, Kleine K, Maierl A, Oliver SG, Pfeiffer F, Zollner A (1997) Overview of the yeast genome. Nature. 387:7–65

Moller EM, Bahnweg G, Sandermann H, Geiger HH (1992) A simple and efficient protocol for isolation of high molecular weight DNA from filament fungi, fruit bodies and infected plant tissue. Nucleic Acids Res 20:6115–6116

Peskan T, Shahollari B, Pham HG, Hehl S, Markent C, Blank V, Kost G, Varma A, Oelmueller R (2004) Association of *Piriformospora indica* with *Arabidopsis thaliana* roots represent a novel system to study beneficial plant-microbe interactions and involve in early plant protein modifications in the endocytoplasmic reticulum and in the plasma membrane. Physiol Plant 122:465–471

Perotto S, Actis-Perino E, Perugini J, Bonfante P (1996) Molecular diversity of fungi from ericoid mycorrhizal roots. Mol Ecol 5:123–131

Pham GH, Kumari R, Singh A, Malla R, Prasad R, Sachdev M, Kaldorf M, Buscot B, Oelmüller R, Hampp R, Saxena AK, Rexer KH, Kost G, Varma A (2004) Axenic Cultures of *Piriformospora indica*. In: Varma A, Abbott L, Werner D, Hampp R (eds) Plant surface microbiology. Springer, Berlin Heidelberg New York, pp 593–613

Podila GK, Lanfranco L (2004) Functional genomic approaches for studies of mycorrhizal symbiosis. In: Varma A, Abbott L, Werner D, Hampp R (eds) Plant surface microbiology, Springer, Berlin Heidelberg New York, pp 567–592

Shahollari B, Varma A, Oelmueller R (2005) Expression of receptor kinase in roots is stimulated by the basidiomycete *Piriformospora indica* and the protein accumulates in Triton-X-100 insoluble plasma membrane microdomains. J Plant Physiol 162:945–958

Solaiman MZ, Abbott LK (2004) Functional diversity of arbuscular mycorrhizal fungi on root surface. In: Varma A, Abbott L, Werner D, Hampp R (eds) Plant surface microbiology, Springer, Berlin Heidelberg New York, pp 331–349

Taylor JW (1986) Fungal evolutionary biology and mitochondrial DNA. Exp Mycol 10:259–269

Varma A, Abbott LK, Werner D, Hampp R (2004) The state of art. Plant surface microbiology. In: Varma A, Abbott L, Werner D, Hampp R (eds) Plant surface microbiology. Springer, Berlin Heidelberg New York, pp 1–11

Verma S, Varma A, Rexer KH, Hassel A, Kost G, Sarabhoy A, Bisen P, Bütenhorn B, Franken P (1998) *Piriformospora indica*, gen. et sp. nov., a new root colonizing fungus. Mycologia 90:896–903

Welsh J, McClelland M (1990) Fingerprinting genomes using PCR with arbitrary primers. Nucleic Acids Res 18:7213–7218

Williams JGK, Kubelik AR, Livak KJ, Rafalski JA, Tingey SV (1990) DNA polymorphism amplified by arbitrary primers are useful as genetic markers. Nucleic Acids Res 18:6531–6535

Weiß M, Selosse MA, Rexer KH, Urban A, Oberwinkler F (2004) Sebacinales: a hitherto overlooked cosm of heterobasidiomycetes with a broad mycorrhizal potential. Mycol Res 108:1–8

Zinno TD, Longree H, Maraite H (1998) Deversity of *Pyrenophora tritici-repentis* isolates from warm wheat growing areas: pathogenicity, toxin production and RAPD analysis. In: Duveiller E, Dubbin HJ, Reeves J, McNab A (eds) *Helminthosporium* blights of wheat, spot blotch and tan spot. CIMMYT, Mexico, pp 302–311

16 Co-Cultivation with Sebacinales

A.C. Kharkwal, R. Prasad, H. Kharkwal, A. Das,
K. Bhatnagar, I. Sherameti, R. Oelmüller, and A. Varma

16.1 Introduction

Mycorrhiza refers to an association or symbiosis between plants and fungi that colonize their roots during periods of active plant growth. The most common and prevalent, arbuscular mycorrhizal (AM) fungi, play an indispensable role in upgrading plant growth, vigour and survival by a positive impact on the nutritional and hydratic status of the plant and on soil health by increasing the reproductive potential, improving root performance and providing a natural defence against invaders including pests and pathogens (Newsham et al. 1995; Auge 2000; Borowicz 2001).

The majority of land plants live in mycorrhizal interaction with fungi, a symbiosis that has a strong impact on ecosystems, agriculture, flori-horticulture and forestry (Sanders 2003; Bidartondo et al. 2004; Koide and Mosse 2004; Pennisi 2004). The benefits of mycorrhizal associations arise from nutrient transport between the plant roots and fungal hyphae. The carbon source is transported from the plant to the fungus, whereas fungal hyphae serve as a fine link between the roots and the rhizosphere and improve the plant's supply of inorganic nutrients (Harrison 1999; Bücking and Heyser 2003; Herrmanns et al. 2004; Koide and Mosse 2004).

Applications of mycorrhizae in micropropagated plantlets are a boon for the micropropagation industry (Varma and Schüepp 1995). The key functions of AM co-cultivation can be summarized as follows: (1) improving root growth and

Amit C. Kharkwal, Ram Prasad, Harsha Kharkwal, Aparajita Das, Kamya Bhatnagar, and Ajit Varma: Amity Institute of Microbial Sciences, Amity University, Sector 125, Noida, Uttar Pradesh, India, email: ajitvarma@aihmr.amity.edu

Irena Sherameti, Ralf Oelmüller: Institute of General Botany, Department of Environmental Sciences, University of Jena, Dornburger Strasse 159, D-07743 Jena, Germany, email: b7oera@hotmail.com

Soil Biology, Volume 11
Advanced Techniques in Soil Microbiology
A. Varma, R. Oelmüller (Eds.)

plant establishment, (2) enhancing plant tolerance to (biotic and abiotic) stresses, (3) improving nutrient cycling, (4) enhancing plant community diversity.

16.2 Sebacinaceae – Novel Fungi

Bandoni (1984) revised the Tremellales and Auriculariales on the basis of ultrastructural, ontogenetic and ecological characters. The Sebacinaceae were transferred to his new concept of Auriculariales that then included taxa with septate basidia and continuous parenthesomes, but without a yeast stage. Weiß and Oberwinkler (2001) validated wide parts of Bandoni's (1984) concept of the Auriculariales in a molecular phylogenetic study using nuclear rDNA coding for the D1/D2 region of the large ribosomal subunit (LSU). Their molecular analysis confirmed the monophyly of the Sebacinaceae (including also *Craterocolla cerasi*, which fits the micromorphological concept of Sebacinaceae); however it also suggested that the Sebacinaceae form a separate lineage of Hymenomycetes that must be excluded from the Auriculariales.

Warcup and Talbot (1967) isolated heterobasidiomycetes that they identified from their sexual stages formed in axenic culture as *Sebacina vermifera* sensu stricto from the roots of Australian terrestrial orchids. Later such fungi were also isolated from pot-cultured ectomycorrhizae and arbuscular mycor-

Table 16.1 Recognized members of the Sebacinaceae

Fungus	Remark
Sebacina incrustans	Non-culturable[a]
S. epigaea	Non-culturable[a]
S. aff. epigaea	Non-culturable[a]
Tremelloscypha gelatinosa	Non-culturable[a]
S. dimitica	Non-culturable[a]
E. bulobasidium rolleyi	Non-culturable[a]
Craterocolla cerasi	Non-culturable[a]
Piriformospora indica	Culturable
Sebacina vermifera sensu stricto	Culturable
Sebacina sp.	Culturable

[a] Scientists have failed to culture these fungi on defined synthetic media

rhizae (Warcup 1988). Recently, using molecular methods like polymerase chain reaction (PCR), molecular cloning and sequencing, members of the Sebacinaceae have been shown to be involved in various mycorrhizal associations in the field: (1) orchid mycorrhizae (McKendrick et al. 2002; Selosse et al. 2002 a, b; Urban et al. 2003), (2) ectomycorrhizae (Berch et al. 2002). Since the remaining taxa of the Auriculariales sensu Bandoni (1984) are likely to be wood decomposers (Wells and Bandoni 2001), the mycorrhizal potential of the Sebacinaceae seems a good ecological features to separate members of this from other, morphologically quite similar heterobasidiomycetes that belong to the Auriculariales. However, sebacinoids were demonstrated recently to be ectomycorrhizal (Selosse et al. 2002a). Observations on ectomycorrhizae and basidiomes suggest that species of Sebacinaceae are fairly common mycobionts in various ectomycorrhizal plant communities (Urban et al. 2003). The phylogenetic position of the Sebacinaceae within the Basidiomycota gives an overview of phylogenetic relationships inside this subgroup of Hymenomycetes for which the new Sebacinales is proposed (Garnica et al. 2003; Michael Weiß, personal communication). Fungal strains included in the Sebacinaceae are given in Table 16.1.

16.3 Host Spectrum

Members of the Sebacinaceae were observed to be associated with a large number of mono- and dicotyledonous plants (Table 16.2), inducing pronounced growth promotional effects (Varma et al. 2001), with the exception of the plants belonging to the Cruciferaceae and some plants belonging to the Chenopodiaceae and Amaranthaceae (Read 1999; Varma et al. 1999, 2001; Singh et al. 2003b). Literature suggests that the members of these groups normally do not form associations with AM fungi (Denison et al. 2003). Under in vitro conditions, *P. indica* and *S. vermifera* sensu stricto were demonstrated to interact with the root system of cruciferous and chenopodaceous plants, viz. mustard (*Brassica junaceae*), cabbage (*Brassica oleracea* var. *capitata*; Kumari et al. 2003), *Arabidopsis thaliana* (Pham et al. 2004a) and spinach (*Spinacia oleracea*). A report indicated the ability of *P. indica* to colonize the rhizoids of a liverwort (bryophyte), and the thalli failed to grow under in situ conditions in the absence of this fungus (Varma et al. 2000, 2001; Pham et al. 2004a). *P. indica* was further shown to form associations with terrestrial orchids such as *Dactylorhiza purpurella* (Stephs.) Soo, *D. incarnate* L. Soo, *D. majalis* (Rchb. F.) Hunt & Summerh. and *D. fuchsia* (Druce) Soo (Blechert et al. 1999; Varma et al. 2001; Pham et al. 2004a; Prasad et al. 2005).

Table 16.2 Plant interactions tested with members of Sebacinaceae. Data is based on the root colonization analysis in vivo and in vitro (c.f. Varma et al. 2001; Singh et al. 2003a, b)

Hosts	
Acacia catechu (L.f.) Willd (black catechu)	*Glycine max* L. Merr. (soybean)
Acacia nilotica (L.) Willd (gum)	*Nicotiana tabaccum* L. (tobacco)
Abrus precatorius L. rosary pea (precatory bean)	*N. attenuata* L. (mountain tobacco)
Adhatoda vasica L. syn. (malabar nut)	*Oryza sativa* L. (rice)
Aneura pinguis L. Dumort. (liverwort)	*Petroselinum crispum* L. (curly parsley)
Arabidopsis thaliana L. Heynh. (mouse ear cress)	*Pisum sativum* L. (pea)
Artemisia annua L. (chinese wormwood)	*Populus tremula* L. (aspen)
Azadirachta indica A. Juss (neem)	*P. tremuloides Michx.* (clone Esch5; quaking)
Bacopa monniera L. Wett. (brahmi)	*Prosopis chilensis* Stuntz sys. (chilean mesquite)
Cassia angustifolia Senna Patti (gallow grass hemp)	*P. juliflora* (Swartz) DC. (honey mesquite)
Chlorophytum borivillianum Baker (musli)	*Quercus robur* L. (clone DF 159; oak)
Ch. tuberosum Baker (mexican orange)	*Setaria italica* L. (thumb millet)
Cicer arietinum L. (chick pea)	*Solanum melongena* L. (egg plant)
Coffea arabica L. (English coffee)	*Sorghum vulgare* L. (millet)
Cymbopogon martinii Staph Van Motia (palmarosa)	*Spilanthes calva* DC (clove)
Dactylorhiza fuchsi Druce (Soo') (spotted orchid)	*Tectona grandis* Linn. f. (teak)
D. incarnata L. Soo' (early marsh orchid)	*Terminalia arjuna* L. (Arjun tree/stembark)
D. maculata L. Verm. (Northern marsh orchid)	*Tephrosia purpurea* L. Pers. (sarphunkha/purpurea)
D. majalis Rchb. f. (broad leaved marsh orchid)	*Withania somnifera* L. Dunal (winter cherry)
D. purpurella (Steph's) Soo' (lady orchid)	*Zea mays* var white (maize)
Daucus carota L. Queen Anne's-lace (carrot)	*Zizyphus nummularia* Burm. fil. (jujube)
Delbergia sisso Roxburg (rosewood)	

16.4 Functions of the Sebacinaceae

Scientists from the Amity University Uttar Pradesh, Noida, have screened a novel endophytic fungus, *Piriformospora indica*, which mimics the capabilities of a typical AM fungus. *P. indica* is a recently isolated root-interacting fungus, related to the Hymenomycetes of the Basidiomycota (Verma et al. 1998). In contrast to AM fungi, it can be easily cultivated in axenic culture where it produces chlamydospores (Peškan-Berghöfer et al. 2004; Oelmüller et al. 2005; Shahollari et al. 2005). The fungus is able to associate with the roots of various plant species in a manner similar to mycorrhiza and promotes plant growth (Varma et al. 1999, 2001; Singh et al. 2002a, b, 2003a; Oelmüller et al. 2004; Pešken-Berghöfer et al. 2004; Pham et al. 2004a; Shahollari et al. 2005). Pronounced growth promotional effects were also seen with terrestrial orchids (Blechert et al. 1999). The fungus can easily be cultivated on a number of synthetic and complex media (Hill and Käfer 2001; Pham et al. 2004b).

P. indica tremendously improves the growth and overall biomass production of diverse hosts, including legumes, medicinal and economically important plants (Varma et al. 1999, 2000). The plants tested in the laboratory conditions as well as in extensive field trials were *Bacopa monieri*, *Nicotiana tobaccum* (Sahay and Varma 1999, 2000), *Artemisia annua*, *Petroselinum crispum* (Varma et al. 1999), *Azadirachta indica* (Singh et al. 2002a, b, 2003a), *Tridex procumbans*, *Abrus precatorius* (Kumari et al. 2004), *Chlorophytum borivilianum* (Pham et al. 2004a), *Withania somnifera* and *Spilanthes calva* (Rai et al. 2001) and *Adhatoda vasica* (Rai and Varma 2005). *P. indica* promotes the antifungal potential of the medicinal plant *Spilanthus calva* due to an increase in spilanthol content after interaction (Rai et al. 2004). *P. indica* promises to be an excellent agent for the biological hardening of tissue culture-raised plants as the fungus rendered more than 90% survival rate of the transferred plantlets of these plants and, by excessive root proliferation and induction of secondary rootlets, protecting them from "transplantation shock" and potent root pathogens (Singh et al. 2002a, b, 2003b; Varma et al. 2000). Therefore, this fungus has promise as a boon for the plant industries (Hazarika 2003; Singh et al. 2003a).

Among the compounds released in root exudates, flavonoids are found to be present in *P. indica*. Flavonoids have been suggested to be involved in the stimulation of pre-contact hyphal growth and branching (Gianinazzi-Pearson et al. 1989; Siqueira et al.1991), which is consistent with their role as signalling molecules in other plant–microbe interactions (Giovannetti and Sbrana 1998). Cell wall-degrading enzymes like CMCase, polygalactouronase and xylanase were found in significant quantities both in the culture filtrate and in roots colonized with *P. indica*.

P. indica showed a profound effect on disease control when challenged with the virulent root and seed pathogen *Gaeumannomyces graminis*. *P. indica* completely blocked growth of this pathogen. This indicates that *P. indica* acts as a potential agent for biological control of root diseases; however the chemical nature of the inhibitory factor is still unknown (Varma et al. 2001).

P. indica has been reported to induce resistance to fungal diseases in the monocotyledonous plant barley, along with tolerance to salt stress without affecting plant productivity (Waller et al. 2005). The beneficial effect on the defence status is detected in distal leaves demonstrating a systemic induction of resistance. The systemically altered "defence readiness" is associated with an elevated antioxidative capacity due to an activation of the glutathione–ascorbate cycle and an overall increase in grain yield. Interaction with *Populus* Esch5 revealed that *P. indica* could be directed in its physiological behaviour from mutualistic to antagonistic by specifically designed cultural conditions (Kaldorf et al. 2005), hence making it a potential model system to study plant–microbe interactions. It provides a promising model organism for the investigation of beneficial plant–microbe interactions, and enables the identification of compounds, which may improve plant growth and productivity and maintain soil productivity. The various multifunctional roles of *P. indica* are outlined in Fig. 16.1.

16.5 Eco-Functional Identity

Members of the Sebacinales, *P. indica* and *S. vermifera* colonize the root cortex and forms inter- and intracellular hyphae. Within the cortical cells, the fungus often forms dense hyphal coils or branched structures, intracellularly. *P. indica* also forms spore- or vesicle-like structures within or between the cortical cells.

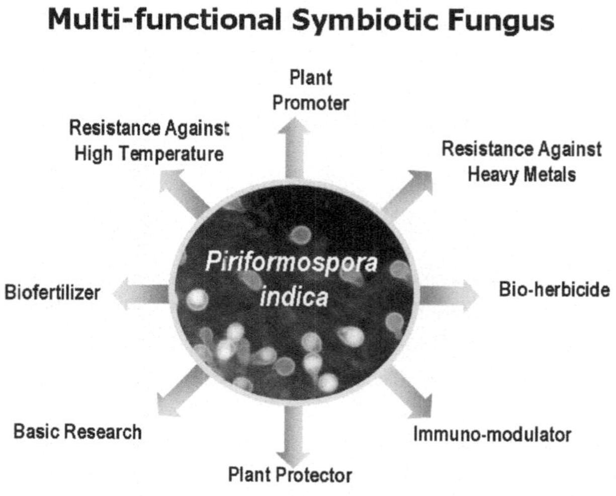

Fig. 16.1 Multifunctional role of *P. indica*

Like AM fungi, hyphae multiply within the host cortical tissues and never traverse through the endodermis. Likewise, they also do not invade the aerial portion of the plant (stem and leaves).

The characteristic features of *P. indica* are the following:

- axenically cultivable on synthetic media,
- no clamp connections,
- anastomosis occurs frequently,
- hypha–hypha aggregation often observed,
- no hyphal knots,
- simple septum with dolipores and continuous, straight parenthosomes (Fig. 16.2 inset),
- chlamydospores 16–25 μm in length, 10–17 μm in width,
- 8–25 nuclei per spore.

The fungus promises to serve as the substitute of AM fungi to overcome the long-standing enigma of science. The functional similarities with AM fungi are the following:

- broad and diverse host spectrum,

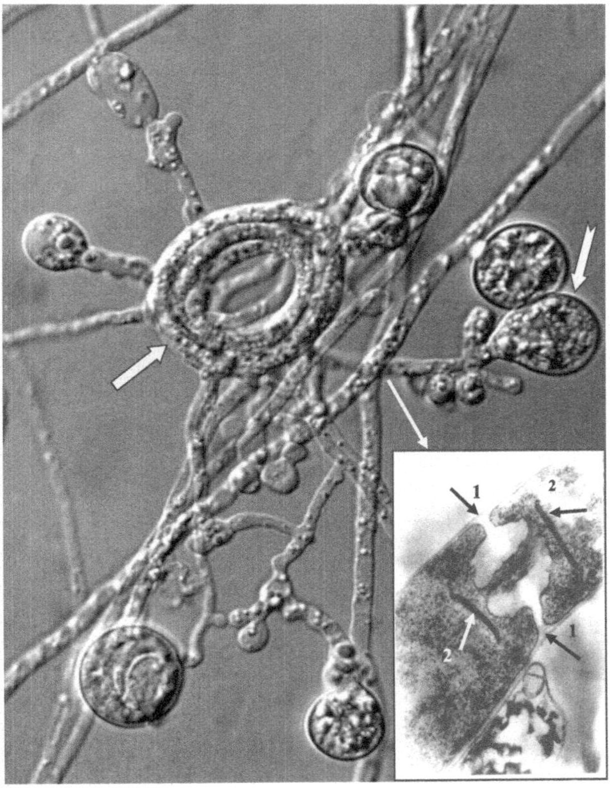

Fig. 16.2 *P. indica*: an overall view of the typical growth and differentiation of hyphae on solidified Käfer medium (the *white arrow* shows the hyphal coil and pear-shaped spore). *Inset*: a magnified view showed the dolipore and parenthosomes of *P. indica* (a section of hypha was observed in electron-transparent material): the *small white arrow* indicates the dolipore and the *black arrows* indicate the continuous parenthosomes. This septal pore is typical for Hymenomycetes

- hyphae extramatrical, inter- and intracellular,
- hyphae never invade the endodermis,
- chlamydospores in soil and within cortical tissues,
- sexual stages not seen,
- positive phytopromotional effects on tested hosts,
- phosphorus mobilizer,
- phosphorus transporter,
- tool for biological hardening of micropropagated plantlets,
- potent biological control agent against root pathogens.

16.6 Axenic Co-Cultivation of Sebacinaceae

Circular agar discs (about 4 mm in diameter) infested with spores and actively growing hyphae of *P. indica* were placed onto Petri dishes (90 mm, disposable; Tarson, India) containing solidified Hill and Kafer medium. Inoculated Petri dishes (90 mm, disposable) were incubated in an inverted position for 7 days at 28±2 °C in the dark. Usually 4–5 fully-grown fungus agar discs (4 mm in diameter) were inoculated into each 500-ml Erlenmeyer flask containing 250 ml of Hill and Kafer broth. Flasks were incubated at 28±2 °C, at constant shaking at 100 rpm on a rotary shaker. The same procedure was performed for *S. vermifera* sensu stricto and *Sebacina* sp.

16.6.1 Procedure

1. Circular agar discs (about 4 mm in diameter) infested with spores and actively growing hyphae of *P. indica* are placed onto Petri dishes (90 mm, disposable) containing solidified Hill and Kafer medium (Fig. 16.3a).
2. Inoculated Petri dishes (90 mm, disposable) are then incubated in an inverted position for 7 days at 28±2 °C in the dark.
3. Four or five fully grown fungus agar discs (4 mm in diameter; Fig. 16.3b) are inoculated into each 500-ml Erlenmeyer flask containing 250 ml of *Aspergillus* broth.
4. Flasks are incubated at 28±2 °C, at constant shaking at 100 rpm on a rotary shaker (Fig. 16.3c).

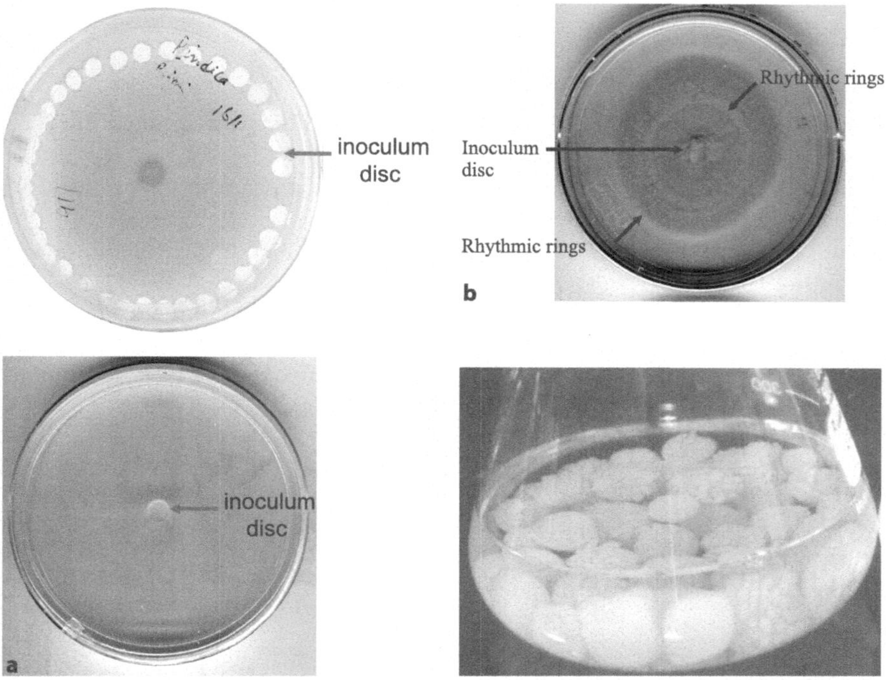

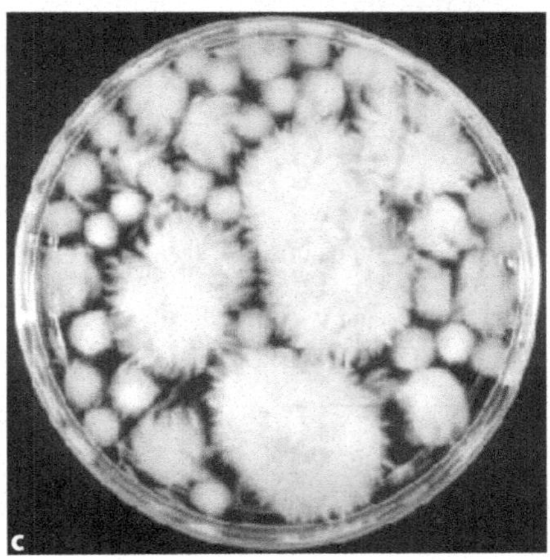

Fig. 16.3 **a** Circular agar discs (about 4 mm in diameter) infested with spores and actively growing hyphae of *P. indica* inoculated onto Petri dishes containing solidified Hill and Kafer medium. **b** Fully grown fungus on solidified Hill and Kafer medium after 7–8 days. **c** Growth of *P. indica* on Käfer liquid medium

16.6.2 Protocol

1. Hold the mother culture of *P. indica* grown on Hill and Kafer medium (Hill and Käfer 2001) inside a laminar flow hood.
2. Make the discs by using the bottom of a sterile glass Pasteur pipette measuring about 4 mm in diameter.
3. Inoculate one disc per Petri plate fortified with Hill and Kafer medium containing 1% agar.
4. Wrap the Petri plates with paraffin tape to avoid any contamination.
5. Incubate the Petri plates at 28±2 °C.
6. Growth normally commences on the third day and, after 12 days, the fungus completely covers the surface of the agar plate (Fig. 16.3b).

16.7 Media Compositions

1. The Hill and Kafer medium composition is given in Table 16.3.
2. For modified Hill and Kafer medium (Varma et al. 2001), the medium composition is the same, except that the quantities of yeast extract, peptone and casein hydrolysate are reduced to one-tenth in quantity.
3. Glucose asparagine agar (for Actinomycetes):

Table 16.3 Composition of Hill and Kafer medium (Hill and Kafer 2001). The pH is adjusted to 6.5 with 1 N HCl/NaOH. All stocks are stored at 4 °C except vitamins, which are stored at −20 °C. In broth culture, agar is excluded

Constituent	Concentration (g/l)
Glucose	20.0
Peptone	2.0
Yeast extract	1.0
Casamino acid	1.0
Vitamin stock solution	1.0 ml
Macroelements from stock	50 ml
Microelements from stock	2.5 ml
Agar	10
$CaCl_2$, 0.1 M	1.0 ml
$FeCl_3$, 0.1 M	1.0 ml

Table 16.3 *(continued)*

Constituent	Concentration (g/l)
Macroelements (major elements) stock (g/l)	
$NaNO_3$	120.0
KCl	10.4
$MgSO_4.7H_2O$	10.4
KH_2PO_4	30.4
Microelements (trace elements) stock (g/l)	
$ZnSO_4.7H_2O$	22.0
H_3BO_3	11.0
$MnCl_2.4H_2O$	5.0
$FeSO_4.7H_2O$	5.0
$CoCl_2.6H_2O$	1.6
$CuSO_4.5H_2O$	1.6
$(NH_4)_6Mo_7O_{27}.4H_2O$	1.1
Na_2EDTA	50.0
Vitamins (%)	
Biotin	0.05
Nicotinamide	0.5
Pyridoxal phosphate	0.1
Amino benzoic acid	0.1
Riboflavin	0.25

Constituent	Concentration (g/l)
Glucose	10
Asparagine	0.5
K_2HPO_4	0.5
Distilled water	1000 ml
Agar	15
pH at 25 °C	6.8±0.2

Directions: ingredients are suspended in 1000 ml of distilled water. Dissolve by boiling completely. Distribute in flasks and sterilize in the autoclave at 15 psi pressure (103 kPa) at 121 °C for 15 min

4. Hoagland solution (Hoagland and Arnon 1938):

Constituent	Concentration (g/l)
Macro-nutrients:	
$MgSO_4.7H_2O$	490
$Ca(NO_3)_2.4H_2O$	492
KNO_3	1002
$CuSO_4 5H_2O$	230
Micro-nutrients:	
$MnCl_2.4H_2O$	1.81
$ZnSO_4.7H_2O$	0.22
H_3BO_3	2.86
$CuSO_4.5H_2O$	0.08
$NaMoO_4$	0.09
Iron source	31.0

Directions: all ingredients are dissolved separately in double-distilled water and then mixed (pH = 6.7)

5. Malt extract medium (Gallowey et al. 1962):

Constituent	Concentration (g/l)
Malt extract	30
Mycological peptone	5
Agar	15
pH	5.4

6. Malt yeast extract medium:

Constituent	Concentration (g/l)
Yeast extract	3
Malt extract	3
Peptone	5
Dextrose	10
pH (25 °C)	6.2±0.2

7. Malt yeast extract agar: add 2% (w/v) agar to the above malt yeast extract medium.

8. Modified Melin–Norkrans (MMN) medium (Johnson et al. 1957):

Constituent	Concentration
NaCl	0.4 mM
KH_2PO_4	3.7 mM
$(NH_4)_2HPO_4$	2.0 mM
$CaCl_2$	0.3 mM
$MgSO_4$	0.6 mM
$FeCl_3$	3.6 mM
Thiamine hydrochloride	0.2 mM
Trypticase peptone	0.1% (w/v)
Glucose monohydrate	1.0% (w/v)
Malt extract	5.0% (w/v)
Trace elements from stock	10 ml/l

a. Stock solution of trace elements:

Constituent	Concentration
KCl	0.2 M
H_3BO_3	0.1 M
$MnSO_4.H_2O$	22.0 mM
$ZnSO_4$	8.0 mM
$CuSO_4$	2.1 mM
pH	5.8

9. MMN agar medium: add 1.2% (w/v) agar to the above MMN medium.
10. Plate count agar (APHA 1978):

Constituent	Concentration (g/l)
Trypton	5.0
Yeast extract	2.5
Dextrose	1.0
Agar	15.0
pH (25 °C)	7.0±0.2

Directions: suspend about 23.5 g of plate count agar in 1000 ml of distilled water. The medium is completely dissolved by boiling and is then sterilized at 15 psi pressure at 121 °C for 15 min.

11. Potato dextrose agar (PDA; APHA 1978):

Constituent	Concentration (g/l)
Potato peel	200
Dextrose	20
Agar	15
Distilled water	1000
pH (25 °C)	5.6±0.2

Directions: the periderm (skin) of potatoes (200 g) is peeled off, cut into small pieces and boiled in 500 ml of water, until a glass rod easily penetrates them. After filtration through cheesecloth, dextrose is added. Agar is dissolved and the required volume (1 l) is made up by the addition of water. The medium is autoclaved at 15 psi pressure at 121 °C for 20 min

12. Water agar (WA):

Constituent	Concentration (g/l)
Daichin agar	7 (0.7%)

13. 20% Knop solution:

Constituent	Concentration (g/l)
Saccharose	20.0 (2.0%)
Daichin agar	8.0 (0.8%)
Vitamin B5 (Gamborg and Phillips 1996)	1.0
Stock solution I	2.0
Stock solution II	2.0
Stock solution III	2.0
Stock solution IV	0.4
Stock solution V	0.2

Adjust pH to 6.4 with 1 N KOH

14. Composition of stock solutions I–V for Knop solution:

Stock solution	Constituent	Concentration (g/l)
Stock solution I	KNO_3	121.32
	$MgSO_4.7H_2O$	19.71
Stock solution II	$Ca(NO_3)_2.4H_2O$	120.0
Stock solution III	KH_2PO_4	27.22
Stock solution IV	FeNaEDTA	7.34
Stock solution V	H_3BO_3	2.86
	$MnCl_2$	1.81
	(or $MnCl_2.4H_2O$)	(2.85)
	$CuSO_4.5H_2O$	0.073
	(or $CuSO_4.2H_2O$)	(0.05)
	$ZnSO_4.7H_2O$	0.36
	$CoCl_2.6H_2O$	0.03
	H_2MoO_4	0.052
	(or $Na_2MoO_4.2H_2O$)	(0.0775)
	NaCl	2.0

16.8 Seed Surface Sterilization and Germination

Maize seeds are soaked in sterile water overnight and surface-sterilized by washing with 90% ethanol for a few seconds and either with 0.01% mercuric chloride solution for 10 min or with 4% (v/v) NaOCl for 15 min, then washed five times with sterile distilled water and finally rinsed with 70% (v/v) ethanol for 30 s. This is followed by a quick treatment with 15% (v/v) NaOCl; chemicals adhered are removed by repeated rinsing with sterile distilled water (or a better alternative method can be used as described in Section 16.8.1; Gamborg and Phillips 1996). The seeds are kept 1 cm apart in a sterile Petri dish layered with germinating paper or aseptically transferred to water agar plates (0.7% agar) and left for germination at 25±2 °C for 4 days in the dark.

16.8.1 Protocol for Seed Surface Sterilization

1. Collect desired quantity of seeds.
2. Soak in sterilized distilled water overnight.
3. Treat with 70% ethanol for 30 s with stirring.
4. Wash three times with sterile distilled water to remove traces of ethanol.
5. Wash with 1.5% NaOCl solution for 20 min with stirring
6. Wash three times with sterilized distilled water.
7. Wash with 15% NaOCl for 20 s.
8. Wash six times with distilled water to remove traces of NaOCl.

Garden soil is sterilized by autoclaving three times at 121 °C at 15 psi pressure (103 kPa), at intervals of 48 h. Sand is acid-treated in 10% HCl overnight and washed in running tapwater until the pH becomes neutral. Sterile soil and acid-washed sand are dried in a hot-air oven. Soil and sand are mixed in the ratio of 3:1 for filling the pots.

16.8.2 Inoculum Placement in the Pots

Live inoculum of *P. indica* is required. This contains spores and fungal hyphae. In the pot, a soil base is added first, up to one-third of the depth of pot. Then live inoculum is layered over it. Above this layer, one layer of soil base is added to sandwich the inoculum between the layers of soil base. For the inoculation of *P. indica*, mycelium is mixed in a small amount of sterile soil and then spread as above, in a sandwich model at the rate of 1%.

Surface-sterilized seeds are transferred to the pots. When the plants reach 2–3 cm, they are then treated with Hoagland solution. The morphological features of each plant are observed and recorded at weeks 2, 4 and 8.

16.8.3 Results

16.8.3.1 *P. indica* – Photobiont Interaction

Fungus-treated plants were compared with untreated control plants in terms of morphological and anatomical characteristics. As an impact of *P. indica*, the treated plants showed early germination in comparison with the uninoculated control. After 30 days, prominent differences were seen. *P. indica*-treated plants become longer with more nodes and leaves than the control.

16.8.3.2
Growth Conditions

Pots were placed in a greenhouse maintained at 30±2 °C, 16 h photoperiod (1000 lux) and 75% relative humidity for four months. The plants were fertilized with 10% strength Hoagland solution (Hoagland and Arnon 1938) on every alternate week, consisting of phosphorus and devoid of phosphorus nutrients. Plants were irrigated with sterile tapwater on every alternate day to maintain a relative moisture of about 60%.

16.8.3.3
Growth Parameters

1. Aerial length: the height of each plant was measured at intervals of 14, 28 and 42 days. Experiments were recorded in triplicate.
2. Aerial biomass: each endophyte-inoculated plant was carefully taken in triplicate from the pots for fresh and dry quantification at intervals of 14, 28 and 42 days. Plants were wiped with tissue paper and air-dried for fresh weight. Later they were dried at 80 °C for 12 h in an air-circulation Memmert-type oven. Samples were desiccated at room temperature before weighing on a Mettler balance (AE 160).
3. Underground length: underground parts were thoroughly washed under running tapwater to remove the adhering soil particles. The length of the underground part was measured in triplicate readings.
4. Underground biomass: after excessive washing, the moisture was blotted out with filter paper, then air-dried and weighed for fresh weight on a Mettler balance (AE 160).
5. Endophyte dependency: the endophyte dependency (ED) of *Zea mays* L. var white was determined using the formula given by Gerdemann (1975), which was modified by Plenchette et al. (1983) to give a percent increase of yield relative to that of mycorrhizal plants. This results in a figure between 0% and 100% rather than an unlimited percent increase:

   ED = (Parameter with mycorrhiza – parameter without mycorrhiza)/(Parameter with mycorrhiza) ×100

   ED was used instead of mycorrhiza dependency (MD) to designate endophyte dependency.

16.9 Comparative Study on Plant Growth with Treated Endosymbionts

Both *P. indica* and *Sebacina vermifera* sensu stricto exhibited the highest positive growth-promoting effect on maize plants, as evidenced by better aerial length (above ground), enhanced and healthier foliage and a well developed rooting system, as compared with other endophytic strains. *S. vermifera* sensu stricto showed a little less growth-promoting effect than *P. indica*.

Mycorrhiza dependency (MD) was used as an index to compare the receptivity of different plant species to AM fungi (Gerdemann 1975; Plenchette et al. 1983). This can also be used for other endophytes, such as *P. indica* and *S. vermifera* sensu stricto. In the present study, ED was used instead of MD, as the test organisms do not develop a typical mycorrhizal association. *P. indica* showed the highest ED over the other related endophytes. The more intense root proliferation in treated plants observed in the present experiments might be due to the synthesis of as yet unidentified extracellular phytohormones by mycobionts (Singh et al. 2000; Varma et al. 2001). The ED value was 211.13 for *Spilanthes calva* and 671.90 for *Withenia somnifera*. These data suggest that *P. indica* has a greater influence on the growth of *W. somnifera* than on that of *S. calva* (Rai et al. 2001). The ED of a host plant can be altered by factors such as soil type and the soil P content of mycorrhizal species (Azcon and Ocampo 1981; Menge et al. 1978). Amongst the reasons proposed for the differences in ED in different plants or varieties of the same species, Baylis (1995) reported that root-hair length and root thickness could determine the ED level. Rajapakse and Miller (1988) observed that the average length of fine roots was negatively correlated with ED in cowpea.

16.10 In Vivo Co-Cultivation of Sebacinales

The Sebacinaceae members were also inoculated into sterile soil in polyethylene or earthenware pots in five replicates, using 0.5 kg capacity pots for mass cultivation. The soil was autoclaved thrice on alternate days and air-dried. Riverbed sand was soaked in 10% HCl overnight and then washed under running tap-water until the pH reached neutrality. An air-dried mixture of soil and sand in the ratio of 3:1 (Feldmann and Idczak 1994) was used as substratum. Pots were also surface-sterilized by 70% ethanol and were then half-filled with this mixture and the inoculum was layered over it. Five holes were made into each pot and into each hole approximately 1 g of endomycorrhizal inoculum was added (80 spores/fungal propagules per 10 g soil). Five germinated seeds of 10 mm length were placed 1–2 cm above the inoculum layer in the marked holes in each pot (Fig. 16.4a, b). The pots were maintained in an environmentally controlled greenhouse at 25±2 °C with 16 h light/8 h dark and relative humidity 60–70%, with a light intensity of 1000 lux. Roots were checked for colonization after 15–20 days. The soil cultures, along with the root propagules obtained after four months, were stored in a cold room for further use. Root pieces with spores and hyphal fragments can be used as a live propagule (inoculum) for experiments or to introduce fungi into soils. Similarly, for comparative photomycobiont growth of *P. indica* and *S. vermifera* sensu stricto, a disc (4 mm diameter) of inoculum infested with hyphae and spores was taken per plant.

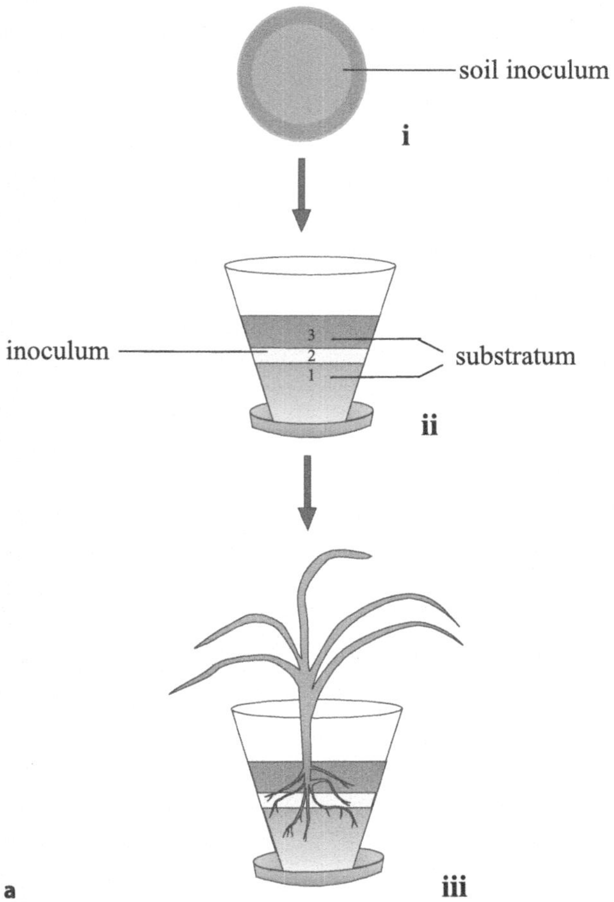

Fig. 16.4 **a** Polyethylene pots (0.5 kg capacity) contained an autoclaved sand and soil mixture (1:3) at pH 7.0. *Stage i* Soil inoculum consisting of spores, hyphae and colonized root propagules. *Stage ii* Sandwich of 1 cm layer of inoculum. *Stage iii* Micropropagated plantlets were plated up to the second layer in an upward direction. A little sterile tap water was gently sprinkled to moisten the upper soil layer **b** *see next page*

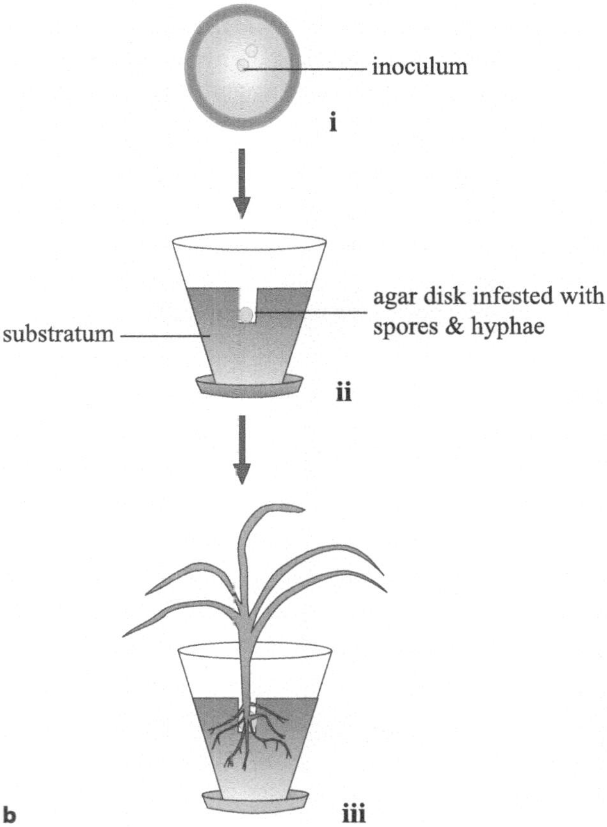

Fig. 16.4 *(continued)* **b** Polyethylene pots (0.5 kg capacity) contained an autoclaved sand and soil mixture (1:3) at pH 7.0. *Stage i* Culture inoculum in Petri dish consisting of spores and hyphae. *Stage ii* A hole was made in the centre of the pot, up to 2 cm deep, with the help of a surface-sterilized, specially designed plastic rod. An agar disc (4 mm diameter) infested with spores and hyphae was placed in the hole. *Stage iii* Micropropagated plantlets were inserted into the hole in an upward direction and the top was covered with the same substratum. A little sterile tap water was gently sprinkled to moisten the upper soil layer

16.11 Conclusions

Members of the Sebacinaceae have been found to be associated with a large number of mono- and dicotyledonous plants. Their interactions have shown growth promotion in diverse plant genera. *P. indica* and *S. vermifera* are root endosymbionts that can be considered as model organisms to study the hidden mystery of mycorrhizal world, since these fungi mimic the AM fungal characters.

The axenic cultivability of Sebacinales members *P. indica* and *S. vermifera* makes them ideal tools for further biotechnological exploitation. They serve as excellent organisms for biotechnological applications in the fields of agriculture, forestry, flori-horticulture, viticulture and arboriculture. They can also be used for the synthesis of herbicides, weedkillers, pesticides and several enzymes of industrial importance. Functionally, co-cultivation with *P. indica* not only promotes plant growth but also increases the plant's active constituents and enhances disease resistant. It is also an excellent biological hardening agent and the fungus renders an above 90% survival rate in tissue culture transplantation plants. Axenic cultivation of *P. indica* is very simple and the fungus can be normally multiplied on a variety of cheap media within a very short time and can be produced on a large scale. The axenically produced fungal inoculum can be directly used for co-cultivation under greenhouse and field conditions.

References

APHA (1978) Standard methods for examination of dairy products, 14th edn. APHA, Washington, D.C.

Auge RM (2000) Stomatal behavior of arbuscular mycorrhizal plants. In: Kapulnik Y, Douds DD (eds) Arbuscular mycorrhizas: physiology and function. pp 201–237

Azcon R, Ocampo JA (1981) Factors affecting the vesicular-arbuscular mycorrhizal infection and mycorrhizal dependency of wheat cultivars. New Phytol 87:677–685

Bandoni RJ (1984) The Tremellales and Auriculariales and alternative classification. Trans Mycol Soc Jpn 25:489–530

Baylis NTJ (1995) Statistical methods in biology. Cambridge University Press, Cambridge

Berch SM, Allen TR, Berbee ML (2002) Molecular detection, community structure and phylogeny of ericoid mycorrhizal fungi. Plant Soil 244:55–66

Bidartondo MI, Burghardt B, Gebauer G, Bruns TD, Read DJ (2004) Changing partners in the dark: isotopic and molecular evidence of ectomycorrhizal liaisons between forest orchids and trees. Proc R Soc Lond 271:1799–1806

Blechert O, Kost G, Hassel A, Rexer R-H, Varma A (1999) First remarks on the symbiotic interactions between *Piriformospora indica* and terrestrial orchids. In: Varma A, Hock B (eds) Mycorrhizae, 2nd edn. Springer, Berlin Heidelberg New York, pp 683–688

Borowicz VA (2001) Do arbuscular mycorrhizal fungi alter plant–pathogen relations? Ecology 82:3057–3068

Bücking H, Heyser W (2003) Uptake and transfer of nutrients in ectomycorhizal associations: interactions between photosynthesis and phosphorus nutrition. Mycorrhiza 13:59–69

Denison RD, Bledsoe C, Kahn M, Gara FO, Simms EL, Thomashow LS (2003) Cooperation in the rhizosphere and the free rider problem. Ecology 84:838–845

Feldmann F, Idczak E (1994) Inoculum production of vescicular–arbuscular mycorrhizal fungi for use in tropical nurseries. In: Norris JR, Read DJ, Varma A (eds) Methods in microbiology, vol 24. Academic, London, pp 339–358

Gallowey LD, Burgess R (1962) Applied mycology and bacteriology, 3rd edn. Hill, London, pp 54–57

Gamborg OL, Phillips GC (1996) Sterile techniques. In: Gamborg OL, Phillips GC (eds) Plant cell tissue and organ culture. Narosa, New Delhi, pp 67–79

Garnica S, Weiß M, Oertel B, Oberwinkler F (2003) Phylogenetic relationships of European *Phlegmacium* species (*Cortinarius*, Agaricales). Mycologia 95:1155–1170
Gerdemann JW (1975) Vesicular–arbuscular mycorrhizae. In: Torrey JG, Clarkson DT (eds) The development and function of roots. Academic, London, pp 575–591
Gianinazzi-Pearson V, Branzanti B, Gianinazzi S (1989) In vitro enhancement of spore germination and early hyphal growth of a vesicular-arbuscular mycorrhizal fungus by host root exudates and plant flavonoids. Symbiosis 7:243–255
Giovannetti M, Sbrana C (1998) Meeting a non-host: the behaviour of AM fungi. Mycorrhiza 8:123–130
Harrison M (1999) Molecular and cellular aspects of the arbuscular mycorrhizal symbiosis. Annu Rev Plant Physiol Plant Mol Biol 50:361–389
Hazarika BN (2003) Acclimatization of tissue-cultured plants. Curr Sci 85:1704–1712
Herrmanns S, Oelmüller R, Buscot F (2004) Manipulation of the onset of ectomycorrhiza formation by indole-3-acetic acid, activated charcoal or relative humidity in the association between oak microcuttings and *Piloderma croceum*: influence on plant development and photosynthesis. J Plant Physiol 161:509–517
Hill TW, Käfer E (2001) Improved protocols for *Aspergillus* minimal medium: trace element and minimal medium salt stock solutions. Fungal Genet Newsl 48:20–21
Hoagland DR, Arnon DI (1938) The water-culture method for growing plants without soil. Univ Calif Agric Exp Stn Circ 1938:346
Johnson CN, Stout PR, Broyer RC, Carlton AB (1957) Comparative chlorine requirements of different plant species. Plant Soil 8:337–353
Kaldorf M, Koch B, Rexer K-H, Kost G, Varma A (2005) Patterns of interaction between *Populus* Esch5 and *Piriformospora indica*: a transition from mutualism to antagonism. Plant Biol 7:210–218
Koide RT, Mosse B (2004) A history of research on arbuscular mycorrhiza. Mycorrhiza 14:145–163
Kumari R, Yadav HK, Bhoon YK, Varma A (2003) Colonization of cruciferous plants by *Piriformospora indica*. Curr Sci 85:1648
Kumari R, Pham GH, Sachdev M, Garg AP, Varma A (2004) Symbiotic fungi for eco friendly environment: a prospective, natural product radiance. CSIR 3:396–400
McKendrick SL, Leake JR, Taylor DL, Reed DJ (2002) Symbiotic germination and development of the myco-heterotrophic orchid *Neottia nidus-avis* in nature and its requirement for locally distributed *Sebacina* spp. New Phytol 154:233–247
Menge JA, Johnson ELV, Platt RG (1978) Mycorrhizal dependency of several citrus cultivars under three nutrient regimes. New Phytol 81:553–559
Newsham KK, Fitter AH, Watkinson AR (1995) Multifunctionality and biodiversity in arbuscular mycorrhizas. Trends Ecol Evol 10:407–411
Oelmüller R, Shahollari B, Peškan-Berghöfer T, Trebicka A, Giang PH, Sherameti I, Oudhoff M, Venus Y, Altschmied L, Varma A (2004) Molecular analyses of the interaction between *Arabidopsis* roots and the growth-promoting fungus *Piriformospora indica*. Endocytobios Cell Res 15:504–517
Oelmüller R, Peškan-Berghöfer T, Shahollari B, Trebicka A, Sherameti I, Varma A (2005) MATH-domain proteins represent a novel protein family in *Arabidopsis thaliana* and at least one member is modified in roots in the course of a plant/microbe interaction. Physiol Plant 124:152–166
Pennisi E (2004) The secret life of fungi. Science 304:1620–1622
Peškan-Berghöfer T, Shahollari B, Giang PH, Hehl S, Markent C, Blank V, Kost G, Varma A, Oelmüeller R (2004) Association of *Piriformospora indica* with *Arabidopsis thaliana* roots represent a novel system to study beneficial plant–microbe interactions and involve in early plant protein modifications in the endocytoplasmic reticulum and in the plasma membrane. Physiol Plant 122:465–471

Pham GH, Singh AN, Malla R, Kumari R, Saxena AK, Rexer K-H, Kost G, Luis P, Kaldorf M, Buscot F, Herrmann S, Peskan T, Oelmüller R, Mittag M, Declerck S, Hehl S, Varma A (2004a) Interaction of *Piriformospora indica* with diverse microorganisms and plants. In: Varma A, Abbott L, Werner D, Hampp R (eds) Plant surface microbiology. Springer, Berlin Heidelberg New York, pp 237–266

Pham GH, Kumari R, Singh An, Kaldorf M, Buscot F, Oelmüller R, Tatjana P, Weiss M, Hampp R, Varma A (2004b) Axenic cultures of *Piriformospora indica*. In: Varma A, Abbott L, Werner D, Hampp R (eds) Plant surface microbiology. Springer, Berlin Heidelberg New York, pp 593–615

Plenchette C, Fortin JA, Furlan V (1983) Growth responses of several plant species to mycorrhizae in a soil of moderate P-fertility 1: mycorrhizal dependency under field conditions. Plant Soil 70:199–209

Prasad R, Pham GH, Kumari R, Singh A, Yadav V, Sachdev M, Peskan T, Hehl S, Oelmüller R, Varma A (2005) Sebacinaceae: culturable mycorrhiza-like endosymbiotic fungi and their interaction with non-transformed and transformed roots. In: Declerck S (ed) Root organ culture of mycorrhizal fungi. Springer, Berlin Heidelberg New York, pp 291–312

Rai M, Varma A (2005) Arbuscular mycorrhiza-like biotechnological potential of *Piriformospora indica*, which promotes the growth of *Adhatpda vasica* Nees. J Biotechnol 8:107–111

Rai MK, Singh A, Arya D, Varma A (2001) Positive growth responses of *Withania somnifera* and *Spilanthes calva* were cultivated with *Piriformospora indica* in field. Mycorrhiza 11:123–128

Rai MK, Varma A, Pandey AK (2004) Enhancement of antimycotic potential in *Spilanthes calva* after inoculation of *Piriformospora indica*, a new growth promoter. Mycoses 47:479–481

Rajapakse S, Miller JC Jr (1988) Relationship between cowpea root systems and mycorrhizal dependency. Hortic Sci 23:568–570

Read DJ (1999) Mycorrhiza: the state of the art. In: Varma A, Hock B (eds) Mycorrhiza: structure, function, molecular biology and biotechnology, 2nd edn. Springer, Berlin Heidelberg New York, pp 3–34

Sahay NS, Varma A (1999) *Piriformospora indica*: a new biological hardening tool for micropropagated plants. FEMS Microbiol Lett 181:297–302

Sahay NS, Varma A (2000) Biological approach towards increasing the survival rates of micropropagated plants. Curr Sci 78:126–129

Sanders I (2003) Preference, specificity and cheating in the arbuscular mycorrhizal symbiosis. Trends Plant Sci 8:143–145

Selosse MA, Bauer R, Moyersoen B (2002a) Basal hymenomycetes belonging to Sebacinaceae are ectomycorrhizal on temperate deciduous trees. New Phytol 155:183–195

Selosse MA, Weiß M, Jany JC, Tillier A (2002b) Communities and populations of sebacinoid basidiomycetes associated with the achlorophyllous orchid *Neottia nidus-avis* (L.) LCM Rich and neighbouring tree ectomycorrhizae. Mol Ecol 11:1831–1844

Shahollari B, Varma A, Oelmüeller R (2005) Expression of receptor kinase in roots is stimulated by the basidiomycete *Piriformospora indica* and the protein accumulates in Triton-X-100 insoluble plasma membrane microdomains. J Plant Physiol 162:945–958

Singh AR, Rexer K-H, Varma A (2000) Plant productivity determinants beyond minerals, water and light. Curr Sci 79:101–106

Singh AN, Singh A, Malla R, Ghosh S, Varma A (2002a) Symbiotic fungi: a boon for plant industry. In: IIT (ed) Conference on biotechnology – the science and the business. IIT/AIBA, New Delhi, pp 52–55

Singh AR, Singh AN, Varma A (2002b) *Piriformospora indica* – in vitro raised leguminous plants: a new dimension in establishment and phytopromotion. Ind J Biotechnol 1:372–376

Singh AN, Singh AR, Kumari M, Kumar S, Rai MK, Sharma AP, Varma A (2003a) AMF-like fungus: *Piriformospora indica* – a boon for plant industry. In: Prasad BN (ed) Biotechnology in sustainable biodiversity and food security, Science Publishers, Enfield, pp 101–124

Singh AN, Singh AR, Kumari M, Rai MK, Varma A (2003b) Biotechnology importance of *Piriformospora indica* – a novel symbiotic mycorrhiza-like fungus: an overview. Ind J Biotechnol 2:65–75

Siqueira JO, Safir GR, Nair MG (1991) Stimulation of vesicular–arbuscular mycorrhiza formation and growth of white clover by flavonoid compounds. New Phytol 118:87–93

Urban A, Weiß M, Bauer R (2003) Ectomycorrhizae involving sebacinoid mycobionts. Mycol Res 107:3–14

Varma A, Schüepp H (1995) Mycorrhizae, their applications in micropropagated plantlets. Crit Rev Biotechnol 15:313–328

Varma A, Verma S, Sudha A, Sahay NS, Franken P (1999) *Piriformospora indica*, a cultivable plant growth promoting root endophyte with similiarities to arbuscular mycorrhizal fungi. Appl Environ Microbiol 65:2741–2744

Varma A, Rai MK, Sudha A, Sahay N (2000) Microbial biotechnology: new paradigms and role in sustainable agriculture. In: Rajak RC (ed) Microbial biotechnology for sustainable development and productivity. Scientific Publishers, New Delhi, pp 22–37

Varma A, Singh A, Sudha A, Sahay NS, Sharma J, Roy A, Kumari M, Rana D, Thakran S, Deka D, Bharti K, Hurek T, Blechert O, Rexer K-H, Kost G, Hahn A, Maier W, Walter M, Strack D, Kranner I (2001) *Piriformospora indica*: an axenically culturable mycorrhiza-like endosymbiotic fungus. In: Hock B (ed) The Mycota IX. Springer, Berlin Heidelberg New York, pp 125–150

Varma A, Singh AR, Sudha A, Sahay NS, Kumari M, Bharati K, Sarbhoy AK, Maier W, Walter MH, Strack D, Franken P, Singh AN, Malla R, Hurek T (2002) *Piriformospora indica*: a plant stimulator and pathogen inhibitor arbuscular mycorrhiza-like fungus. In: Markandey DK, Markandey NR (eds) Microorganisms in bioremediation. Capital Book, New Delhi, pp 71–89

Verma S, Varma A, Rexer K-H, Hassel A, Kost G, Sarbhoy A, Bisen P, Buetehorn P, Franken P (1998) *Piriformospora indica* gen. nov., a new root-colonizing fungus. Mycologia 90:895–909

Waller F, Achatz B, Baltruschat H, Fodor J, Becker K, Fischer M, Heier T, Hückelhoven R, Neumann C, Von Wettstein D, Franken P, Kogel K-H (2005) The endophytic fungus *Piriformospora indica* reprograms barley to salt-stress tolerance, disease resistance, and higher yield. Proc Natl Acad Sci USA 102:13386–13391

Warcup JH (1988) Mycorrhizal associations of isolates of *Sebacina vermifera*. New Phytol 110:227–231

Warcup JH, Talbot PHB (1967) Perfect states of Rhizoctonias associated with orchids. New Phytol 66:631–641

Weiß M, Oberwinkler F (2001) Phylogenetic relationships in *Auriculariales* and related groups – hypotheses derived from nuclear ribosomal DNA sequences. Mycol Res 105:403–415

Wells K, Bandoni RJ (2001) Heterobasidiomycetes. In: Mclaughlin DJ, McLaughlin EG, Lemke PA (eds) The Mycota VII. Systematics and evolution, part B. Springer, Berlin Heidelberg New York, pp 85–120

17 Quantitative Histochemistry: a Forgotten Tool with New Applications

R. Hampp and S. Haag

17.1 Introduction

Biological tissues are not homogenous; instead they consist of cells having specific functions. A typical bifacial leaf, for example, contains not only photosynthetic mesophyll cells (palisade parenchma, spongy parenchyma) but also epidermal, guard and bundle sheath cells, as well as conducting elements. They all have defined functions, which indicate profound biochemical differences between adjacent cells. Such differences are obliterated by tissue homogenation, which precedes most analytical biochemistry. This is even more a problem when different organisms come into close vicinity such as in symbiotic interactions. The roots of most plants form such symbiotic structures with soil fungi (mycorrhiza). Here, fungal hyphae either grow along the surface of fine roots or penetrate root cortex cells, forming structures which extend their surface area for solute exchange (arbuscular mycorrhiza), or produce a hyphal mantle covering the surface of the fine root and connected with hyphae penetrating the cell wall of root cortex cells, thereby forming finger-like structures (Hartig net). This net also greatly increases the surface area available for solute exchange between fungus and host (the ectomycorrhiza). Here also, simple tissue homogenation does not reveal the biochemical properties of the respective organisms. Biologists have thus been challenged to develop methods that allow for selective sampling of specific cell types and analysis of the resultant small amounts of material.

A general problem is the conservation of the metabolic state of the intact organ. Ideally, biochemical analysis would be non-invasive, but this goal is elusive. Unaltered cells are difficult to isolate, and even when they can be isolated, intact membranes limit the uptake of reaction components. Thus, histochemical approaches include various implementations of chemical fixation, embedding,

University of Tübingen, Physiological Ecology of Plants, Auf der Morgenstelle 1
72076 Tübingen, Germany, email: ruediger.hampp@uni-tuebingen.de

Soil Biology, Volume 11
Advanced Techniques in Soil Microbiology
A. Varma, R. Oelmüller (Eds.)

or freeze stop. Because of its wider utility, we generally use the last method in our research. Although written more than 30 years ago "*A flexible system of enzymatic analysis*" (Lowry and Passonneau 1972) is still the "bible" in this respect. In this publication, the reader can find most of the important details of procedures which are not specifically referenced in this chapter.

17.2 Sample Preparation and Handling

1. Freeze stop of mycorrhiza:
   We generally develop mycorrhized roots in Petri dishes, in which fungal suspensions are added to already developed sterile roots of seedlings, the shoots of which are outside the Petri dish (Hampp et al. 1996). When mycorrhizae are well developed, the Petri dish is opened and flooded with liquid N_2. This approach is the preferred method of quenching tissue because it stops endogenous reactions immediately. Under regular conditions, liquid N_2 is at its boiling temperature. The resulting gas layer between liquid N_2 and the sample insulates it and thus slows freezing. This effect can be reduced by precooling liquid N_2 to its freezing point (–210 °C) by evacuation (a Dewar flask is sealed with a rubber and connected to vacuum pump by an insulated copper tubing). Safety note: due to the condensation of O_2 from the air into the liquid N_2 (which can form an explosive mixture), the Dewar flasks should not be open for extended periods of time.
2. Storage of frozen tissue:
   Storage temperatures must be low enough to prevent ice crystal growth, metabolic activities and diffusion of solutes. In general, frozen tissues can be stored without significant losses or metabolic alterations for several months at –50 °C or below. We routinely keep our samples at –80 °C.
3. Freeze-drying:
   The principles and equipment of freeze-drying are simple. In outline, the samples are transferred to a –35 °C to –40 °C compartment (commercial kitchen freezer at "super frost") and dried by reducting the pressure to around 10^{-3} mbar (100 Pa). The vacuum pump and the sample compartment are separated by a cold trap. This can either be a container with dry ice (cheap) or a freezer working at around –100 °C.
4. Storage of freeze-dried material:
   Dried tissue is stable at –20 °C, if stored under vacuum. To avoid the entry of water into the sample container upon admitting air, the orifice of the sample container has to be plugged before storage in a frost-free freezer (a "no frost" freezer). If this precaution is not taken, water condensing inside the orifice while the sample container is warming up could be sucked in upon releas-

ing the vacuum. This can ruin the samples. Samples should only be removed when they have reached ambient temperature. Exposure to ambient conditions (more than 40% relative humidity, more than 20 °C) should be kept to a minimum.

5. Dissection of tissue:
The room used for tissue dissection should be air-conditioned and not exceed 40% relative humidity or 20 °C stored samples are taken from the respective container under the precautions mentioned above. For handling, the samples are transferred to a piece of translucent Plexiglass, which rests on the stage of a stereomicroscope. As small samples are subject to static electricity, the latter can make handling a pain. Charges can be eliminated by spraying surfaces and tools with ionized air (ionizing compressed-air "guns" are commercially available). Alternatively, suitable radiation sources can be used (discs or bars containing ^{241}Americium or ^{210}Polonium).
   a. Sample transfer and dissection require special tools. Large samples (>1 μg; a freeze-dried poplar mycorrhiza weighs between 10 μg and 15 μg) are handled with commercial preparatory needles. Smaller samples are handled with hair points. Such tools are made from Pasteur pipettes, the capillary tips of which are cut off. After fire-polishing the end, a curved hair is epoxyed onto it. Glueing a fine quartz fiber (2–5 μm diameter) to the hair points yields a small tip (for the production of glass fibers, see Lowry and Passonneau 1972).
   b. Different sizes of knives are needed for different operations. Larger samples such as whole mycorrhizas can be cut into smaller sections with a scalpel or an ordinary razor blade. Smaller knives are made as follows: the cutting edge from a ordinary razor blade (about 2 mm wide) is cut from the rest of the blade by a paper cutter (a pair of heavy scissors will also do). Each sliver is then cut into pieces of 1–2 mm length. A fragment is epoxyed onto a nylon bristle (taken from a tooth brush), which is then glued to the trimmed metal end of a preparation needle. Care has to be taken to keep the blades free of rust (storage in a box with dry pearls) and of grease. The latter can be removed by successive washes with ethanol and acetone.
6. Collection and transfer of samples:
Dissected samples are collected on a transfer platform from the microscope stage (see Lowry and Passonneau 1972). In principle the platform consists of a wooden handle to which a 3- to 6-mm wide strip of a glass cover slip is glued (Fig. 17.1). Under a dissecting microscope, the samples are arranged in a single row parallel to the leading edge of the cover slip. To keep them clean and to prevent sample loss, the platforms are kept in (glass) Petri dishes (polystyrene dishes build up electrical charges which can dissipate the samples).
7. Determination of sample mass
As the entire sample is used for analysis, sample mass is the only feasible reference. Mass, however, can vary considerably between samples due to changes

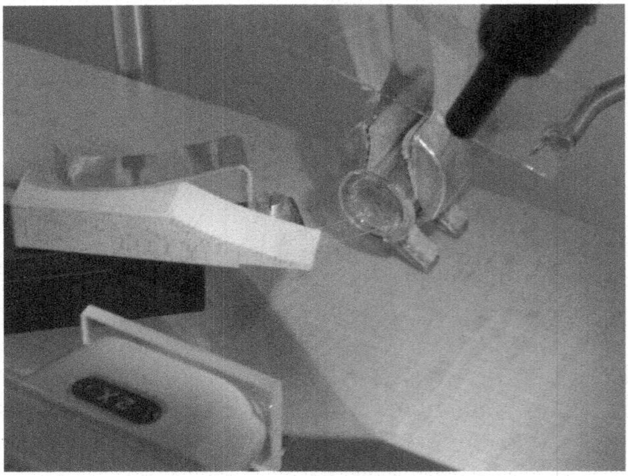

Fig. 17.1 Transfer platform for small samples fixed in front of the quartz fiber balance housing

in cell wall thickness. It is thus advisable to establish mass per protein, etc., conversion factors with comparable samples on a larger-scale sample (compare Outlaw et al. 1981). Owing to the generally small size of sample, conventional balances are not suitable. Instead, quartz fibre balances are used. These consist of a quartz fibre with a diameter in the lower micron range, contained in a glass syringe housing. For fabrication, the lower end of the syringe barrel is cut off. At the lower end of the plunger, a short piece of copper wire is attached with epoxy resin. Then the plunger is inserted all the way into the barrel until the copper wire is exposed at the other end. The quartz fiber is then expoxyed to the copper wire, and the fiber is withdrawn into the body of the syringe for protection and to avoid turbulence from air currents. Details on balance construction, ranges of sensitivity and calibration can be found in Lowry and Passonneau (1972). For our mycorrhiza samples, we use balance capacities of about 1 µg. Figure 17.2 details some steps about sample handling. Samples (Fig. 17.2a) arranged on a glass cover slip (compare Fig. 17.1) are transferred by means of a pointed tip to the end of the quartz fiber (Fig. 17.2b) contained in the glass syringe housing (Fig. 17.2c). The whole process is viewed by a horizontal stereomicroscope (Fig. 17.2d). Deflection of the fiber tip upon sample transfer is monitored by a calibrated microscale contained in one of the oculars (for calibration, see Lowry and Passonneau 1972).

17.3 Microphotometry

The sample chambers consist of a 5 mm thick Teflon tray with holes of 3 mm diameter. The holes are closed by a thin Teflon film stretched across the lower

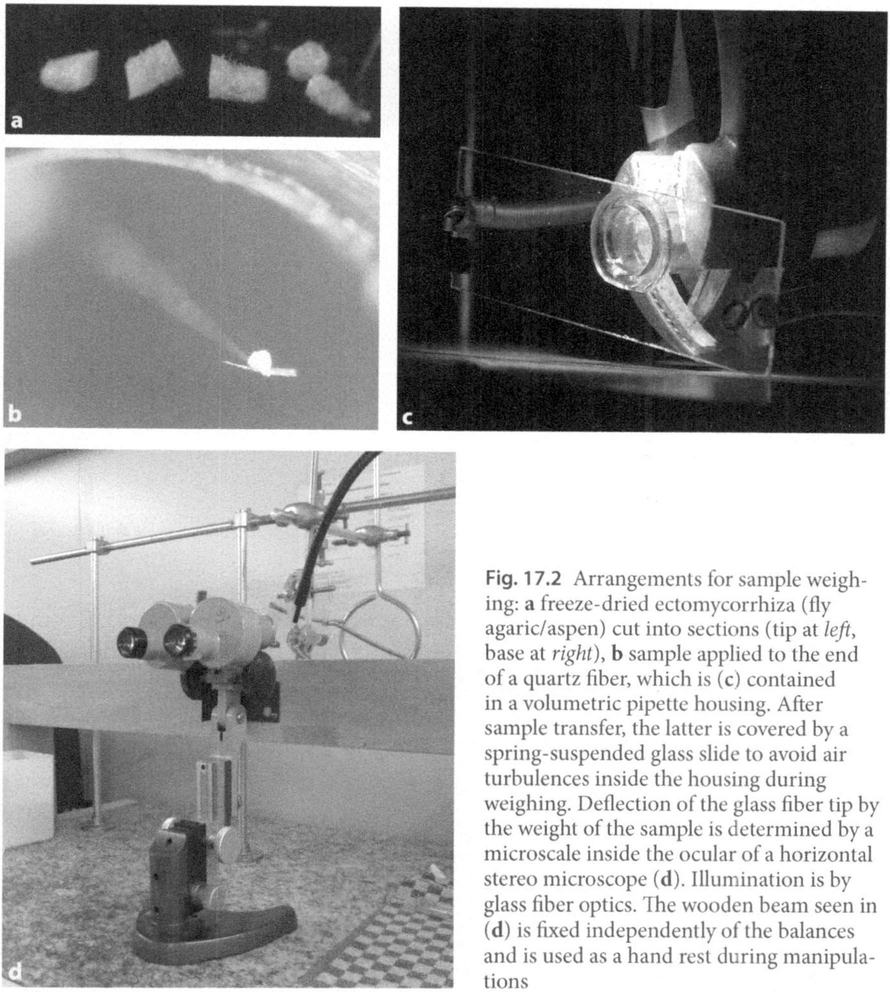

Fig. 17.2 Arrangements for sample weighing: **a** freeze-dried ectomycorrhiza (fly agaric/aspen) cut into sections (tip at *left*, base at *right*), **b** sample applied to the end of a quartz fiber, which is (**c**) contained in a volumetric pipette housing. After sample transfer, the latter is covered by a spring-suspended glass slide to avoid air turbulences inside the housing during weighing. Deflection of the glass fiber tip by the weight of the sample is determined by a microscale inside the ocular of a horizontal stereo microscope (**d**). Illumination is by glass fiber optics. The wooden beam seen in (**d**) is fixed independently of the balances and is used as a hand rest during manipulations

end and fixed by insertion of Teflon tubing, with an outer diameter identical to the hole diameter. The Teflon membrane used is selected according to minimal fluorescence. The membrane supplied by Hansa Tech (UK) for use with their oxygen electrodes meets the requirements (Sauer, Reutlingen, Germany).

In a practical assay, to avoid evaporation, the wells were filled with 10 μl of purified light mineral oil (Sigma, Taufkirchen, Germany). Using glass constriction pipettes, 2 μl of the assay cocktail were submerged in the oil. Subsequently, the tissue sample was pushed through the oil into the assay droplet by means of a tiny quartz fibre glued to a glass or wooden handle. Contact with the assay cocktail was indicated by a color change due to wetting of the sample. The whole Teflon tray was then transferred to the stage of the inverted microscope. The objective lens (PL Fluotar, Leitz; 40×/0.70 EF) was focused to a layer above the

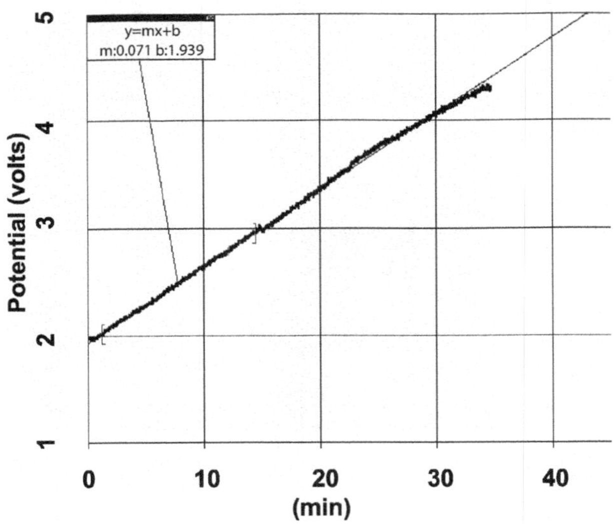

Fig. 17.3 Example of the kinetics of neutral trehalase after digitizing of the photomultiplier signals

Teflon membrane within the brightest area of the droplet, avoiding any shadowing by the sample. The excitation light (Hg lamp HBO 103W/2; Leitz, Bensheim, Germany) was passed to the assay droplet by an excitation filter (330–380 nm) and reflected by a dichroic mirror (<400 nm). For exact details, see Outlaw et al. (1985). The analog signal resulting from the photomultiplier tube was digitized by an AD converter (Logger Pro, Vernier, Canada) and made visible as a kinetics graph on a PC screen (Fig. 17.3).

17.4 Biochemical Analysis: Real Time Microassays

Due to the small amounts of sample material, usually about 1 µg dry weight, photometer signals resulting from NADH fluorescence in microdroplets were recorded. For this purpose a photomultiplier connected to an inverted microscope was used, principally as outlined by Outlaw et al. (1985).

1. Sample preparation:
   Freeze-dried mycorrhizae (Fig. 17.4) were cut into four or five pieces of about equal length with microknives and weighed with a glass fiber balance in a conditioned room (40% RH, 20 °C).
2. Enzyme assays.
   a. Trehalose phosphate synthase:
      The assay was carried out according to Vanderkammen et al. (1989; method 2). A total volume of 2 µl contained HEPES (50 mM, pH 7.6), glucose 6-phosphate (40 mM), $MgCl_2$ (2 mM), NADH (0.6 mM), phos-

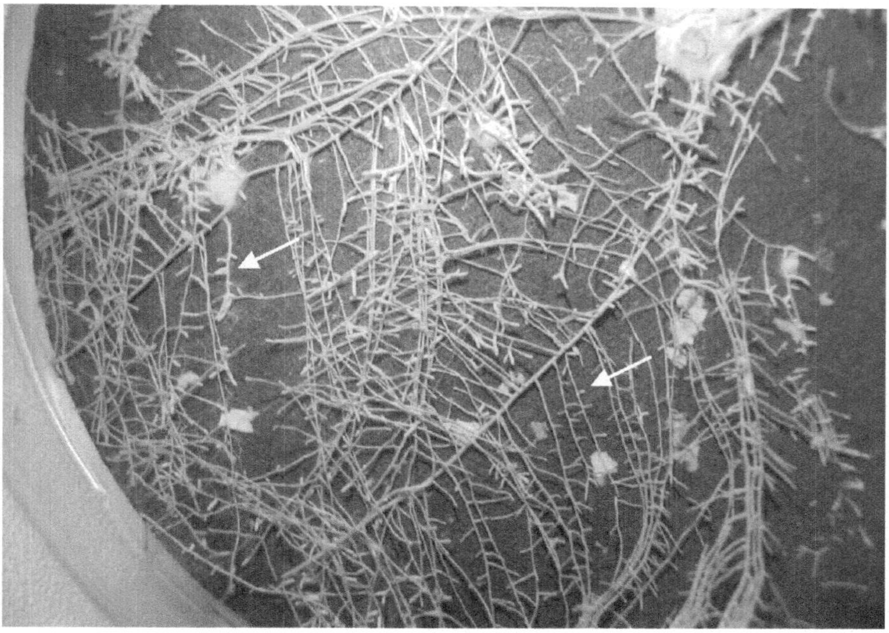

Fig. 17.4 Source of the ectomycorrhizas investigated. The ectomycorrhizas (swollen fine roots, *arrows*) were obtained by inoculation of the roots of aspen seedlings developed inside a Petri dish with a cell suspension of fly agaric. The photograph shows the root system after lyophilization

phoe*nol*pyruvate (1.5 mM), lactate dehydrogenase (2 units), pyruvate kinase (2 units), and UDP glucose (1.7 mM).

b. Neutral trehalase:
Enzyme activity was assayed as glucose produced from trehalose hydrolysis. Glucose was quantified enzymatically via phosphorylation by hexokinase and subsequent oxidation by NADP-dependent glucose 6-phosphate dehydrogenase (Jones et al. 1981). The assay volume was 2 μl, submersed below light mineral oil by means of a 2-μl-constriction pipette.

17.5 Spatial Resolution of Basic Steps of Fungal Trehalose Metabolism in Symbiosis

Trehalose, a non-reducing disaccharide consisting of two molecules of glucose, is present in most organisms except vertebrates (Benaroudj et al. 2001). It plays an important role as a protectant against abiotic stress. In mycorrhiza-form-

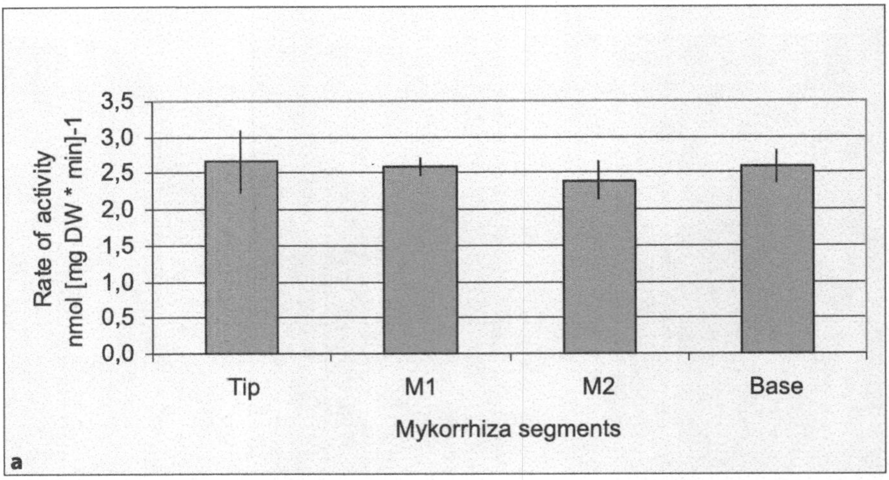

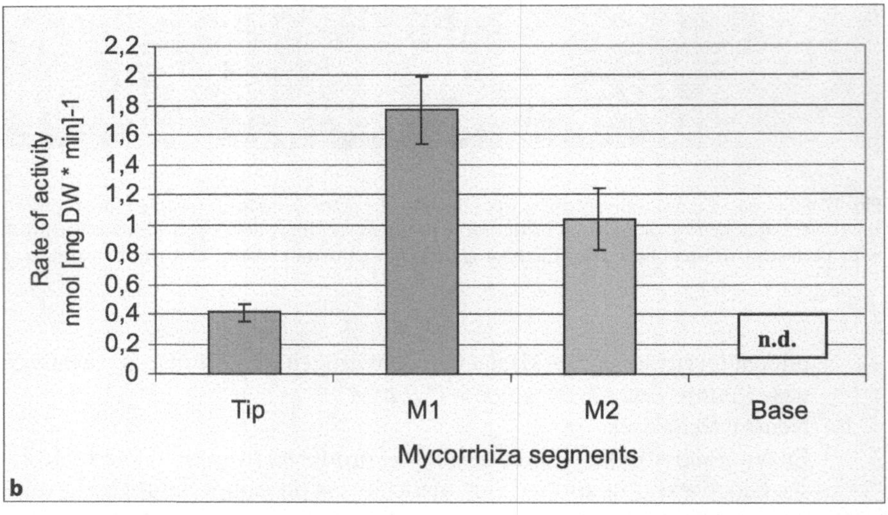

Fig. 17.5 Rates of activity of enzymes along four sections of single mycorrhiza of about 2.5 µg each. **a** Neutral trehalase, **b** trehalose phosphate synthase. The values given are means for 12 individual mycorrhizas (±SE). *n.d.* Not detectable

ing fungi, trehalose constitutes an important intermediary form of carbohydrate storage (Smith and Read 1997).

Key enzymes of turnover are trehalose phosphate synthase and neutral trehalase (for a recent review, see Bonini et al. 2004). The balanced acivity of both enzymes determines the amount of trehalose within fungal cells. In an ectomycorrhiza, as formed between fly agaric and fine roots of poplar, there is a distinct distribution of both enzyme activities (Fig. 17.5). While, on a total dry weight basis, neutral trehalase is nearly equally distributed across all sections (Fig. 17.5a), trehalose phosphate synthase is most prominently present closely behind the growing tip of the fine root. The activity of this enzyme declines towards the base of the fine root, where it is no longer detectable (Fig. 17.5b).

Other data obtained for the same sections (amounts of partner-specific carbohydrates, enzyme activities; Hampp et al., unpublished data) clearly indicate highest fungal activities in zone M1.

The examples given are based on "macro dissection" as we only used relatively large parts of mycorrhizas. Spatial resolution can, however, largely be extended by cutting smaller parts (e.g. separation of the hyphal mantle from the Hartig net) and concomitantly decreasing the assay volume (see Outlaw et al. 1985).

Acknowledgements

Part of this work presented was financed by a grant from the Deutsche Forschungsgemeinschaft.

References

Benaroudj N, Lee DH, Goldberg AL (2001) Trehalose accumulation during cellular stress protects cells and cellular proteins from damage by oxygen radicals. J Biol Chem 276:24261–24267

Bonini BM, Van Dijck P, Thevelein JM (2004) Trehalose metabolism: enzymatic pathways and physiological functions. In: Brambl R, Marzluf GA (eds) The mycota III, biochemistry and molecular biology. Springer, Berlin Heidelberg New York, pp 291–332

Hampp R, Ecke M, Schaeffer C, Wallenda T, Wingler A, Kottke I, Sundberg B (1996) Axenic mycorrhization of wild type and transgenic hybrid aspen expressing T-DNA indoleacetic acid-biosynthetic genes. Trees 11:59–64

Jones MGK, Outlaw WH Jr, Lowry OH (1981) Enzymic assays of 10^{-7} to 10^{-14} moles of sucrose in plant tissues. Plant Physiol 60:379–383

Lowry OH, Passonneau JV (1972) A flexible system of enzymatic analysis. Academic, New York

Outlaw WH Jr, Manchester J, Zenger V (1981) The relationship between protein content and dry weight of guard cells and other single cell samples of *Vicia faba* L. leaflet. Histochem J 13:329–336

Outlaw WH Jr, Springer SA, Tarczynski MC (1985) Histochemical technique. A general method for quantitative enzyme assays on single cell extracts with a time resolution of seconds and a reading precision of femtomoles. Plant Physiol 77:659–666

Smith SE, Read DJ (1997) Mycorrhizal symbiosis. Academic, London

Vanderkammen A, Francois J, Hers H-G (1989) Characterization of trehalose-6-phosphate synthase and trehalose-6-phosphate phosphatase of *Saccharomyces cerevisiae*. Eur J Biochem 182:613–620

18 Ion Cyclotron Resonance Fourier Transform Mass Spectrometry for Non-Targeted Metabolomics of Molecular Interactions in the Rhizosphere

P. Schmitt-Kopplin, N. Hertkorn, M. Frommberger, M. Lucio, M. Englmann, A. Fekete, and I. Gebefugi

18.1 Introduction

Plant health and quality is challenged by the attack of soil-borne pathogens and increased environmental stress. Therefore, a detailed understanding of mechanisms and processes in the rhizosphere is essential to ensure, e.g. safe food production of high quality. The rhizosphere (the root zone of plants) is a very special environment and a region of extremely intense interaction of organisms of all taxonomic levels, and as such a chemical environment with a large number of substances with different origin; either excreted by the organisms to their neighbourhood (bacteria, fungi and many other organisms as well as natural organic matter) or vice versa (Fig. 18.1; see also http://rhizosphere.gsf.de).

Metabolites of various interacting partners (plants, microbes, soil) in the rhizosphere constitute the chemical scenario of activation or inhibition (gene activation and control of metabolic and signalling pathways). Known signalling compounds of the plants, like salicylic acid, jasmonic acids and flavonoids and novel components of early response mediators or S-containing metabolites may have protective or signalling functions inside and/or outside the roots. Microbes excrete a variety of secondary compounds, like antibiotics of different chemical nature, siderophores or specific signalling molecules (i.e. phytohormones or *N*-acylhomoserine lactones; AHLs) to enable interacting with other microbes or the plant (Eberl 1999). AHL-mediated cross-talk is reported to be effective also across species borders within microbial communities (Pierson et al. 1998) and even between prokaryotes and eukaryotes (Mathesius et al. 2003; Schuhegger et al. 2006).

GSF – Research Center for Environment and Health, Institute of Ecological Chemistry, Molecular BioGeoanalysis / BioGeomics, Ingolstädter Landstraße 1, D-85764 Neuherberg, Germany, Fax: ++49-89-3187-3358, email: schmitt-kopplin@gsf.de

Soil Biology, Volume 11
Advanced Techniques in Soil Microbiology
A. Varma, R. Oelmüller (Eds.)

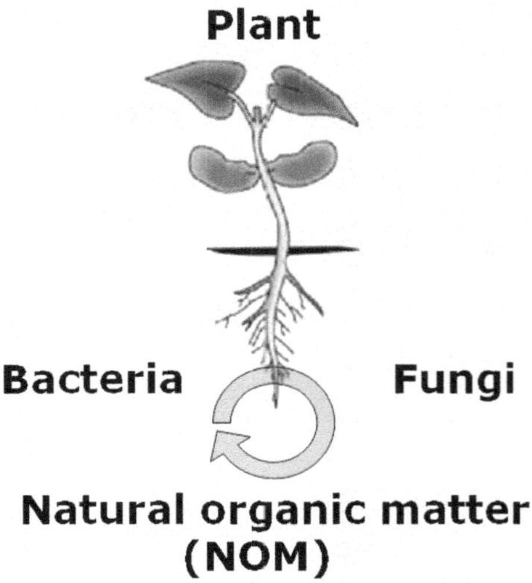

Fig. 18.1 Some important interacting partners studied in the rhizosphere

These compounds induce specific gene expression in neighbouring microbes and the root, initiating specific responses. The soil environment also contributes to the metabolic scenario of the root/soil interface through water-soluble compounds released from the organic soil matrices. The analysis of these signalling molecules and the related soluble natural organic matter in the different ecological environments requires a combination of analytical approaches complementary in their resolution and sensitivity for a qualitative and quantitative approach.

The big challenge is in analytical chemistry to find the best tools to enable description of the chemical space and to understand the chemical regulations in living systems in general. Here we present our conceptual analytical approach to assess information on molecular interactions in the rhizosphere using *non-targeted metabolomics* with an integrated *chemical biology* approach; this implies the combination of complementary disciplines such as *biology* and *chemistry*, supported with analytical tools for a quantitative and qualitative assessment of selected metabolites and signalling molecules in the rhizosphere.

18.2 The Chemical Biology Approach

Chemical biology is a modern approach that utilizes the full spectrum and concepts of organic, physical, analytical and inorganic chemistry and mathematical analysis for the examination of biological processes (Schreiber 2005). Small molecules play key roles at the core of life sciences, health and environment, including topics related to the origin of life, memory and cognition, sensing and signalling, modulation and regulation, cell circuiting and, along these lines, open novel avenues in the understanding and treatment of diseases. Protein interactions mediate the formation of specific small molecules, that themselves modulate many of the multiple individual functions of protein networks. Additional distinct small and macromolecules such as peptides, RNA, carbohydrates, lipids and their covalent and non-covalent adducts (glycoproteins, liposaccharides, lipoproteins, etc.) are involved in interactions and modulations of molecular processes of life. Different disciplines have become merged under the umbrella of chemical biology, involving *proteomics, glycobiology* (Bertozzi and Kiessling 2001) and *lipidomics* (Wenk 2005); *metabolomics* represents a tool to describe the complement of small molecules and biomarkers involved in biological processes with time and spatial resolution.

Chemical biology has matured into an interdisciplinary approach with impulses from bioorganic chemistry (Kadereit et al 2000) and medical chemistry (Wess et al 2001), organic synthesis, structural biology, molecular and cell biology as well as biotechnology, microarray techniques and molecular informatics. The goal is to understand the effect of (de novo) *small molecules in biological systems* and to use this knowledge to trigger these systems (Stockwell 2004). Therefore, chemical biology can also be regarded as an extension of *medical chemistry* and *drug discovery* to natural and living systems in general. The chemical biology approach is ideally applicable to study the complexity of the interactions between the different living and nonliving compartments occurring in the rhizosphere on a molecular level.

18.3 Complementary Analytical Approaches

Complementary approaches of analytical chemists need to be used and adapted in close interaction with soil scientists, soil microbiologists and biologists to reach the goals of the studying interactions in the rhizosphere on a molecular level. Adequate sampling and sample preparation (cleanup or concentration of target compounds) is followed with a separation of the analytes based on their structural properties (charge, size, hydrophilicity/hydrophobicity, etc.) and a detection strategy offering best sensitivity or best profiling properties (Fig. 18.2). The analytical platform is complemented with nuclear magnetic resonance spectroscopy and ultrahigh resolution mass spectrometry for structural description on the molecular level.

Different levels of the analytical approaches must be distinguished (Fig. 18.3): targeted analysis, metabolite profiling and non-targeted analysis.

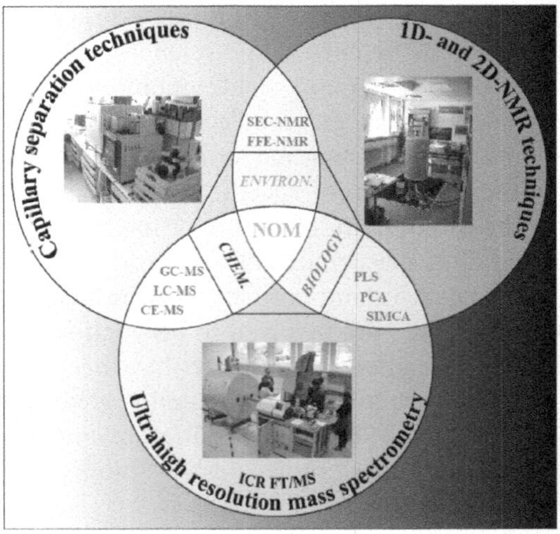

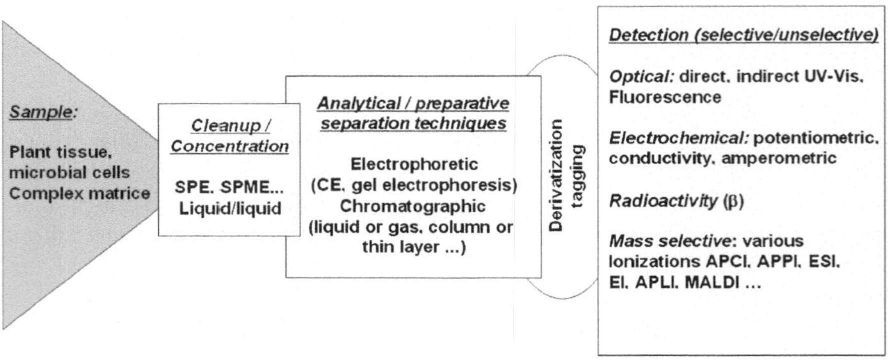

Fig. 18.2 Sample preparation, cleanup/concentration, separation and detection as a classic approach in analytical chemistry

18.3.1 Targeted Analysis

Quantitative evaluation of concentrations of chemicals (organic and inorganic, natural and anthropogenic, ambient to trace amounts) from various matrices after precise and adapted sample preparation (cleanup, concentration).

The substances are of known chemical structures and of known or in-silico estimated physico-chemical properties. Each analytical approach allows the analysis of a few to a hundred components (multi-residue approach) per run. The use of standard chemicals (commercially available or synthesized) leads to the possible quantification of the analytes. Cleanup and sample pre-concentration [e.g. Solid phase (micro)extraction; SPE] are the first steps in the analysis,

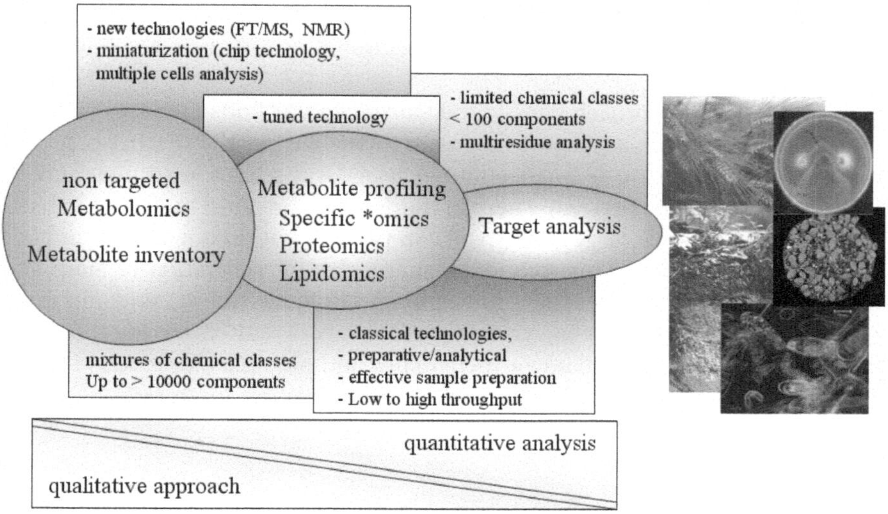

Fig. 18.3 Illustration of the different levels in the analytical approach

followed by electrophoretic or chromatographic separation (capillary electrophoresis, electrochromatography, liquid chromatography, gas chromatography) and adapted detection (UV-Vis, fluorescence, mass spectrometry).

18.3.2 Metabolite Profiling

Here the targeted components are more in number and one focuses on classes of metabolites (i.e. lipids, sugars, peptides, proteins, etc.). The cleanup methods and instrumentation are tuned for these classes of compounds, based on their physico-chemical properties, and form the basis of the *Omics fields developed recently (lipidomics, glycomics, peptidomics, proteomics, etc.). The information is of a structural basis with further possible target quantification. Very often miniaturization of the separation techniques is a goal to enable high throughput (HTP) analysis for screening purposes.

18.3.3 Non-Targeted Analysis

Here new technologies are developed and optimized for a qualitative/semi-quantitative evaluation of the presence of chemical classes in complex mixtures – the molecular inventory needed to allow process descriptions or the discovery of new biomarkers. Cleanup is done in such a way as to alter the sample as little as possible, trying to keep as much as structural information as possible in the complex samples.

Within this last approach *non-targeted metabolomics* finds its place and high-end technologies such as nuclear magnetic resonance (NMR) spectroscopy and especially ion cyclotron resonance fourier transform mass spectrometry (ICR-FT/MS; which will be presented here in more detail) are ideal tools to get structural information on a large quantity of components within one single sample.

The combined use of microfluidic technologies for separation/cleanup and liquid microhandling, with new ionization techniques (APPI) and ultrahigh resolution mass spectrometry certainly are uniquely suited for identification of known and hitherto unknown metabolites and possible biomarkers in very complex samples and for the analysis of often unseparable mixtures.

18.4 Resolving Structural Information from Molecular Complexity with ICR-FT/MS

Because of its ultrahigh resolution in excess of 200 000 full width at half maximum (FWHM) in broad band measurements (or up to 2 000 000 FWHM in ultrahigh resolution mode) and a mass accuracy of routinely less than 0.2 ppm, high field (12 Tesla) FTICR-MS allows for an advanced chemical characterization of metabolites of known and hitherto unknown structure in complex and heterogenous samples of biological origin like plant or bacteria cell extracts, natural or artificial oligo- and multispecies systems, or in complex mixtures derived from biological precursors: "*in particular, systems in which very high mass measurement accuracy is required, very complex mixtures are to be analyzed, or very limited amounts of sample are available may be uniquely suited to interrogation by FTICR mass spectrometry*" (Hofstadler et al. 2005). The use of newly available on-chip nanoelectrospray ionization systems enables in addition another significant increase in sensitivity, a drastic reduction in sample amount and a more efficient ionization of low-abundant ions in the presence of highly abundant (matrix) species. The improvement in mass resolution is exemplified in Fig. 18.4 by comparing the mass profiles and details of a surface water natural organic matter (Suwannee river fulvic acids) as analysed with classic ion-trap technology and ICR-FT/MS. This profile is representative of the hydrophobic fraction that can be extracted from soil percolation water and shows already the complexity of the chemistry in the root zone.

By setting sensible chemical constraints, FTICR-MS allows for the assignment of individual elemental compositions to most of up to 10 000 peaks from one single measurement across a sizable mass range. This information and the informative order of (changing) *elemental composition patterns* gives *unprecedented* insight into, e.g. the nature of metabolites, their possible origin and their

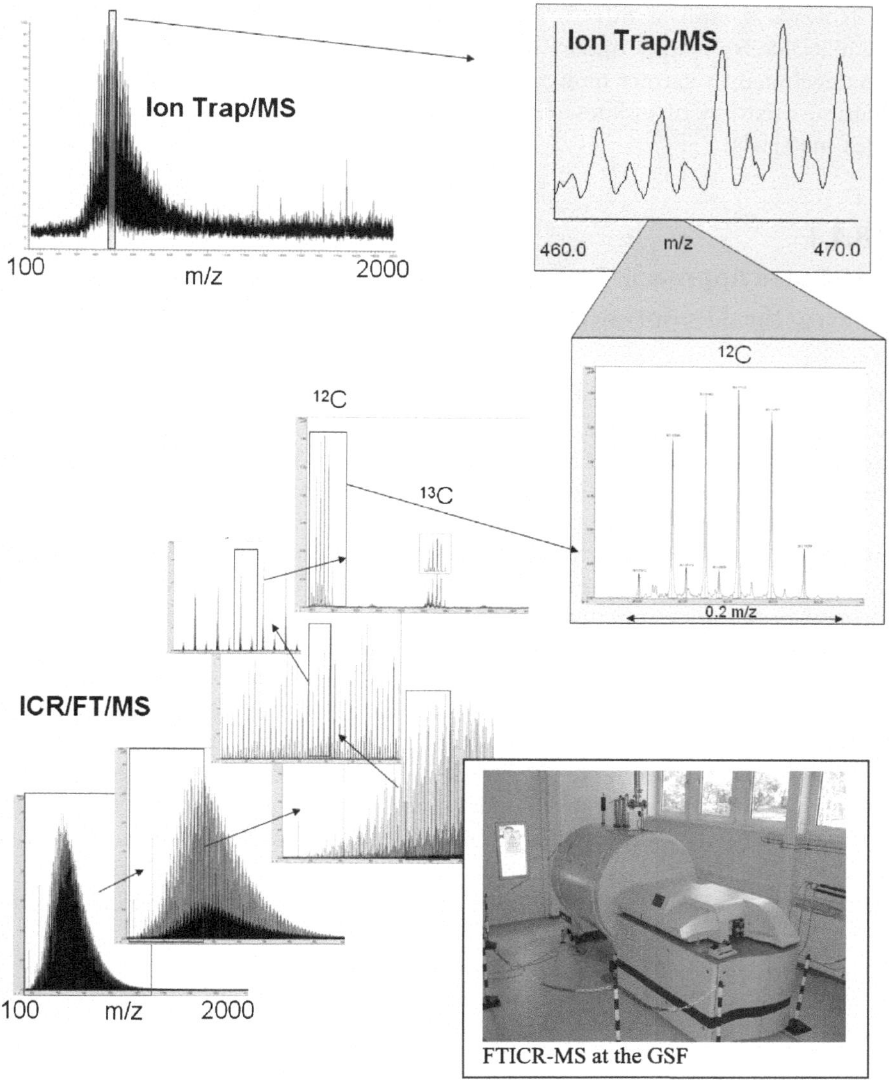

Fig. 18.4 Comparison of ion trap MS and ICR-FT/MS of a natural organic matter (fulvic acid) showing the high resolving power of ICR-FT/MS

function in an organism. Unlike in classic mass spectrometry, which in practice seldomly exceeds a resolution of 10 000 FWHM or a mass accuracy of 5 ppm, and even in comparison with low-field FTMS instruments (7 T), where typical values may be 100 000 FWHM and 1 ppm, respectively, strategies to compare different datasets are completely different in high-field FTMS. New ways of peak alignment and detecting similarity/dissimilarity in samples through new software algorithms and statistical methods are necessary.

Nowadays, only a tiny fraction of the enormous datasets acquired with FTICR-MS are readily understood, and new and intuitive procedures have to be developed to extract molecular level structural information from complex natural mixtures of unknowns (metabolites, peptides, macromolecules, complex materials).

18.4.1 Top-Down Approach: From ICR-FT/MS-Profiling Analysis to Structural Hypothesis

For thousands up to tens of thousands of peaks from a single FTMS spectrum of a *highly complex* sample, a detailed description of the complexity itself and analysis of the presence (or absence) and of the consequences of the *informative order* in different visualization approaches is indispensable.

The crucial point of meaningful data interpretation and the first step of the analytical process towards the above-mentioned *identification* of, e.g. signalling molecules, biomarkers or metabolites reacting (i.e. increasing, decreasing, appearing or disappearing) in relation to environmental parameters, or existing in different taxonomic groups, is a detailed comparison of the datasets.

A possible approach is illustrated in Fig. 18.5 with the profile of a complex growth medium based on agar (Fig. 18.5a) and that of the corresponding bacterial extract on this medium (similar at first glance but full of differences in details, Fig. 18.5b). From the profiles, these peaks appearing in Fig. 18.5b can be picked out and assigned in elementary composition (CHONS) from their exact mass. The analysis of a sufficient number of samples is needed for a statistical approach, as illustrated in Fig. 18.6, and these tools can be used to answer a number of different open questions, e. g. in chemical taxonomy of micro-organ-

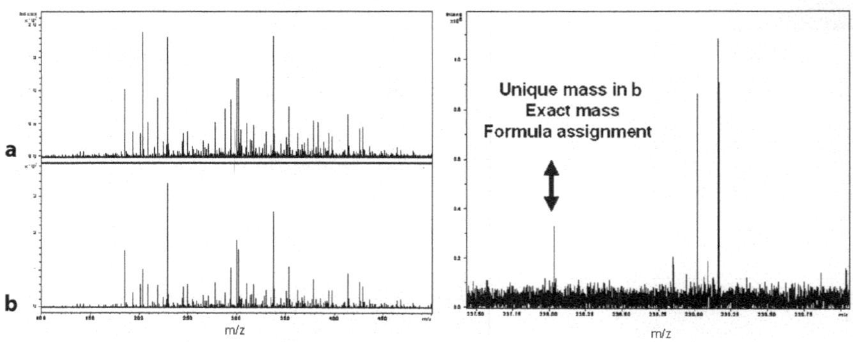

Fig. 18.5 Positive electrospray ionization ICR-FT/MS of agar growth medium (**a**) and a bacterial groth (Burcholderia *sepacia* LA3) grown on agar (**b**). On the *right* is a possible comparison of the two complex mixtures and assignment of specific bacterial components

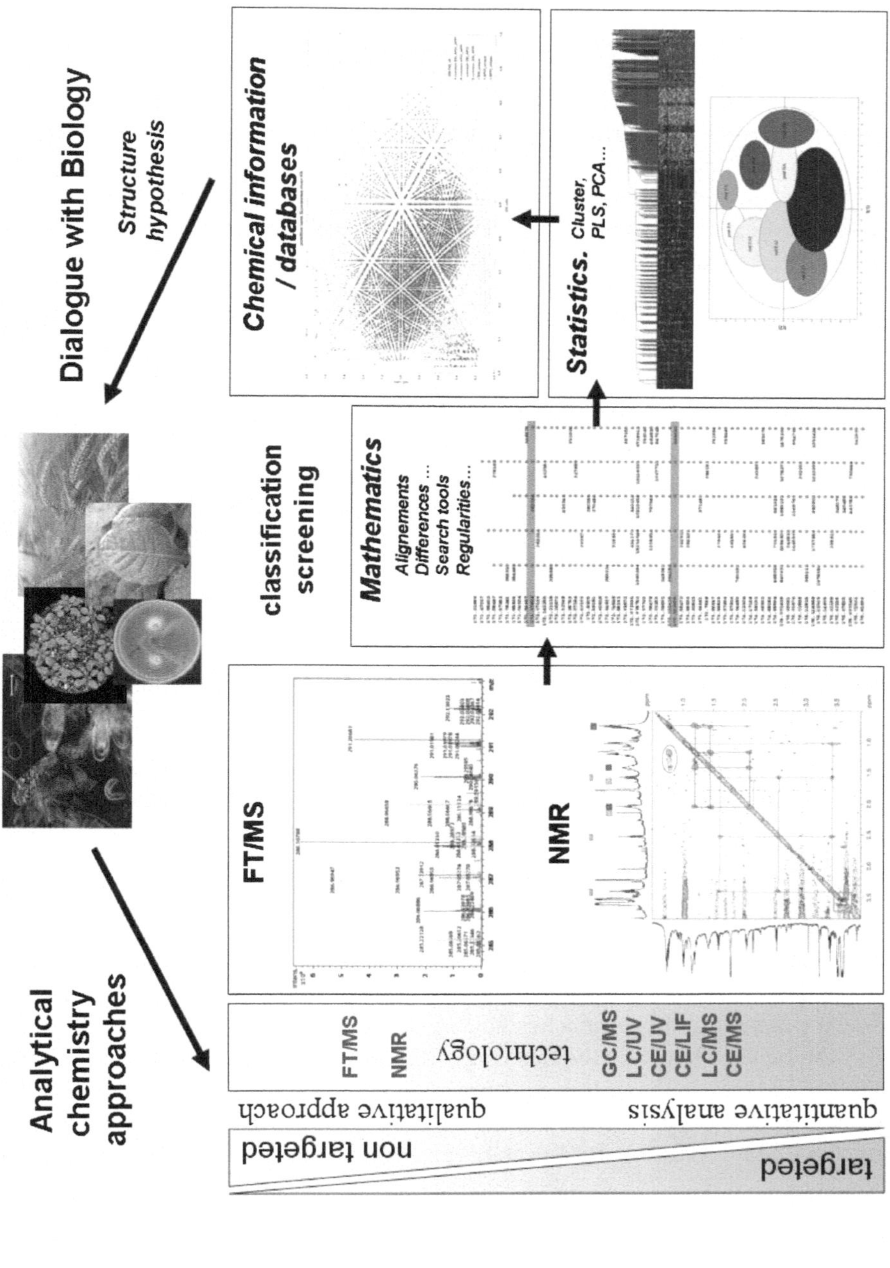

Fig. 18.6 Representation of the non-targeted approach in the analysis of space-, time- and interaction-dependent studies. This platform approach is available for studies in the rhizosphere

isms, plant mutant analysis, metabolite detection in plant organs and their time-dependent translocation.

18.4.2 Complementary Analytical Tools

The ICR-FT/MS and NMR technologies generate enormous amounts of data (in comparison with other analytical approaches that rather average the structural information one can derive from them), that first need to go through a series of mathematical and statistical data mining and visualization algorithms prior squeezing out the essence of information.

This is illustrated in Fig. 18.6. When choosing a non targeted approach, both ICR-FT/MS and NMR generate thousands of experimental values (chemical shift related to functional group and respectively exact mass assignable to an elemental composition in C, H, O, N) per analysed variable (interaction partners, bacterial strain, interaction partner, time and space scales, etc.). For exploratory visualization, here hierarchical clustering analysis and the correlation matrix were used. The patterns or trends related to class separation could be detected by PLS discriminant analysis (PLS-DA). It was performed in order to sharpen the separation between groups of observations, by rotating principal components analysis (PCA) components such that a maximum separation among classes is obtained, and to understand which variables carry the class separating information.

18.4.3 Bottom-Up Approach: From Hypothesis-Driven Experiments Upwards to ICR-FT/MS

Certain micro-organisms within the diverse soil microflora are able to interact specifically with plant roots. In the genera *Burkholderia* and *Pseudomonas*, e.g. many root-associated strains (plant protective or pathogenic) but also human pathogens are known. It has recently been found that signalling molecules of the *N*-acylhomoserine lactone (AHL) type also trigger responses similar to induced systemic resistance in plant roots. Looking for AHL signalling molecules in complex media involves target analysis strategies described above using classic chromatographic approaches and in-silico separation simulation. ICR-FT/MS can be used here as a tool for confirmation of the presence of the one or the other AHL molecule through exact mass determination. As already pointed out by others (Kind and Fiehn 2006), a mass accuracy of even less than 1 ppm *alone* may not be high enough for the assignment of *one* unique elemental formula to

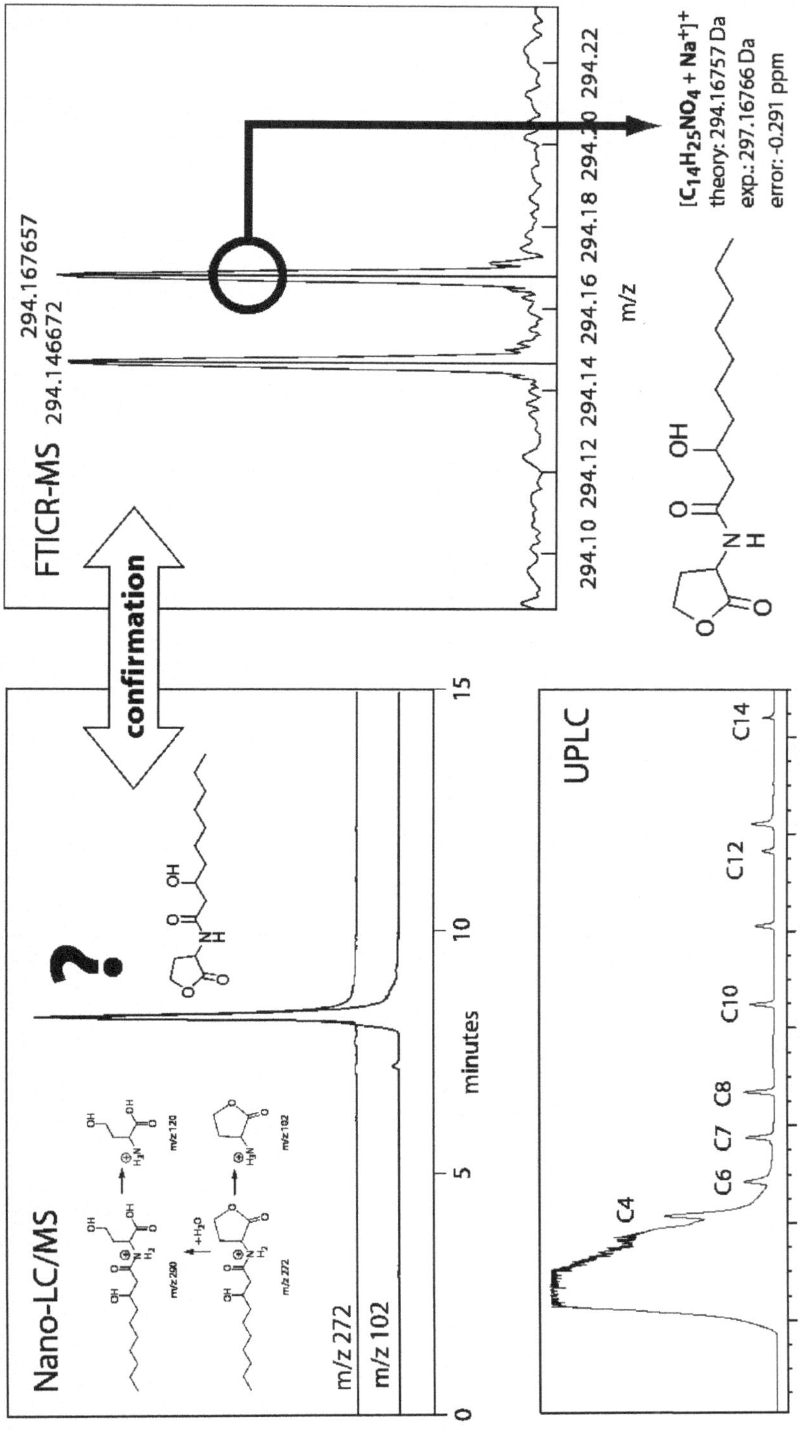

Fig. 18.7 Nano-LC/MS, UPLC and FTMS for the analysis of known and unknown signalling molecules in selected rhizosphere bacteria

an exact mass, when, like in the present bacterial extracts, a plethora of possible structures can be expected. Thus especially the *combination* of exact mass measurements by ultrahigh resolution mass spectrometry with chromatographic data from ultrahigh resolution separation gives unprecedented insight into the nature of metabolites, their possible origin and their function in an organism confronted with pathogens or symbiotic partners (Fig. 18.7).

18.5 Conclusion

As a result of the considerations above, it should become clear that chemical analysis of the rhizosphere, in which FTICR-MS can play a decisive role, should always be part of an interdisciplinary approach involving biologists, chemists and biomathematicians, as it requires the development of new sampling strategies, new ways of sample preparation, new strategies for the analysis itself, and, maybe as the most important part, new ways of data interpretation in a meaningful way.

References

Bertozzi C, Kiessling L (2001) Chemical Glycobiology. Science 291:2357–2364

Eberl L (1999) N-Acyl homoserinelactone-mediated gene regulation in Gram-negative bacteria. Syst Appl Microbiol 22:493–506

Hofstadler SA, Sampath R, Blyn LB, Eshoo MW, Hall TA, Jiang Y, Drader JJ, Hannis JC, Sannes-Lowery KA, Cummins LL, Libby B, Walcott DJ, Schink A, Massire C, Ranken R, Gutierrez J, Manalili S, Ivy C, Melton R, Levene H, Barrett-Wilt G, Li F, Zapp V, White N, Samant V, McNeil JA, Knize D, Robbins D, Rudnick K, Desai A, Moradi E, Eckera DJ (2005) TIGER: the universal biosensor. Int J Mass Spectrom 242:23–41

Kadereit D, Kuhlmann J, Waldmann H (2000) Linking the field – the interplay of organic synthesis, biophysical chemistry, and cell biology in the chemical biology of protein lipidation. Chem Biochem 1:144–169

Kind T, Fiehn T (2006) Metabolomic database annotations via query of elemental compositions: mass accuracy is insufficient even at less than 1 ppm. BMC Bioinformatics 7:234

Mathesius U, Mulders S, Gao M, Teplitski M, Caetano-Anolles G, Rolfe BG, Bauer WD (2003) Extensive and specific responses of a eukaryote to bacterial quorum-sensing signals. Proc Natl Acad Sci USA 100:1444–1449

Pierson EA, Wood DW, Cannon JA, Blachere FM, Pierson LS (1998). Interpopulation signaling via N-acyl-homoserine lactones among bacteria in the wheat rhizosphere. Mol Plant Microbe Interact 11:1078–1084

Schreiber SL (2005) Small molecules: the missing link in the central dogma. Nat Chem Biol 1:64–66

Schuhegger R, Ihring A, Gantner S, Bahnweg G, Knappe C, Vogg G, Hutzler P, Schmid M, Van Breusegem F, Eberl L, Hartmann A, Langebartels C (2006) Induction of systemic resistance in tomato plants by N-acylhomoserine lactone-producing rhizosphere bacteria. Plant Cell Environ 29:909–918

Stockwell B (2004) Exploring biology with small organic molecules. Nature 432:846–854

Wenk MR (2005) The emerging field of lipidomics. Nature 4:594–610

Wess G, Urmann M, Sickenberger B (2001) Medizinische Chemie: Herausforderung und Chancen. Angew Chem 113:3443–3453

19 Application of Terminal-Restriction Fragment Length Polymorphism for MolecularAnalysis of Soil Bacterial Communities

A. Mengoni, E. Giuntini, and M. Bazzicalupo

19.1 Introduction

DNA-based techniques have become a powerful tool for studying the diversity and the composition of soil bacterial communities in cultivation-independent ways (Torsvik and Øvreås 2002). One of the most important methods for the surveys of soil bacteria is the analysis of a 16SrDNA clone library (Giovannoni et al. 1990; Hugenholtz et al. 1998) or, more and more promising, the analysis of a metagenomic library (Rondon et al. 2000). However, due to the complexity of soil communities and the effort required for this type of analysis, clone libraries have been restricted to the analysis of a single or a few samples in an environment. To circumvent the limitations of the clone library approach, several PCR-based methods exist which allow rapid fingerprinting and monitoring of many samples. Terminal-restriction fragment length polymorphism (T-RFLP) is a PCR-based tool which has been introduced for specifically studying the genetic diversity of bacterial communities (Liu et al. 1997; Marsh 1999; Kitts 2001). T-RFLP analysis is based on the detection of a single restriction fragment in each sequence amplified directly from the environmental sample of DNA and is capable of surveying dominant members comprising at least 1% of the total community (Dunbar et al. 2000). Community fingerprintings obtained are well correlated with those obtained with other methods like denaturing gradient gel electrophoresis (DGGE) or ribosomal intergenic spacer analysis (RISA; Hartmann et al. 2005). T-RFLP has been widely used in recent years for the analysis of bacterial communities in different conditions (for examples, see Moeseneder et al. 1999; Osborn et al. 2000; Richardson et al. 2002; Sakano et al. 2002; Fierer et al. 2003; Lueders and Friedrich 2003; Nagashima et al. 2003) and to assess

Dipartimento di Biologia Animale e Genetica 'Leo Pardi', Via Romana 17
I-50125, Firenze, Italy, Tel: +390552288242, Fax: +390552288250,
email: marco.bazzicalupo@dbag.unifi.it

Soil Biology, Volume 11
Advanced Techniques in Soil Microbiology
A. Varma, R. Oelmüller (Eds.)

spatial and temporal heterogeneity and dynamics of bacterial communities in soil (Kuske et al. 2002; Mengoni et al. 2004, 2006), sediments and water environments (Scala and Kerkhof 2000; Braker et al., 2001; Casamayor et al. 2002; Konstantinidis et al. 2003). Moreover, T-RFLP has recently been proposed as a standard methodology for assessing soil fertility in comparison to fatty acid methyl ester (FAME) analysis (Suzuki et al. 2005).

Terminal restriction fragment (TRF) patterns obtained by using the T-RFLP technique are generated and analyzed in a series of steps that combine PCR, restriction enzyme digestion and gel electrophoresis. Extracted DNA is subjected to PCR amplification using primers homologous to conserved regions in a target gene. One primer is labeled on the 5' end, usually with a fluorescent dye. Amplicons are then digested with a restriction enzyme and the restricted products subjected to electrophoresis in either a polyacrylamide gel or a capillary gel electrophoresis apparatus. The obtained TRF patterns are then compared among different samples to depict similarities and dissimilarities among different communities. For the phylogenetic description of a bacterial community, PCR is performed using primers which anneal to conserved sequences of the 16S rRNA gene and the TRF pattern obtained represents an estimate of the number of different 16S rRNA genes present in the community, i.e. different bacterial groups (Fig. 19.1). Obtained bands can then be taxonomically interpreted either by their direct cloning (Mengoni et al. 2002), or by comparing the length of the restriction fragments with sequences present in a library of 16S

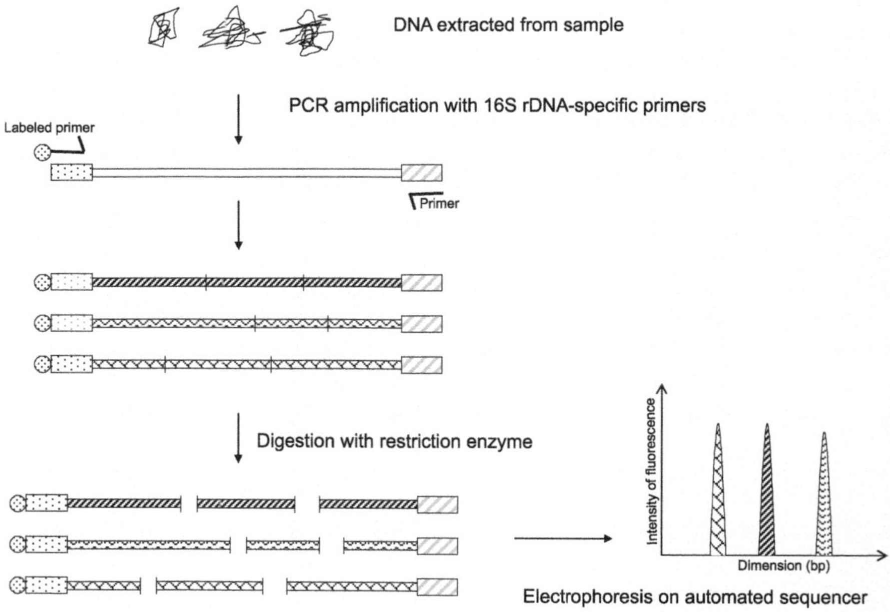

Fig. 19.1 Schematic representation of T-RFLP technique. See text for details

rDNA previously prepared (Dunbar et al. 2000; Urakawa et al. 2000; Fey et al. 2001; Grant and Ogilvie 2004), or by an "in-silico" approach on a 16S rDNA database (Marsh et al. 2000; Kent et al. 2003).

19.2
A General Protocol for Taxonomic T-RFLP Profiling of Soil Bacterial Communities

19.2.1
Materials

19.2.1.1
Equipment

Thermal cycler, gel electrophoresis apparatus with power supply, agarose, automated sequencer for capillary electrophoresis equipped with discrete band analysis software, UV transilluminator and gel documentation system.

Caution: UV rays are dangerous. Protect eyes with a plastic shield.

19.2.1.2
Reagents and Solutions

- Double distilled water (ddH_2O) sterilized by autoclaving or filtering. Prepare 100 µl aliquots before sterilization and keep at −20 °C. Discard the aliquot after each use.
- 50 mM $MgCl_2$ stock solution (usually supplied with the *Taq* DNA polymerase).
- Stock solution of a mixture of desoxyribonucleotide triphosphates (dNTPs): 2 mM of each dNTP in ddH_2O.
- *Taq* DNA polymerase or other thermostable DNA polymerase.
- Restriction enzymes with 4-base recognition sequence (i.e. *Hin*fI, *Msp*I, *Hha*I, *Taq*I, *Rsa*I) and their specific buffers.
- Primer for the amplification of the gene of interest. For instance, on 16SrDNA, 27f primer (5' GAGAGTTTGATCCTGGCTCAG) and 1495r primer (5' CTACGGCTACCTTGTTACGA) give good results. 27f primer is labeled at the 5' end with a fluorescent dye (6-FAM). Prepare stock solutions of primers at 10 µM.

- DNA size marker: good examples are a 100-bp or 1-kbp ladder for agarose gel electrophoresis and TAMRA 500 (Applied Biosystems) for capillary electrophoresis.
- Genomic DNA: use concentrations of 10 ng/μl. For extracting DNA from soil we routinely use the FASTDNA kit for soil (BIO101). Alternatively, good results are obtained using the extraction method developed by Bürgmann et al. (2003) for extracting both DNA and RNA from soil.
- Kit for the purification of PCR products from unincorporated primers and salts. We usually obtain good results with the MinElute PCR purification kit (Qiagen).
  Note: All the above reagents should be kept at –20 °C.
- TAE buffer: 40 mM Tris/acetate, 1 mM EDTA, pH 8. Prepare a 50× stock solution.
- Agarose.
- 10× loading buffer: 70% (w/v) glycerol, 0.5% (w/v) bromophenol blue; store at 4 °C.
- Ethidium bromide stock solution: 10 mg/μl; store in a dark bottle.
  Caution: Ethidium bromide is a powerful mutagen: wear gloves when handling this compound; wear mask when weighing it.

19.2.2 Experimental Procedure

Extract the DNA from soil by using an extraction procedure which works both on gram-negative and on gram-positive bacteria (for references, see Section 19.2.1 Materials). Perform the PCR amplification in a total volume of 50 μl containing: the diluted buffer of *Taq* DNA polymerase, 1.5 mM $MgCl_2$, 10 mM dNTPs, 2 units of *Taq* DNA polymerase, 10 pmol of each primer, 10–20 ng of template DNA.

For instance for ten samples, consider a Master Mix solution for 11 reactions using the following volumes: add 1 μl of template DNA (from solutions previously prepared at 10 ng/μl) in 0.2-ml PCR tube; prepare a Master Mix , adding 485.1 μl of dH_2O, 11 μl of each primer solution, 55 μl of 10× PCR buffer (provided with *Taq* DNA polymerase), 16.5 μl of 50 mM $MgCl_2$ and 4.4 μl of *Taq* DNA polymerase (5 units/μl); mix and aliquot 49 μl of Master Mix solution in the 0.2-ml PCR tubes.

The described PCR conditions have been optimized in a Perkin-Elmer 9600 thermal cycler (Perkin-Elmer). The reaction mixtures, after incubation at 95 °C for 1.5 min, are cycled through the following temperature profile: denaturation at 94 °C for 30 s, annealing temperature for 30 s and extension at 72 °C for 2 min. For the first five cycles, the annealing temperature is set at 60 °C, for the following five cycles 55 °C and for the last 25 cycles 50 °C. Finally, the reaction mixtures are incubated at 72 °C for 10 min.

1. Check 5 µl of each amplification mixture by agarose gel (1.0% w/v) electrophoresis in TAE buffer (0.04 M Tris-acetate, 0.001 M EDTA) containing 1 µg/ml (w/v) of ethidium bromide.
2. Purify (and concentrate if necessary) the amplification products from primers and salts by using a dedicated kit.
3. Digest 600 ng of amplified 16S rDNA in a total volume of 15 µl with 20 units of the chosen restriction enzyme. Incubate for 3 h at the optimal incubation temperature for the restriction enzyme (37 °C or 65 °C for *Taq*I). Heat-inactivate the enzyme by incubation the mixture at 70 °C for 15 min (80 °C for 20 min for *Taq*I).
4. Separate the digested products by capillary electrophoresis on an automated sequencer (ABI 310 genetic analyzer, Applied Biosystems). Inject 5 µl of digestion product with 0.5 µl of molecular weight standard TAMRA-500 (Applied Biosystems). Run the electrophoresis as indicated by the instrument's manufacturer.

19.2.3 Troubleshooting

1. Low intensity of T-RFLP bands or weak amplification of 16S rDNA:
   Check purity and quantity of extracted DNA. Try a different extraction method and a range of different template DNA concentrations. Check reagents and procedure with a control primer pair and DNA. Verify that primers are labeled correctly and load higher quantities of digestion product on the automated sequencer to optimize signal fluorescence.
2. Too many or too few bands:
   Assay different restriction enzymes. A good theoretical estimation of the level of polymorphism shown by the enzyme can be obtained from the in-silico digestion of the database present in the Ribosomal Database Project by using the TAP software (Marsh et al. 2000; http://35.8.164.52/html/TAP-trflp.html#program).

19.3 Standardization of T-RFLP Profiles

The T-RFLP technique usually produces a profile of a bacterial community, which shows 20–40 peaks (bands or TRFs) for each restriction enzyme used (Fig. 19.2). The information gained by the single experiment can be increased by digesting the amplified 16S rDNA with other restriction enzymes and then combining the obtained profile to reach 100–200 peaks per sample.

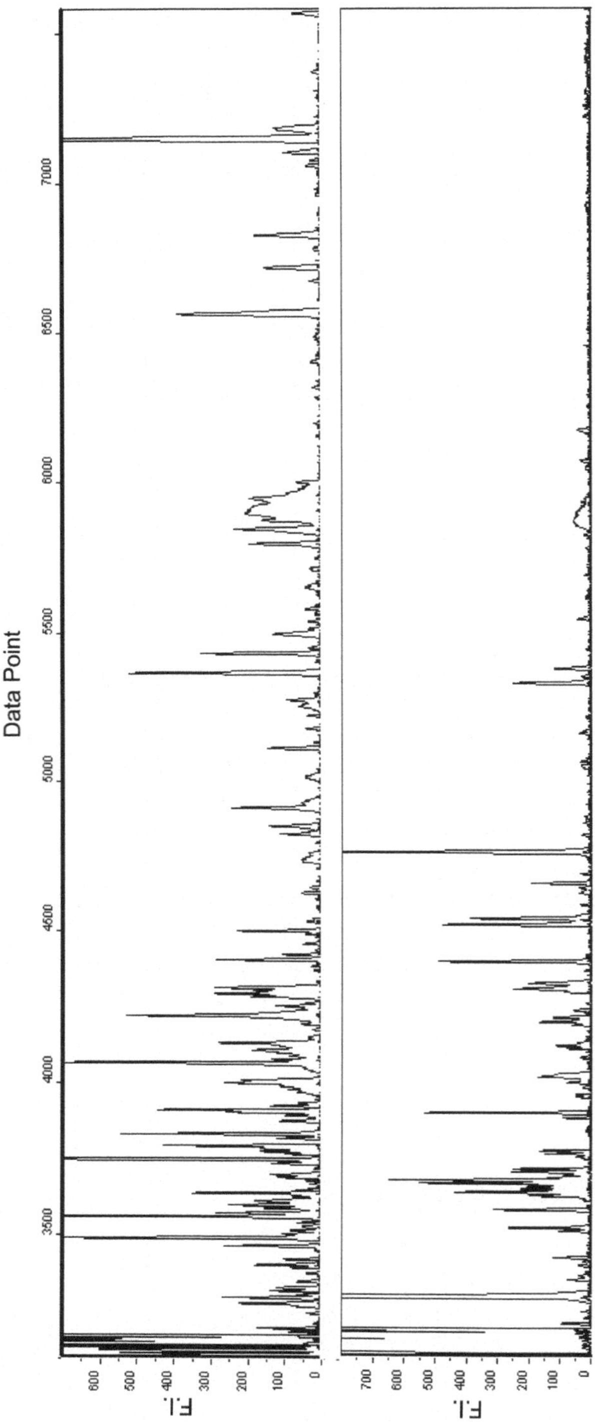

Fig. 19.2 Examples of T-RFLP profiles obtained after *Hin*fI digestion of amplified 16S rDNA from an agricultural soil (*upper* chromatogram) and a heavy-metal rich (serpentine) soil (*lower* chromatogram). *Data Point* indicates size range between 20 bp and 500 bp. *F.I.* Fluorescence intensity

One of the most frequent problems in T-RFLP pattern analysis is the presence of very small peaks resulting from either artefacts or differences in DNA loading which can skew similarity profiles that are based on presence/absence data (Liu et al. 1997; Dunbar et al. 2000). To avoid the presence of nonreproducible peaks derived from artefacts, at least two independently extracted DNA and three PCR reactions for each extracted DNA can be performed. The resulting three digestion products from the same extracted DNA are then mixed and run in a single capillary electrophoresis. In this way nonreproducible peaks, which are often due to 16S rDNA species present in a concentration at the limit of the detection threshold or to artefacts generated during PCR amplification tend to disappear because of the dilution performed with the replicate reactions. The possibility to compare profiles obtained from different analyses then allow consideration of only those peaks present in both DNA extraction, thus reconstructing a "synthetic profile" of the community. The software GelComparII (Applied Maths) is a good platform for the analysis of chromatograms obtained after capillary electrophoresis and includes useful tools for a first statistical analysis of data (cluster and principal component analyses). In general, for the analysis of chromatograms, peaks below 50–100 units of fluorescence are not included because of their low level of reproducibility. However, differences in DNA loading can also generate slightly different profiles and it could be useful to standardize the DNA quantities loaded into the capillary. Kaplan et al. (2001) present a method for standardizing T-RFLP patterns based on TRF peak area. The amount of DNA loaded onto a gel or a capillary is estimated as the sum of all TRF peak areas in a pattern (total peak area). Dunbar et al. (2001) propose a method for standardizing TRF patterns based on peak height. The sum-of-peak-height values are then standardized between samples by proportionally decreasing the height of each peak in the profiles until the sum of peak heights (total fluorescence) for each profile equals the lowest value represented among the samples.

19.4 Other Applications of T-RFLP to Soil Bacterial Communities

In addition to taxonomic profiling, T-RFLP can also be used to characterize functional diversity in a bacterial community. In fact, in principle one can make use of primers anchored to conserved sequences present in functional genes and generate amplicons and TRFs from a DNA sample which reflect the functional genetic diversity present in the community. This approach has been used to explore the diversity of genes involved in nitrogen fixation (*nif*H; Noda 1999), nitrification (*amo*A; Horz et al. 2000), denitrification (*nos*Z; Scala et al. 2000), nitrite reduction (*nir*S; Braker et al. 2001), methane oxidation (*pmo*A; Pester et al. 2004) and mercury resistance (*mer*R; Bruce 1997). Moreover, T-RFLP can also be used to type the genetic diversity of retro-transcribed RNA extracted from soil. This approach can be useful when the diversity of expressed functional genes or the taxonomic diversity (16S rRNA) of metabolically active cells has to be targeted (Rogers et al. 2005; Mengoni et al. 2006). Actually, because RNA is labile and ribosome numbers have been correlated with cellular activity, total rRNA might reflect the diversity of the metabolically active members of the community. After RNA extraction from soil, reverse transcription is performed to produce cDNA molecules, which are subsequently PCR-amplified with the selected primer pair following the standard T-RFLP protocol. In this way it is possible to compare, within a soil community, the genetic diversity of the most metabolically active bacterial groups (T-RFLP on 16S rRNA), with that of the bacterial community present (T-RFLP on 16S rDNA).

19.5 Conclusions

Analysis of soil bacterial communities with T-RFLP patterns provides a rapid and reproducible way to compare communities and assess community dynamics. However, some precautions must be taken when preparing the data for analysis to minimize artifacts and to produce robust profiles. T-RFLP profiling has the advantage of being simply and rapidly produced on existing standard DNA sequencing equipment. T-RFLP patterns are then automatically digitized and easily analyzed with a variety of clustering and multivariate statistical techniques (Rees et al. 2004; Blackwood et al. 2003). Due to this easy and automated processing, many samples can be analyzed at the same time, allowing (as the large and increasing amount of literature shows) far unprecedented opportunities to correlate the structure and diversity of soil bacterial communities to the environmental parameters.

References

Blackwood CB, Marsh T, Kim SH, Paul EA (2003) Terminal restriction fragment length polymorphism data analysis for quantitative comparison of microbial communities. Appl Environ Microbiol 69:926–932

Braker G, Ayala-del-Rio HL, Devol AH, Fesefeldt A, Tiedje JM (2001) Community structure of denitrifiers, Bacteria, and Archaea along redox gradients in Pacific Northwest marine sediments by terminal restriction fragment length polymorphism analysis of amplified nitrite reductase (*nir*S) and 16S rRNA genes. Appl Environ Microbiol 67:1893–1901

Bruce KD (1997) Analysis of the *mer* gene subclass within bacterial communities in soils and sediments resolved by fluorescent-PCR restriction fragment length polymorphism profiling. Appl Environ Microbiol 63:4914–4919

Bürgmann H, Widmer F, Sigler WV, Zeyer J (2003) mRNA extraction and reverse transcription-PCR protocol for detection of *nifH* gene expression by *Azotobacter vinelandii* in soil. Appl Environ Microbiol 69:1928–1935

Casamayor EO, Massana R, Benlloch S, Øvreås L, Díez B, Goddard VJ, Gasol JM, Joint I, Rodríguez-Valera F, Pedrós-Alió C (2002) Changes in archaeal, bacterial and eukaryal assemblages along a salinity gradient by comparison of genetic fingerprinting methods in a multipond solar saltern. Environ Microbiol 4:338–348

Dunbar J, Ticknor LO, Kuske CR (2000) Assessment of microbial diversity in four Southwestern United States soils by 16S rRNA gene terminal restriction fragment analysis. Appl Environ Microbiol 66:2943–2950

Dunbar J, Ticknor O, Kuske CR (2001) Phylogenetic speciticity and reproducibilily and new method for analysis of terminal restriction fragment profiles of 16S rRNA genes from bacterial communities. Appl Environ Microbiol 67:190–197

Fey A, Chin KJ, Conrad R (2001) Thermophilic methanogens in rice field soil. Environ Microbiol 3:295–303

Fierer N, Schimel JP, Holden PA (2003) Influence of drying-rewetting frequency on soil bacterial community structure. Microb Ecol 45:63–71

Giovannoni SJ, Britschgi TB, Moyer CL, Field KG (1990) Genetic diversity in Sargasso Sea bacterioplankton. Nature 345:60–63

Grant A, Ogilvie LA (2004) Name that microbe: rapid identification of taxa responsible for individual fragments in fingerprints of microbial community structure. Mol Ecol Notes 4:133–136

Hartmann M, Frey B, Kolliker R, Widmer F (2005) Semi-automated genetic analyses of soil microbial communities: comparison of T-RFLP and RISA based on descriptive and discriminative statistical approaches. J Microbiol Methods 61:349–360

Horz H, Rotthauwe J, Lukow T, Liesack W (2000) Identification of major subgroups of arnmonia-oxidizing bacteria in environmental sampies by T-RFLP analysis of *amo*A PCR products. J Microbiol Methods 39:197–204

Hugenholtz P, Goebel BM, Pace NR (1998) Impact of culture-independent studies on the emerging phylogenetic view of bacterial diversity. J Bacteriol 180:4765–4774

Kaplan CW, Astaire JC, Sanders ME, Reddy BS, Kitts CL (2001). 16S ribosomal DNA terminal restriction fragment pattern analysis of bacterial communities in feces of rats fed *Lactobacillus acidophilus* NCFM. Appl Environ Microbiol 67:1935–1939

Kent AD, Smith DJ, Benson BJ, Triplett EW (2003) Web-based phylogenetic assignment tool for analysis of terminal restriction fragment length polymorphism profiles of microbial communities. Appl Environ Microbiol 69:6768–6776

Kitts CL (2001) Terminal restriction fragment patterns: a tool for comparing microbial communities an assessing community dynamics. Curr Iss Int Microbiol 2:17–25

Konstantinidis KT, Isaaca N, Fett J, Simpson S, Long DT, Marsh TL (2003) Microbial diversity and resistance to copper in metal-contaminated lake sediment. Microb Ecol 45:191–202

Kuske CR, Ticknor LO, Miller ME, Dunbar JM, Davis JA, Barns SM, Belnap J (2002) Comparison of soil bacterial communities in rhizospheres of three plant species and the interspaces in an arid grassland. Appl Environ Microbiol 68:1854–1863

Liu W-T, Marsh TL, Cheng H, Forney LJ (1997) Characterization of microbial diversity by determining terminal restriction fragment length polymorphism of genes encoding 16 rRNA. Appl Environ Microbiol 63:4516–4522

Lueders T, Friedrich MW (2003) Evaluation of PCR amplification bias by terminal restriction fragment length polymorphism analysis of small-subunit rRNA and *mcrA* genes by using defined template mixtures of methanogenic pure cultures and soil DNA extracts. Appl Environ Microbiol 69:320–326

Marsh TL (1999) Terminal restriction fragment length polymorphism (T-RFLP): an emerging method for characterizing diversity among homologous populations of amplification products. Curr Opin Microbiol 2:323–327

Marsh TL, Saxman P, Cole J, Tiedje J (2000) Terminal restriction fragment length polymorphism analysis program, a Web-based research tool for microbial community analysis. Appl Environ Microbiol 66:3616–3620

Mengoni A, Grassi E, Bazzicalupo M (2002) A cloning method for the taxonomic interpretation of T-RFLP patterns. Biotechniques 33:990–992

Mengoni A, Grassi E, Barzanti R, Biondi EG, Gonnelli C, Kim C-K, Bazzicalupo M (2004) Genetic diversity of bacterial communities of serpentine soil and of rhizosphere of the nickel-hyperaccumulator plant *Alyssum bertolonii*. Microb Ecol 48:209–217

Mengoni A, Tatti E, Decorosi F, Viti C, Bazzicalupo M, Giovannetti L (2006) Comparison of 16S rRNA and 16S rDNA T-RFLP approaches to study bacterial communities in soil microcosms treated with chromate as perturbing agent. Microb Ecol (in press)

Moeseneder MM, Arrieta JM, Muyzer G, Winter C, Herndl GJ (1999) Optimization of terminal-restriction fragment length polymorphism analysis for complex marine bacterioplankton communities and comparison with denaturing gradient gel electrophoresis. Appl Environ Microbiol 65:3518–3525

Nagashima K, Hisada T, Sato M, Mochizuki J (2003) Application of new primer–enzyme combinations to terminal restriction fragment length polymorphism profiling of bacterial populations in human feces. Appl Environ Microbiol 69:1251–1262

Noda S, Ohkuma M, Usami R, Horikoshi K, Kuda T (1999) Culture independent charactemization of a gene responsible for nitrogen fixation in the symbiotic microbial community in the gut of the termite *Neotermes koshunensis*. Appl Environ Microbiol 65:4935–4942

Osborn AM, Moore ER, Timmis KN (2000) An evaluation of terminal-restriction fragment length polymorphism (T-RFLP) analysis for the study of microbial community structure and dynamics. Environ Microbiol 2:39–50

Pester M, Friedrich MW, Schink B, Brune A (2004) *pmoA*-based analysis of methanotrophs in a littoral lake sediment reveals a diverse and stable community in a dynamic environment. Appl Environ Microbiol 70:3138–3142

Rees GN, Baldwin DS, Watson GO, Perryman S, Nielsen DL (2004) Ordination and significance testing of microbial community composition derived from terminal restriction fragment length polymorphisms: application of multivariate statistics. A Van Leeuw Int J Microbiol 86:339–347

Richardson RE, Bhupathiraju VK, Song DL, Goulet TA, Alvarez-Cohen L (2002) Phylogenetic characterization of microbial communities that reductively dechlorinate TCE based upon a combination of molecular techniques. Environ Sci Technol 36:2652–2662

Rogers GB, Carroll MP, Serisier DJ, Hockey PM, Kehagia V, Jones GR, Bruce KD (2005) Bacterial activity in cystic fibrosis lung infections. Respir Res 6:49

Rondon MR, August FR, Bettermann AD, Brady SF, Grossman TH, Liles MR, Loiacono KA, Lynch BA, MacNeil IA, Minor C, Tiong CL, Gilman M, Osburne MS, Clardy J, Handelsman J, Goodman RM (2000) Cloning the soil metagenome: a strategy for accessing the genetic and functional diversity of uncultured microorganisms. Appl Environ Microbiol 66:2541–2547

Sakano Y, Pickering KD, Strom PF, Kerkhof LJ (2002) Spatial distribution of total, ammonia-oxidizing, and denitrifying bacteria in biological waste water treatment reactors for bioregenerative life support. Appl Environ Microbiol 68:2285–2293

Scala DJ, Kerkhof LJ (2000) Horizontal heterogeneity of denitrifying bacterial communities in marine sediments by terminal restriction fragment length polymorphism analysis. Appl Environ Microbiol 66:1980–1986

Suzuki C, Kunito T, Aono T, Liu CT, Oyaizu H (2005) Microbial indices of soil fertility. J Appl Microbiol 98:1062–1074

Torsvik V, Øvreås L (2002) Microbial diversity and function in soil: from genes to ecosystems. Curr Opin Microbiol 5:240–245

Urakawa H, Yoshida T, Nishimura M, Ohwada K (2000) Characterization of depth-related population variation in microbial communities of a coastal marine sediment using 16S rDNA-based approaches and quinone profiling. Environ Microbiol 2:542–554

20 Molecular Symbiotic Analysis Between *Arabiopsis thaliana* and *Piriformospora indica*

B. Shahollari, K. Bhatnagar, I. Sherameti,
A. Varma, and R. Oelmüller

20.1 Introduction

The molecular analysis of beneficial interactions of plants and fungi is often difficult, either because one or both of the symbiotic partners are not well characterized at the molecular level or because they can only grow together in symbiosis. We study the interaction of an endophytic fungus, *Piriformospora indica*, with *Arabdiopsis thaliana*. *P. indica* can interact with many different plant species including *A. thaliana* and promotes their growth, development and seed production. We use the model organism *A. thaliana* as the plant partner to understand the molecular basis for this beneficial plant/microbe interaction. The availability of a large number of well characterized mutants, knock-out lines and molecular tools for genetic analysis in *A. thaliana* allows a rapid identification of mechanisms and molecules involved in this interaction. This information can then be used to analyze the molecular basis of the interaction of *P. indica* with other economically important plant species. Since *A. thaliana* is normally not the host found in nature, the type of interaction and the molecules identified might also differ from the interaction of *P. indica* with other plant species. However, since the endophytic fungus interacts with many different plant species and since the plant responses are very similar, it is likely that the basic mechanisms are comparable in all organisms. Thus, the *A. thaliana*/*P. indica* interaction is an attractive model system to study beneficial plant/microbe interaction at the molecular level.

B. Shahollari, I. Sherameti, R. Oelmüller: Institute of General Botany, University of Jena, Dornburger Strasse 159, D-07743 Jena, Germany,
email: b7oera@hotmail.com; bationa@hotmail.com

K. Bhatnagar, A. Varma: Amity Institute of Microbial Sciences, Amity University, Uttar Pradesh, Sector 125, Noida 201303, India, email: ajitvarma@aihmr.amiity.edu

Soil Biology, Volume 11
Advanced Techniques in Soil Microbiology
A. Varma, R. Oelmüller (Eds.)

20.2 Beneficial Interaction Between Plants and Fungi: *Piriformospora indica* and *Arabidopsis thaliana* as a Model System

We study the interaction of *P. indica* with *A. thaliana*. *P. indica* was originally isolated from the Indian desert and is a wide-host root-colonizing fungus, which allows the plants to grow under extreme physical and nutrient stress. The fungus can be cultivated on complex and minimal substrates and belongs to the Sebacinales in the Basidiomycota (Varma et al. 1999; Weiss et al. 2004). *P. indica* has a vast geographical distribution and is reported from Asia, South America and Australia. The fungus is interesting for basic research as well as biotechnological applications because it functions: (1) as a plant biofertilizer in nutrient-deficient soils, (2) as a bioprotector against root pathogens, insects and heavy metals, (3) as a bioregulator for plant growth development, early flowering, enhanced seed production, and stimulation of active ingredients in medicinal plants and (4) as a bio-agent for the hardening of tissue culture-raised plants. Positive interaction has been established for many plants of economic importance in agriculture, forestry and flori-horticulture, including orchids (Bhatnagar and Varma 2006) and those utilized for biodiesel production. *P. indica* also interacts with a bryophyte, *Aneura pinguis*, and a pterdophyte, *Pteris ensiormis*. Similar to arbuscular mycorrhizal fungi, *P. indica* stimulates nitrate assimilation in the roots and solubilizes insoluble phosphatic components in the soil. The interaction of *P. indica*

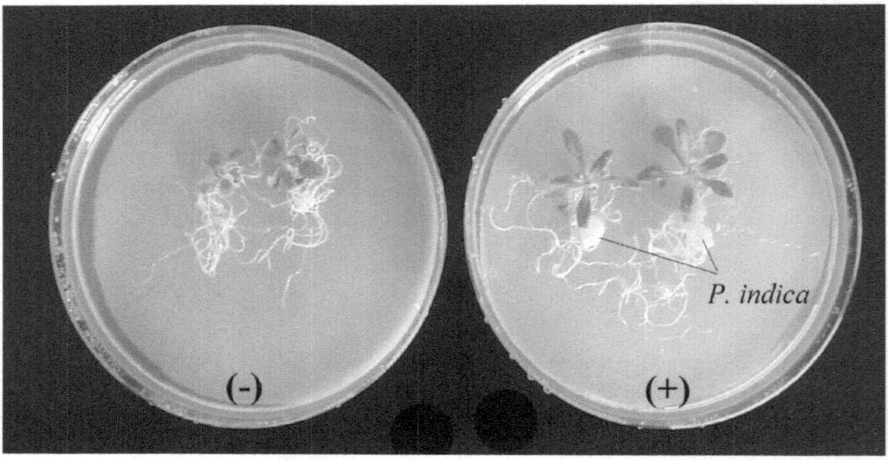

Fig. 20.1 Wild-type *Arabidopsis* seedlings, which were grown in the absence (-) or presence (+) of *P. indica* for 6 days. For a better demonstration, two seedlings were grown in one Petri dish

with the model plant *Arabidopsis* is being used to understand the molecular basis of this beneficial plant/microbe interaction.

The lack of any specificity in the host plant species suggests that the fungus utilizes well conserved recognition and signaling molecules which are present in all plant species. Furthermore, since an interaction can be seen already at the level of bryophytes, one might hypothesize that the symbiosis is based on ancient phylogenetic routes in plants.

Our strategy is based on the observation that *Arabdiopsis* in etalic seedlings (as well as adult plants) grow taller in the presence of a defined amount of fungal hyphae (Fig. 20.1). We have established conditions which allow us to follow the growth promotion mediated by *P. indica* over a period of 14 days in Petri dishes. If colonized seedlings are then transferred to soil, growth promotion is easily visible and is accompanied by an increase in seed production.

20.3 Co-Cultivation of *P. indica* and *Arabidopsis* under Standardized Growth Conditions

1. Co-cultivation of the two symbiotic partners:
   P. indica promotes growth of *Arabidopsis* seedlings in nature, in the greenhouse and under sterile conditions in Petri dishes. *A. thaliana* seeds [from wild-type, ecotype Columbia, EMS mutant lines (Lehle, San Diego, USA) or homozygote T-DNA insertion lines (http://www.arabidopsis.org/portals/mutants/worldwide.jsp)] are surface-sterilized and placed on Petri dishes containing MS nutrient medium (Murashige and Skoog 1962). After cold treatment at 4 °C for 48 h, plates are incubated for 7 days at 22 °C under continuous illumination (100 μmol m^{-2} s^{-1}). Simultaneously, *P. indica* is cultured as described previously (Verma et al. 1998; Peškan-Berghöfer et al. 2004) on Hill and Kafer medium, solidified with 1% (w/v) agar (Hill and Kafer 2001). For the co-cultivation experiments, 9 day-old *A. thaliana* seedlings are transferred to nylon disks (mesh size 70 μm) placed on top of a modified PNM culture medium (5 mM KNO_3, 2 mM $MgSO_4$, 2 mM $Ca(NO_3)_2$, 0.01 μM $FeSO_4$, 70 μM H_3BO_3, 14 μM $MnCl_2$, 0.5 μM $CuSO_4$, 1 μM $ZnSO_4$, 0,2 μM Na_2MoO_4, 0.01 μM $CoCl_2$, 10.5 g l^{-1} agar, pH 5.6), in 90 mm Petri dishes. One seedling is used per Petri dish. Fungal plugs of approximately 5 mm in diameter are placed at a distance of 1 cm from the roots. Plates were incubated at 22 °C under continuous illumination from the side (max. 80 μmol m^{-2} s^{-1}). Growth of the seedlings is monitored by taking pictures every day, or by determining the fresh weight of the roots and aerial parts over a period of 14 days of co-cultivation. Figure 20.1 shows seedlings 6 days after co-cultivation.

Thereafter the plants are transferred to soil and cultivated in multi-trays with Aracon tubes in a temperature-controlled growth chamber at 22 °C under long-day conditions (light intensity: max. 80 µmol m^{-2} s^{-1}). Before transfer to soil, the roots of the seedlings grown in the presence of the fungus are examined under the microscope to test whether hyphae and spores have developed within and around the roots (Fig. 20.2). In addition, for plants which are growing in the presence of *P. indica*, the soil is mixed carefully with the fungus (1%, w/v). The fungal mycelium is obtained from liquid cultures after removal of the medium and washed with an excess of distilled water. Control seedlings without the fungus are grown in soil without the fungus. Uninoculated control plants and those infected by the fungus are kept in Aracon tubes and their growth is monitored until the collection of seeds (Fig. 20.3). Seeds are collected in the Aracon tubes and quantified as grams of seed per plant. The weight of a seed is not altered by the fungus.

2. Isolation of mutants:
   We isolated (ethyl-methanesulfonate, EMS or knock-out) mutants which failed to respond to the fungus with regard to growth promotion under the above-described conditions. Originally, mutations were induced by EMS (0.3%, Sigma-Aldrich Chemie) in the Columbia ecotype, as described by Sommerville and Ogren (1982). EMS is an alkylating agent that produces point mutations by adding an ethyl group to a nucleic acid, resulting in a GC to AT transition. Mutagenized seeds were sown on soil and after 12 weeks, seeds from each M1 plant were collected. Approximately 10 000 individual seedlings were then used to test for their response to *P. indica*, as described above. Positive candidates were further analyzed for other responses which are induced in *Arabidopsis* roots in response to *P. indica*.

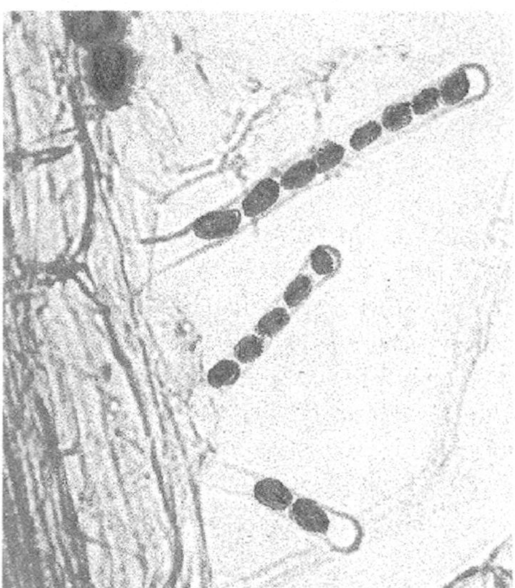

Fig. 20.2 Root hair colonized by fungal spores. Parts of the roots were removed from the seedling prior to transfer to soil to test whether the fungus had colonized the roots. Sections were stained with cotton blue and examined under the light microscope (Zeiss Axioplan model MC 100)

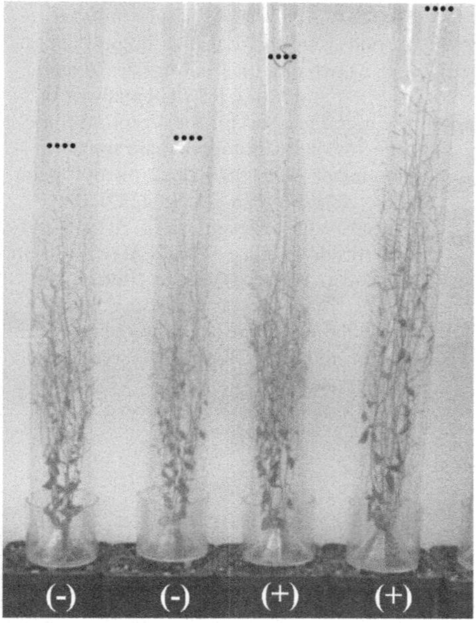

Fig. 20.3 *Arabidopsis* plants grown in the absence (–) or presence (+) of *P. indica* in a multi-tray with Aracon tubes. The picture was taken while the seeds were ripening. The line of dots (••••) indicates the height of each plant

3. Seed production:
   Putative mutants obtained in the primary screen were transferred to soil to collect seeds; 50 seedlings of the next generation were then co-cultivated with *P. indica* (and 50 wild-type seedlings were used as controls) to confirm that the mutants did not respond to *P. indica*. Positive candidates were then again transferred to soil to collect seeds of the third generation.
4. Analysis of gene expression in the mutants:
   In addition to the growth-promoting response, we tested whether several independent responses normally induced by the fungus were not induced in the mutants. Based on microarray analyses and suppression substractive hybridization techniques (cf. below) we identified genes which responded very early to co-cultivation with the fungus in *A. thaliana* roots. We used some of these genes as markers to test whether they were not regulated in the mutants co-cultivated with *P. indica* (cf. Fig. 20.4).
5. Analysis of a post-translational modification of a MATH protein in the plasma membrane of roots:
   The interaction of *P. indica* with *Arabidopsis* roots is also detectable at the protein level. We identified a MATH protein in the plasma membrane of the roots which is transiently modified in response to the fungus (Peškan-Berghöfer et al. 2004). This modification is not detectable in a mutant which does not recognize the fungus (Oelmüller et al. 2005). Although the function of the MATH protein and the kind of the mutation is not known at present, it represents a characteristic and a highly specific post-translational response which is independent from those involved in gene expression.

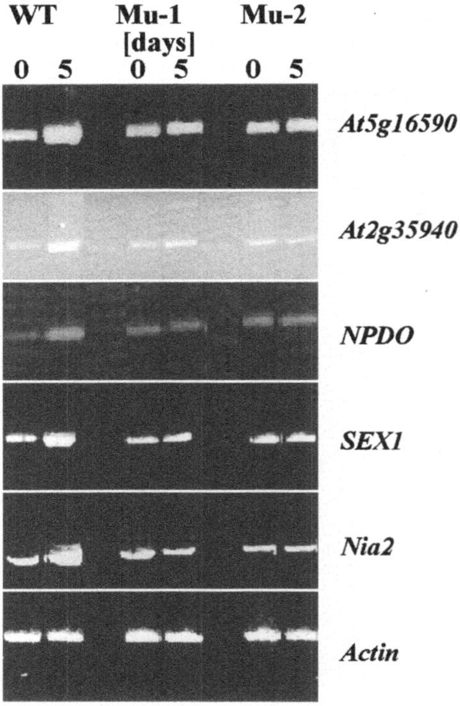

Fig. 20.4 Analysis of *P. indica* induced responses in wild-type seedlings (*WT*) as well as two mutant seedlings designated as Mu-1 and Mu-2. RT-PCR analyses of the message levels for the receptor kinase At5g16590, the homeodomain transcription factor At2g35940, the 2-nitropropane dioxygenase (*NPDO*, At5g64250), the glucan-water dikinase (*SEX1*, At1g10760) and nitrate reductase (*Nia2*, At1g37130) in the roots. Actin was used as control. The RT-PCR analysis was performed 0 days and 5 days after the beginning of co-cultivation, and the PCR products were run on an agarose gel

20.4 Map-Based Cloning of a Mutated Gene

The strategy of map-based cloning tries to identify molecular markers which are closely linked to the gene of interest. Since the complete *Arabidopsis* genome is sequenced, many of these markers can be found on the *Arabidopsis* homepage (www.arabidopsis.org.). The mutations were mapped using restriction fragment length polymorphism (RFLP)- and polymerase chain reaction (PCR)-based markers on a F2 progeny and F3 families deriving from a single cross between the male donor plants of homozygous lines of the ecotype *Columbia* and the female recipient plants of the ecotype *Landsberg errecta*. The markers display different patterns for plants that are homozygous or heterozygous for an appropriate genomic locus. DNA for PCR based mapping was prepared from a single leaf of a plant using a rapid DNA isolation procedure (see below). The markers which were generally used are available under www.arabidopsis.org. (cf. also Bell and Ecker 1994; Klimyuk and Jones 1997; Neff et al. 1998). To assign the mutant locus to one of the *Arabidopsis* chromosomes, a population of at least 28 plants homozygous at the gene of interest was analyzed (cf. Konieczny and Ausubel 1993). Furthermore, the distance between two markers on the same chromosome should be approximately 50 cM before a linkage can be detected.

20.5 Rapid DNA Extraction

For PCR-based mapping purposes, DNA was isolated from individual leaves of adult plants. One has to make sure that removal of the leaf does not cause severe damage to the plant, since seeds need to be collected from the adult plants. The leaf was homogenized with a metal pestle in a 1.5-ml tube after addition of 0.5 ml extraction buffer (7 M urea, 300 mM NaCl, 50 mM Tris-HCl, pH 8.0, 1% N-laurylsacrosinate, 20 mM EDTA). After incubation at 37 °C for 5 min, 500 µlof phenol/chloroform/isopropanol (25:24:1) was added, centrifugal for 10 min, and the DNA was precipitated with isopropanol from the aeqnous phase. After washing with 80% ethanol, the DNA was resuspended in 30 µl TE buffer.

20.6 Confirmation of a Mutated Phenotype of an EMS Mutant by the Analysis of an Independent T-DNA Insertion Line

T-DNA knock out lines are available from: http://signal.salk.edu/cgi-bin/tdnaexpress. The *Arabidopsis* mutants can be obtained from the Salk, Sail, Gabi, Riken or other collections. In many cases, the seed material obtained from the stock center is heterozygote with regard to the T-DNA insertion and might also contain more than one insertion in the genome. Thus, seeds of the next generation need to be screened for homozygote knock-out lines. Therefore, the seeds obtained from the stock center are surface-sterilized and placed on Petri dishes containing MS nutrient medium (Murashige and Skoog 1962). After cold treatment at 4 °C for 48 h, plates are incubated for 3–4 weeks at 22 °C under continuous illumination (100 µmol m^{-2} s^{-1}). Plants are then transferred to the soil. After 2 weeks growing in soil, DNA is isolated from one leaf of each plant, as described. This DNA is used for PCR. The primers for PCR are available from: http://signal.salk.edu/tdnaprimers.2.html. The PCR reactions for the determination of the insertion sites are performed as described by: http://signal.salk.edu/tdnaprimers.html.

20.7 Differntial Display to Identify Genes which are Regulated in Response to *P. indica*

1. Isolation of lateral roots and lateral root RNA: *Arabidopsis* seedlings were grown in the presence or absence of *P. indica*. The seedlings were removed from the nylon membrane with a forcept and dipped into liquid nitrogen for approximately 5 s. Most of the lateral roots remained in the liquid nitrogen, while main roots stayed with the seedlings. Approximately 0.4 g of lateral roots were collected from at least five independent experiments and ground in liquid nitrogen. The resulting powders were used for RNA extraction and cDNA syntheses. The tester cDNA was obtained from the root material co-cultivated with the fungus, while the driver cDNA derived from the control material.
   The cDNA from control roots can be mixed with cDNA from the fungus. Normally, we used 10% fungal cDNA and 90% plant cDNA as driver cDNA. We performed this protocol with different plant species. When *Arabidopsis* was used as a host, we omitted this step since *Arabidopsis* sequences were easily detectable after sequencing of clones from the suppression substracted fraction.
   RNA extraction was performed with the RNA extraction kit (RNeasy Quiagen, Hilden, Germany). The quality of the RNA was checked spectrophotometrically by measuring the absorbance (A) at 260, 280 and 300 nm. The A_{260}/A_{280} ratio should be at least 1.95, the A_{260}/A_{300} ratio should be at least 0.5.
2. Generation of cDNAs for a suppression substraction library: an equal amount of RNA (0.1 μg) was used for cDNA synthesis using the SMART PCR cDNA-synthesis kit from Clontech (Palo Alto). The cDNAs were used for the creation of a substracted library with the help of the PCR-select cDNA substraction kit (BD Bioscience Clontech, Palo Alto). A detailed protocol is available under: http://www.clontech.com/clontech/archive/OCT01UPD/pdf/PCR-SelectProducts.pdf.
   In brief:
   a. The resulting tester and driver cDNAs are digested with *Rsa*I to obtain small blunt-end cDNA fragments.
   b. The tester cDNA are separated into two fractions, and ligated to two different types of adapters.
   c. The two tester cDNA pools are hybridized two times to an excess of driver cDNA to enrich differentially expressed sequences among single-stranded tester cDNAs.
   d. Differentially expressed cDNAs are amplified by PCR.
   e. Depending on the co-cultivation time of the roots with *P. indica*, the substracted libraries contains between 20 (14 days of co-cultivation) and 350 (3 days of co-cultivation) clones.

3. Differential screening
   a. Individual bacteria are grown to isolate plasmid DNA.
   b. The insertions are amplified by PCR with M13 forwards and reverse primers.
   c. The PCR products are separated on 1% agarose gels and blotted onto nylon membranes (Hybond N; Amershan Biosciences, Freiburg, Germany).
   d. cDNA from control and inoculated lateral roots are radiolabeled with ^{32}P ($>5\times10^6$ cpm ml^{-1}) and Superscript II reverse transcriptase (Invitrogen, Karlsruhe, Germany).
   e. Hybridization is performed to two identical nylon membranes and the signals are quantified using a phosphorimager (cf. Fig. 20.5).

20.8 Activation Tagged Lines

Activation tagging with constitutive enhancer elements was developed by Rick Walden at the Max Planck Institute in Cologne, Germany. A T-DNA vector with

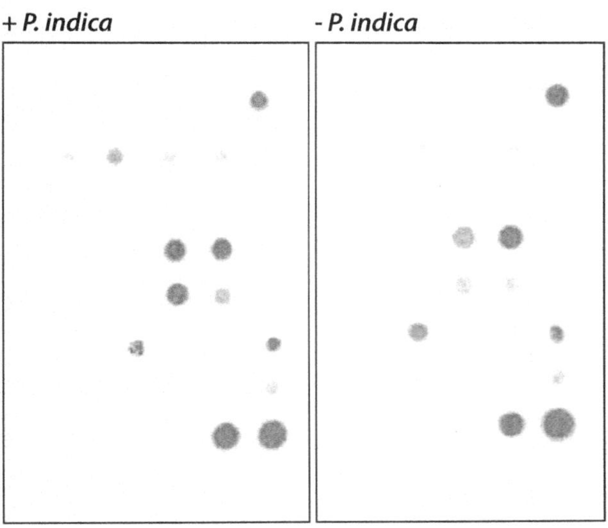

Fig. 20.5 Dot blot hybridization of radiolabled cDNA from *Arabidopsis* roots, which were grown in the presence (+ *P. indica*) or absence (– *P. indica*) of the fungus. Note that the message levels for some of the genes are upregulated by *P. indica*, while others are identical (controls). Hybridization occurred to two identical filters, on which cDNA fragments from genes were spotted which were identified in the suppression substractive hybridization or which were used as controls

enhancer elements from the highly active cauliflower mosaic virus 35S promoter can cause transcriptional activation of nearby genes, because activated genes can be associated with a T-DNA insertion (cf. Weigel et al. 2000; http://arabidopsis.org/abrc/weigel.jsp).

We screened activation tagged lines from various sources for their response to *P. indica* and found that a few lines responded much more sensitive to the fungus than the wild type (cf. Fig. 20.6). When grown in the absence of the fungus, no difference from the wild type could be detected. Thus, this approach is very helpful for the identification of genes and proteins which can stimulate the response to *P. indica* when present at higher levels in the plants.

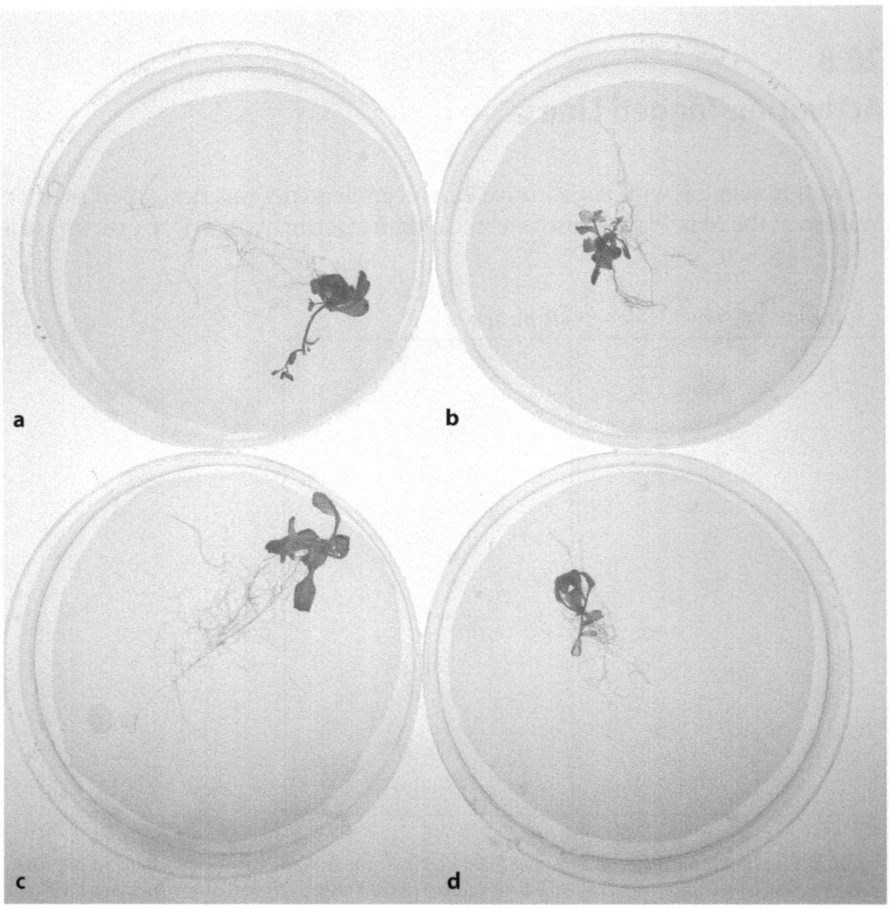

Fig. 20.6 An activation tagged line is more sensitive to *P. indica* than the wild type. **a** Wild type with the fungus, **b** wild type without the fungus, **c** activation tagged line with the fungus, **d** activation-tagged line without the fungus

20.9 Identification of Biochemical Pathways in *A. thaliana* which are Regulated by *P. indica*

The available microarray data as well as differentially expressed gene data can be used to understand (signaling) pathways which are targets for *P. indica*. To this end, the regulatory network can be best analyzed by superimposing the proteins identified at the level of their mRNAs on the Kyoto encyclopedia of genes and genomes (KEGG; www.genome.jp/kegg/pathway.html). The KEGG pathway database is a collection of pathway maps representing our knowledge on the molecular interaction and reaction networks for metabolism, carbohydrate energy, lipid, nucleotide and amino acid metabolisms, glucan, cofactor and vitamins biosynthesis, enzymes involved in secondary metabolites, etc. Such an analysis should finally lead to the understanding of signaling processes, crucial signaling molecules as well as metabolic pathways which can be manipulated in *A. thaliana* by the endophytic fungus *P. indica*.

References

Bell CJ, Ecker JR (1994) Assignment of 30 microsatellite loci to the linkage map of *Arabidopsis*. *Genomics* 19:137–144

Bhatnagar K, Varma A (2006) The healthy marriage between terrestrial orchids and fungi. J Hill Res 19:1–12

Hill TW, Käfer E (2001) Improved protocols for *Aspergillus* medium: trace elements and minimum medium salt stock solutions. Fungal Genet Newsl 48:20–21

Klimyuk VI, Jones JD (1997) *AtDMC1*, the *Arabidopsis* homologue of the yeast *DMC1* gene: characterisation, transposon-induced allelic variation and meiosis-associated expression. Plant J 11:1–14

Konieczny A, Ausubel FM (1993) A procedure for mapping *Arabidopsis* mutations using co-dominant ecotype-specific PCR-based markers. Plant J 4:403–410

Murashige T, Skoog F (1962) A revised medium for rapid growth and bioassays with tobacco tissue cultures. Physiol Plant 15:473–497

Neff MM, Neff DJ, Chory J, Pepper AE (1998) dCAPS, a simple technique for the genetic analysis of single nucleotide polymorphisms: experimental applications in *Arabidopsis thaliana* genetics. Plant J 14:387–392

Oelmüller R, Peškan-Berghöfer T, Shahollari B, Sherameti I, Varma A (2005) MATH-domain containing proteins represent a novel gene family in *Arabidopsis thaliana* and are involved in plant/microbe interactions. Physiol Plant 124:152–166

Peškan-Berghöfer T, Shahollari B, Giang PH, Hehl S, Markert C, Blanke V, Varma AK, Oelmüller R (2004) Association of *Piriformospora indica* with *Arabidopsis thaliana* roots represents a novel system to study beneficial plant–microbe interactions and involves early plant protein modifications in the endoplasmatic reticulum and at the plasma membrane. Physiol Plant 122:465–477

Sommerville CR, Ogren WL (1982) Isolation of photorespiration mutants in *Arabidopsis thaliana*. In: Edelman M, Hallick RB, Chua N-H (eds) Methods in chloroplast molecular biology. Elsevier Biomedicinal, New York, pp 129–139

Varma A, Verma S, Sudha V, Sahay N, Britta B, Franken P (1999) *Piriformospora indica* – a cultivable plant growth promoting root endophyte with similarities to arbuscular mycorrhizal fungi. Appl Environ Microbiol 65:2741–2744

Verma SA, Varma A, Rexer K-H, Hassel A, Kost G, Sarbhoy A, Bisen P, Bütehorn B, Franken P (1998) *Piriformospora indica*, gen. et sp. nov., a new root-colonizing fungus. Mycologia 90:898–905

Weigel D, Ahn JH, Blazquez MA, Borevitz JO, Christensen SK, Fankhauser C, Ferrandiz C, Kardailsky I, Malancharuvil EJ, Neff MM, Nguyen JT, Sato S, Wang ZY, Xia Y, Dixon RA, Harrison MJ, Lamb CJ, Yanofsky MF, Chory J (2000) Activation tagging in *Arabidopsis*. Plant Physiol 122:1003–1013

Weiss M, Selosse MA, Rexer KH, Urban A (2004) Sebacinales: a hitherto overlooked cosm of heterobasidiomycetes with a broad mycorrhizal potential. Mycol Res 108:1–8

21 Biophysical Phenomics Reveals Functional Building Blocks of Plants Systems Biology: a Case Study for the Evaluation of the Impact of Mycorrhization with *Piriformospora indica* *

R.J. Strasser, M. Tsimilli-Michael, D. Dangre, and M. Rai

21.1 Introduction

Soil microbial activity is a main parameter in ecosystem functions. Arbuscular mycorrhiza fungi are mutualistic microsymbionts of about 90% of higher plants in natural, semi-natural and agricultural plant communities. As mycorrhizosphere systems can be tailored to help plants to survive in nutrient-deficient, degraded habitats or during stress periods, they are highly advantageous in sustainable agriculture. However, the success of this practise, as for any microbial inoculation, depends strongly on the effectiveness of mycorrhization, which depends on complex interactions between plant and symbiont.

Mycorrhization has multiple effects on the physiology of the plant at different levels. We focus our interest on the responses of the photosynthetic apparatus and especially on photosystem (PS) II (see e.g. Tsimilli-Michael et al. 2000) which is well known to be a component of the plant system highly sensitive to any stress.

Our approach for the evaluation of the effectiveness of mycorrhization, which we term *biophysical phenomics*, is based on the description of an in vivo vitality analysis (behaviour/performance) of PSII, i.e. the description of a biophysical phenotype. The tools that provide access to this phenotyping, termed the "JIP-test", are based on the analysis of the fast fluorescence kinetics O-J-I-P exhibited

Reto J. Strasser, Merope Tsimilli-Michael: Laboratory of Bioenergetics, University of Geneva, Chemin des Embrouchis 10, CH-1254 Jussy-Geneva, Switzerland,
email: Reto.Strasser@bioveg.unige.ch

Merope Tsimilli-Michael: Ath. Phylactou 3, CY-1100, Nicosia, Cyprus,
email: tsimicha@spidernet.com.cy

Devanand Dangre, Mahendra Rai: Department of Biotechnology, SGB Amravati University, Amravati-444602, Maharashtra, India,
email: mkrai123@rediffmail.com; pmkrai@hotmail.com

* Dedicated to the memory of Hannes Schuepp, a great scientist and a wonderful person.

Soil Biology, Volume 11
Advanced Techniques in Soil Microbiology
A. Varma, R. Oelmüller (Eds.)

by all oxygenic photosynthetic organisms upon illumination (for reviews, see Strasser et al. 2000, 2004).

Moreover, in our analysis, we show that biophysical phenotyping, which refers to the system macrostate, allows us to recognise and evaluate impacts on the function and (re)distribution of the heterogeneous microstates – functional building blocks – whose balance defines the macrostate (Strasser and Tsimilli-Michael 2005).

We also present, as an application of our approach, a case study of the beneficial role of the emerging growth booster and in vitro-cultivable *Piriformospora indica* (Varma et al. 1999) on chick peas (*Cicer arietinum* L. Chafa variety) exposed to cadmium stress, which we further compare with the impact of typical arbuscular mycorrhiza fungi (*Glomus mosseae*, *Glomus caledonium*).

21.2 Biophysical Phenomics of the Fast Fluorescence Rise O-J-I-P

21.2.1 The Energy Cascade in the Photosynthetic Apparatus

A simplified scheme for the energy cascade in photosystem II of the photosynthetic apparatus is presented in Fig. 21.1 (modified after Epitalawage et al. 2003). Not only solar energy but any light energy of suitable wavelengths, i.e. wavelengths that can be absorbed by chlorophyll and accessory pigments, has the same fate. In the energy conservation pathway, the flux of photons is transformed sequentially to a flux of excitons, a flux of electrons and a flux of molecules. The electron flow is coupled to the formation of adenosine triphosphate (ATP), which is a high-energy compound. The end-products are molecular oxygen (O_2), evolved by water (H_2O) splitting, and sugars, formed from carbon dioxide (CO_2). Part of the excitation is not conserved; it is dissipated, mainly as heat and less as fluorescence emission by chlorophyll (Chl) *a*. As the kinetics of Chl *a* fluorescence reflect changes in the function and structure of PSII and, concomitantly, changes in the whole electron transport chain, they provide a very useful, non-invasive tool for the investigation of the behaviour and performance of the photosynthetic apparatus.

21.2.2 Microstates – Functional Building Blocks of Photosynthesis

According to open system thermodynamics, the Gibb's energy, linked to biochemical activity (quantity term), and the entropy-related energy component,

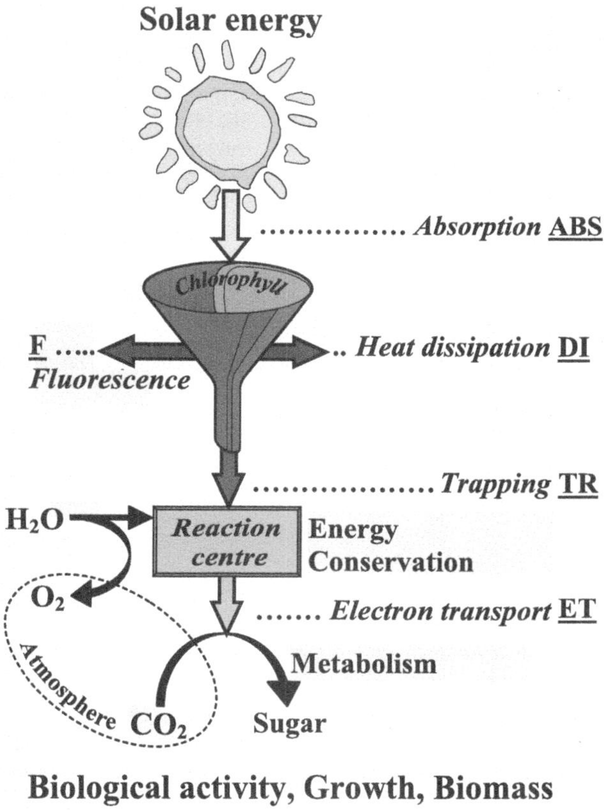

Fig. 21.1 A simplified scheme for the energy cascade in photosystem II (PSII) of the photosynthetic apparatus (modified after Epitalawage et al. 2003). Light absorption (ABS) creates excited chlorophyll. Part of the excitation energy is dissipated, mainly as heat (heat dissipation, *DI*) and less as fluorescence emission (F); another part is channelled to the reaction centre (trapping, TR) to be converted to redox energy (*Energy Conservation*), with the simultaneous evolution of oxygen (O2) by water (H2O) splitting. The redox energy creates electron transport (ET), which, via PSI (not shown), leads ultimately to CO_2 fixation into sugars (*Metabolism*)

linked to the structure, complexity and organization of the system (quality term) follow optimization strategies, potentially establishing steady-states (stability term) under given conditions (see Strasser and Tsimilli-Michael 2005).

Recognizing the high complexity and heterogeneity of the photosynthetic system in nature (see also Strasser 1985; Strasser and Tsimilli-Michael 1998), we propose that its apparent state is a heterogeneous macrostate, determined by the statistical distribution of microstates, as listed in the model shown in Fig. 21.2 (modified after Strasser and Tsimilli-Michael 2005): architecture of PSII and PSI antenna, i.e. size and connectivity among units (grouped/separate), light-harvesting complexes (LHC II and I), kinase-catalyzed migration from PSII to PSI of an LHC component when phosphorylated (LHC~P), spill-over from PSII to PSI antenna, types of electron donation to PSII raction centers (from wa-

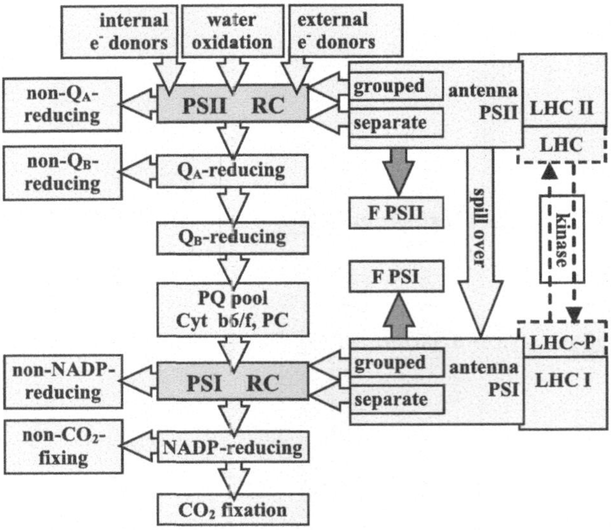

Fig. 21.2 Heterogeneity of microstates/functional units, whose balance determines the macrostate of the photosynthetic system (modified after Strasser and Tsimilli-Michael 2005). For details, see text

ter oxidation or internal/external donors), Q_A^- reducing reaction centers (RC) or non-Q_A^- reducing (heat sinks), Q_B^- reducing or non-Q_B^- reducing (slow Q_A^- reoxidizing) units, states of intermediate electron carriers (PQ pool, Cytb6/f, PC), splitting of PSI acceptor side, in non-NADP-reducing pathways and NADP-reducing pathways, the latter further split in non-CO_2-fixing and towards CO_2 fixation. For any steady-macrostate (optimal/adapted), the balance of mechanisms governing the distribution of microstates is equivalent to optimizations of quantity, quality and stability, which are interrelated, governed by the genetics of the system, its resources and its environment. Concomitantly, stress is any disturbance of the achieved balance, upon which the system undergoes microstate changes towards a new optimal balance, i.e. a new macrostate (Strasser and Tsimilli-Michael 2005).

21.2.3 Measuring Fluorescence Transients with PEA, Handy-PEA and FIM- Fluorimeters

Chl *a* fluorescence transients exhibited by any photosynthetic material are measured by a PEA (Plant Efficiency Analyser) or Handy-PEA fluorimeter (Hansatech Instruments, King's Lynn, UK; Fig. 21.3) or FIM fluorimeter (Fluorescence Induction Meter 1500; ADC, Hoddesdon, UK). The transients are induced by a red light (peak at 650 nm) of 600 W m^{-2} (equivalent to 3200 μE s^{-1} m^{-2}) provided by an array of six (PEA and FIM fluorimeters) or three (Handy-PEA fluorimeter) light-emitting diodes, and recorded for 1 s with 12 bit resolution. The data

Fig. 21.3 The PEA (Handy-PEA) fluorimeter used for our studies. The photo is from in situ measurements, with the clips already put on the leaves to dark-adapt them. The insert shows the sensor of the instrument, which provides, by LEDs, the red actinic light (650 nm) and collects the fluorescence signals

acquisition in the PEA and FIM fluorimeters is every 10 μs for the first 2 ms, every 1 ms between 2 ms and 1000 ms and every 100 ms thereafter (for further details, see Strasser et al. 1995; for reviews, see Strasser et al. 2000, 2004), while in the Handy-PEA fluorimeter it is every 10 μs (in the interval 10 μs to 0.3 ms), every 0.1 ms (0.3–3.0 ms), every 1 ms (3–30 ms), every 10 ms (30–300 ms), etc. (see e.g. Tóth et al. 2005). The first reliable measurement with PEA and FIM fluorimeters is at 50 μs, while with the Handy-PEA fluorimeter it is at 20 μs.

21.2.4 How Fluorescence Kinetics Provide an Insight to the Microstates – Functional Blocks of PSII

21.2.4.1 Qualitative Screening of Many Samples

Screening of many samples in situ by recording Chl *a* fluorescence transients is a very simple task. Figure 21.4 depicts Chl *a* fluorescence transients of dark-adapted leaves of *Hedera* (left panel) and *Shefflera* (right panel), measured with

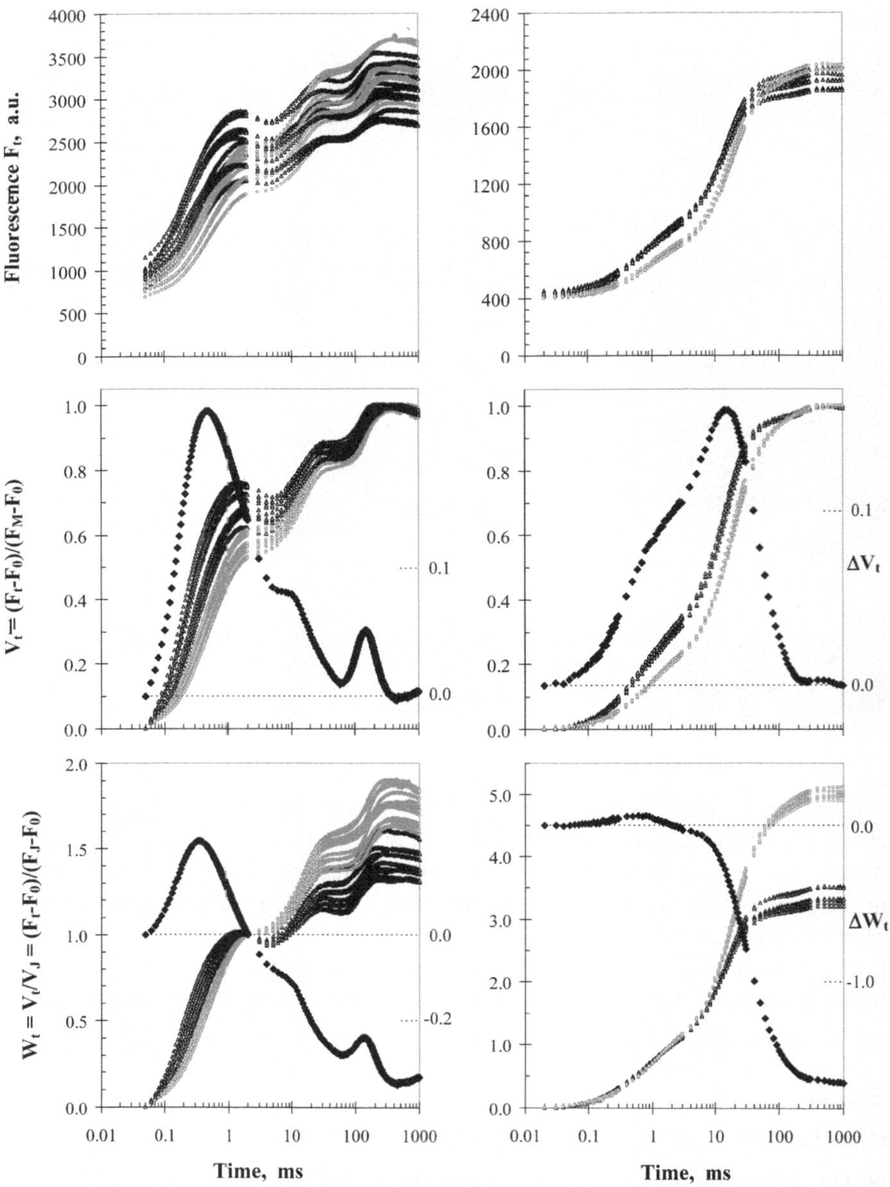

Fig. 21.4 Chl *a* fluorescence transients of dark adapted leaves of *Hedera* (*left panels*) and *Shefflera* (*right panels*), measured with PEA and Handy-PEA fluorimeters respectively. Young leaves (*black open triangles*) and mature leaves (*open grey circles*) from both plants were measured. The transients, induced by saturating red actinic light (peak at 650 nm) of 600 W m^{-2} (equivalent to about 3000 μE s^{-1} m^{-2}), are plotted on a logarithmic time-scale. The plots in the *upper panels* depict the kinetics of the raw fluorescence data, F_t. The kinetics of the relative variable fluorescence V_t and W_t, calculated from the raw data as $V_t = (F_t - F_0)/(F_M - F_0)$ and $W_t = V_t/V_J = (F_t - F_0)/(F_J - F_0)$, are depicted in the plots of the *middle and lower panels* respectively (*left axis*). In the plots of V_t and W_t, the corresponding differences ΔV_t and ΔW_t, between the average transients of young and mature leaves (young minus mature) are also plotted (black closed diamonds; *right axis*). For other details, see text

PEA and Handy-PEA fluorimeters, respectively. Young leaves (black open triangles) and mature leaves (open grey circles) from both plants were measured. The transients, induced by saturating red actinic light (peak at 650 nm) of 600 W m^{-2} (equivalent to about 3000 μE s^{-1} m^{-2}), are plotted on a logarithmic time-scale.

The plots in the upper panels depict the kinetics of the raw fluorescence data, F_t. The first to observe is that all the transients are polyphasic with, however, differences between the two species. We can also clearly observe that *Shefflera* (right panel) exhibits a high homogeneity among samples, which permits the distinction between young and mature leaves, while in *Hedera* possible differences are hidden under the wide heterogeneity between samples.

However, when we transform the kinetics of the raw data to the kinetics of the relative variable fluorescence $V_t = (F_t - F_0)/(F_M - F_0)$ or $W_t = V_t/V_J = (F_t - F_0)/(F_J - F_0)$ [for the definition of terms and symbols, see text below as well as Fig. 21.5 and Table 21.1], we can see in the respective plots of the middle and lower panels a clear distinction between young and mature leaves for both species. In these plots, the corresponding differences ΔV_t and ΔW_t, between the average transients of young and mature leaves (young minus mature) are also plotted (black closed diamonds; secondary vertical axis).

21.2.4.2 The Typical O-J-I-P Fluorescence Transient: Definition of Steps and Selection of Fluorescence Data for the JIP-Test

The Chl *a* fluorescence transient, known as the Kautsky transient (Kautsky and Hirsh 1931), consists of a rise completed in less than one second and a subsequent slower decline towards a steady state. Our method presented here utilises only the fast rise that is generally accepted to reflect the accumulation of the reduced form of the primary quinone acceptor Q_A, otherwise the closure of the reaction centres (RCs), which is the net result of Q_A reduction due to PSII activity and Q_A^- reoxidation due to photosystem I (PSI) activity. When the photosynthetic sample is kept for a few minutes in the dark, Q_A is fully oxidised, hence the RCs are all open, and the fluorescence yield at the onset of illumination is denoted as F_0 (minimal fluorescence). The maximum yield F_P at the end of the fast rise, depending on the achieved reduction–oxidation balance, acquires its maximum possible value – denoted as F_M – if the illumination is strong enough to ensure the closure of all RCs. A lot of information has been driven during the past 70 years from the fluorescence transient (for reviews, see Papageorgiou 1975; Briantais et al. 1986; Govindjee et al. 1986; Krause and Weiss 1991; Dau 1994; Govindjee 1995; Strasser et al. 2000, 2004).

Transients recorded with high time-resolution fluorimeters, e.g. the PEA- or Handy-PEA instrument that we use, have provided additional and/or more accurate information (Strasser and Govindjee 1992; Strasser et al. 1995; for reviews, see Strasser et al. 2000, 2004). The fluorescence rise kinetics was shown to be polyphasic, clearly exhibiting, when plotted on a logarithmic time-scale,

Table 21.1 Summary of terms, definitions and formulae used by the JIP-test

Experimental signals	Symbol	Formula
Minimal fluorescence intensity[a]	F_0	
Maximal fluorescence intensity	F_M	
Fluorescence intensity at 2 ms (J-step)	F_J	
Fluorescence intensity at 30 ms (I-step)	F_I	
Fluorescence intensity at 300 μs	$F_{300\mu s}$	
Normalised signals		
Maximum variable fluorescence	F_V	$= F_M - F_0$
Relative variable fluorescence (from F_0 to F_M)	V_t	$= (F_t - F_0) / (F_M - F_0)$
Relative variable fluorescence (from F_0 to F_J)	W_t	$= (F_t - F_0) / (F_J - F_0)$
Relative variable fluorescence at the J-step	V_J	$= (F_J - F_0) / (F_M - F_0)$
Relative variable fluorescence at the I-step	V_I	$= (F_I - F_0) / (F_M - F_0)$
Initial slope[a] of the $V = f(t)$ transient: $M_0 = (\Delta V/\Delta t)_0 \cong (dV/dt)_0 = [dQ_A^- / Q_{A,total})/dt]_0 =$ initial rate of primary photochemistry	M_0	$= 4\times (F_{300\mu s} - F_0) / (F_M - F_0)$
Specific fluxes: energy fluxes per reaction centre		
Specific flux for absorption	ABS/RC	$= (M_0/V_J) / [1 - (F_0/F_M)]$
Specific flux for trapping[a]	TR_0/RC	$= (M_0/V_J)$
Specific flux for dissipation[a]	DI_0/RC	$= (ABS/RC) - (TR_0/RC)$
Specific flux for electron transport[a]	ET_0/RC	$= (M_0/V_J) \times (1 - V_J)$
Yields or ratios of fluxes		
Maximum quantum yield[a] of primary photochemistry: $\varphi_{Po} \equiv TR_0/ABS$	φ_{Po}	$= [1 - (F_0/F_M)]$
Maximum yield[a] of electron transport: $\varphi_{Eo} \equiv ET_0/ABS$	φ_{Eo}	$= [1 - (F_0/F_M)] \times (1 - V_J)$
Efficiency[a] of a trapped exciton to move an electron into the electron transport chain further than Q_A^- : $\psi_0 \equiv ET_0/TR_0$	ψ_0	$= (1 - V_J)$

[a] At the onset of illumination (at 50 μs for PEA and FIM, or at 20 μs for Handy-PEA, in which case the initial slope must be calculated between 20 μs and 270 μs)

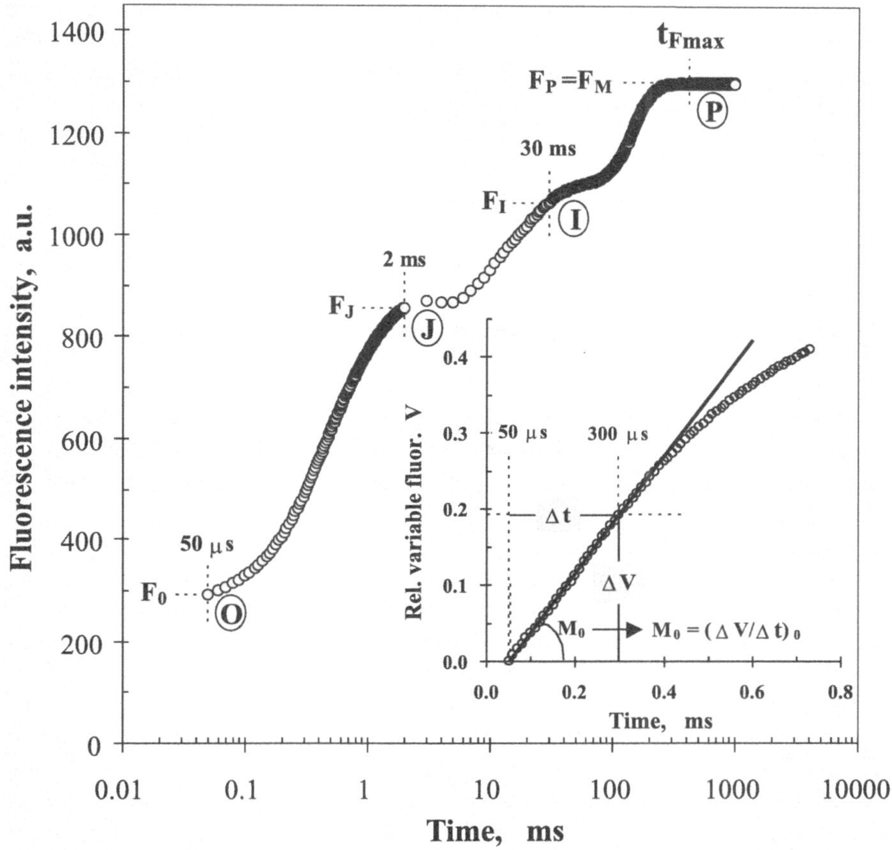

Fig. 21.5 A typical Chl *a* polyphasic fluorescence rise O-J-I-P, exhibited by higher plants. The transient is plotted on a logarithmic time-scale from 50 μs to 1 s. The marks refer to the selected fluorescence data used by the JIP-test for the calculation of structural and functional parameters. The signals are: the fluorescence intensity F_0 (at 50 μs), the fluorescence intensities F_J (at 2 ms) and F_I (at 30 ms) and the maximal fluorescence intensity $F_P = F_M$ (at t_{Fmax}). The insert presents the transient expressed as the relative variable fluorescence $V = (F{-}F_0)/(F_M{-}F_0)$ vs time, from 50 μs to 0.75 ms on a linear time-scale, demonstrating how the initial slope, also used by the JIP-test, is calculated: $M_0 = (dV/dt)_0 \cong (\Delta V/\Delta t)_0 = (V_{300\mu s})/(0.25\ \text{ms})$

the steps J (at 2 ms) and I (at 30 ms) between the initial O (F_0) and maximum P level (F_P). Moreover, a much more precise detection of F_0 is achieved, as well as the detection of the initial slope, which offers a link to the maximum rate of photochemical reaction.

Despite differences among species (as e.g. shown in Fig. 21.4), all oxygenic photosynthetic material investigated so far using this method show this polyphasic rise, labelled O-J-I-P. A typical Chl *a* fluorescence transient O-J-I-P is shown in Fig. 21.5, plotted on a logarithmic time-scale. The following original data are utilised by the JIP-test: the maximal measured fluorescence intensity,

F_P, equal here to F_M since the excitation intensity is high enough to ensure the closure of all RCs of PSII; the fluorescence intensity at 50 μs considered as the intensity F_0 when all RCs are open; the fluorescence intensity at 300 μs ($F_{300\mu s}$) required for the calculation of the initial slope $M_0 = (dV/dt)_0 \cong (\Delta V/\Delta t)_0$ of the relative variable fluorescence (V) kinetics (see insert in Fig. 21.5); the fluorescence intensities at 2 ms (J-step) denoted as F_J, and at 30 ms (I-step) denoted as F_I (for reviews, see Strasser et al. 2000, 2004).

21.2.4.3 The O-L-K-J-I-H-G-P Fluorescence Transient: From Steps to Bands

As shown in Fig. 21.4, the kinetics of ΔV_t and ΔW_t (where V_t and W_t are different expressions of relative variable fluorescence) reveal bands hidden in the J- and I-steps of the fluorescence kinetics F_t, which are much richer in information than the original O-J-I-P.

Figure 21.6, which utilises the results of a stress study – nitrogen deficiency in cowpea plants (*Vigna unguiculata* L) – reported by Schmitz et al. (2001), offers a detailed presentation of these bands, regarding their position and labelling and their relation with the main steps. (Note: the transients are here presented as kinetics of ΔV_t; choosing ΔW_t would lead to the resolution of the same bands, as can be seen in Fig. 21.4). The sequence of events is distinguished in single-turnover and multiple-turnover events. Moreover, the main information that can be derived from each band is depicted, namely information concerning: Q_A, the oxidised primary quinone acceptor; p_{2G}, the overall grouping probability within PSII antenna; OEC, the oxygen-evolving complex; Q_B^-, the reduced (one electron) secondary quinone acceptor; Q_B^{2-}, the reduced (two electrons) secondary quinone acceptor; Q_BH_2, the protonated secondary quinone acceptor; EC_{red}, the fully reduced electron carriers.

One can easily see that this list of derivable information provides an insight to microstates – functional building blocks of photosynthesis, to which we deconvolute (see Fig. 21.2) the macrostate – biophysical phenotype.

21.2.4.4 The JIP-Test: Conversion of Experimental Signals to Biophysical Parameters and the Performance Index

For the evaluation of the impact of any stress and, similarly, of mycorrhizosphere activity on plants, we apply the "JIP-test", which provides a quantitative analysis of the in vivo vitality – behaviour/performance – of PSII, i.e. a quantitative description of the biophysical phenotype – macrostate, by accessing the different microstates – functional building blocks. The "JIP-test" is an analysis of the fast fluorescence kinetics O-J-I-P exhibited by all oxygenic photosynthetic

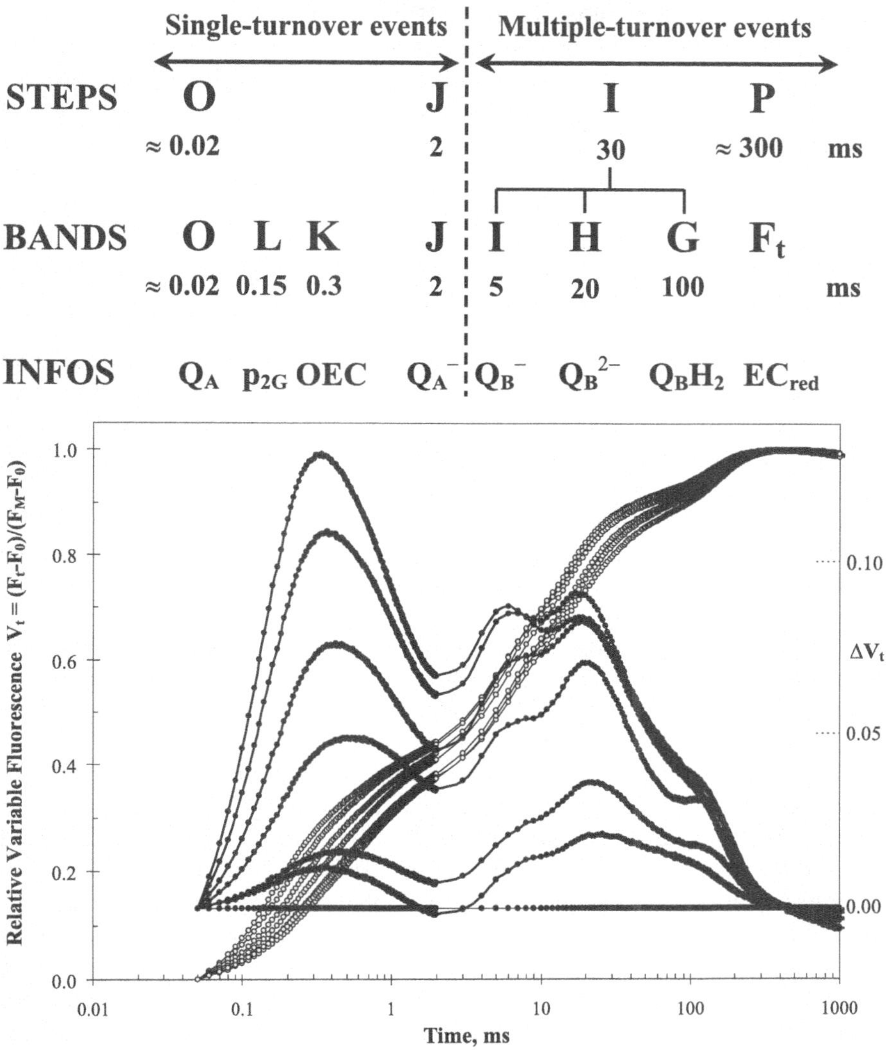

Fig. 21.6 Chlorophyll *a* fluorescence transients exhibited by dark-adapted leaves of cowpea plants (*Vigna unguiculata* L) grown at different KNO_3 concentrations (based on data from Schmitz et al. 2001) are presented as kinetics of relative variable fluorescence $V_t = (F_t - F_0)/(F_M - F_0)$, *open circles, left axis*, and as difference kinetics, ΔV_t, i.e. nitrogen deficient minus control (*closed circles, right axis*; see also legend of Fig. 21.4), where the extent of deficiency increases form bottom to top. The transients show the typical basic *STEPS* O-J-I-P (see also Fig. 21.5) while, as demonstrated in the *upper panel*, the difference transients reveal the full sequence of *BANDS* O-L-K-J-I-H-G-P (P or any F_t), which is much richer in information (for the L-band, see Fig. 21.8). *INFOS* refer to the main information that can be derived by each band, i.e. information concerning Q_A (oxidised primary quinone acceptor), p_{2G} (overall grouping probability within PSII antenna), OEC (oxygen evolving complex), Q_B^- -(reduced secondary quinone acceptor, one electron), Q_B^{2-} (reduced secondary quinone acceptor, two electrons), Q_BH_2 (protonated secondary quinone acceptor) and EC_{red} (fully reduced electron carriers). The *vertical dashed line* separates the phase of single-turnover events of primary photochemistry from that of multiple turnover events

organisms upon illumination, based on a simple model and the Theory of Energy Fluxes in Biomembranes (Strasser 1978, 1981). It is well documented that the shape of the O-J-I-P transient and its analysis by the JIP-test are efficient biophysical tools, not only in the recognition and evaluation of the beneficial role of mycorrhiza symbiosis on PSII activity (which we also approach as stress – see Tsimilli-Michael and Strasser 2002; see also Tsimilli-Michael et al. 2000) but, much more generally, in the biophysical phenotyping of the photosynthetic apparatus of a plant under any stress, biotic (e.g. Tsimilli-Michael et al. 2000) or abiotic, i.e. any stress caused by changes in different environmental conditions, e.g. light intensity, temperature, drought, atmospheric CO_2 or ozone elevation, chemical influences (Srivastava and Strasser 1995, 1996; Srivastava et al. 1997; Tsimilli-Michael et al. 1995, 1996, 1999, 2000; Tsimilli-Michael and Strasser 2001; Van Rensburg et al. 1996; Krüger et al. 1997; Ouzounidou et al. 1997; Clark et al. 1998, 2000; Van Heerden et al. 2003), even by diurnal changes (Strasser and Tsimilli-Michael 2001), as well as by senescence (Prakash et al. 2003). For a review, see Strasser et al. (2004).

The "JIP-test", which is a conversion of fluorescence data (selected as described above in 21.2.4.2; see also Fig. 21.5) to biophysical parameters and the performance index, is schematically presented in Fig. 21.7 and analytically in Table 21.1.

The biophysical parameters, all referring to time zero (onset of fluorescence induction) are: (a) the specific energy fluxes (per reaction centre, RC) for absorption (ABS/RC), trapping (TR_0/RC), dissipation (DI_0/RC = ABS/RC – TR_0/RC; not shown in Fig. 21.7, but see Table 21.1) and electron transport (ET_0/RC), (b) the flux ratios or yields, namely, the maximum quantum yield of primary photochemistry ($\varphi_{Po} = TR_0/ABS$), the efficiency ($\psi_0 = ET_0/TR_0$) with which a trapped exciton can move an electron into the electron transport chain further than Q_A^- and the quantum yield of electron transport ($\varphi_{Eo} = ET_0/ABS = \varphi_{Po} \times \psi_0$), (c) the phenomenological energy fluxes (per excited cross-section, CS) for absorption (ABS/CS), trapping (TR_0/CS), dissipation (DI_0/CS) and electron transport (ET_0/CS). The amount of active PSII reaction centres per excited cross-section (RC/CS) is also derived by the JIP-test. [Note: the calculation of the phenomenological fluxes is based on the ABS/CS flux, which can be directly determined (as Chl/area or Chl/dry weight or Chl/protein) or approximated by F_0 or F_M].

Figure 21.7 depicts also the definition of the performance index on an absorption basis, PI_{ABS}, which compiles all basic biophysical parameters and, as well

► **Fig. 21.7** Conversion of fluorescence data selected from an O-J-I-P fluorescence transient to biophysical parameters with the JIP-test, which is based on the Theory of Energy Fluxes in Biomembranes. The figure distinguishes experimental signals, normalised signals and biophysical parameters, the latter being further distinguished in specific and phenomenological fluxes and yields (or flux ratios). The definition of the performance index on absorption basis, PI_{ABS}, is also depicted, together with the formulae which link the PI_{ABS} with the experimental and normalised signals and the biophysical parameters. (Chl_{RC} and Chl_{ant} stand for the chlorophyll content of the reaction centers and of the antenna respectively, while $Chl_{total} = Chl_{ant} + Chl_{RC}$ stands for the total chlorophyll; for the definition of other symbols, see text, Table 21.1 and Fig. 21.5)

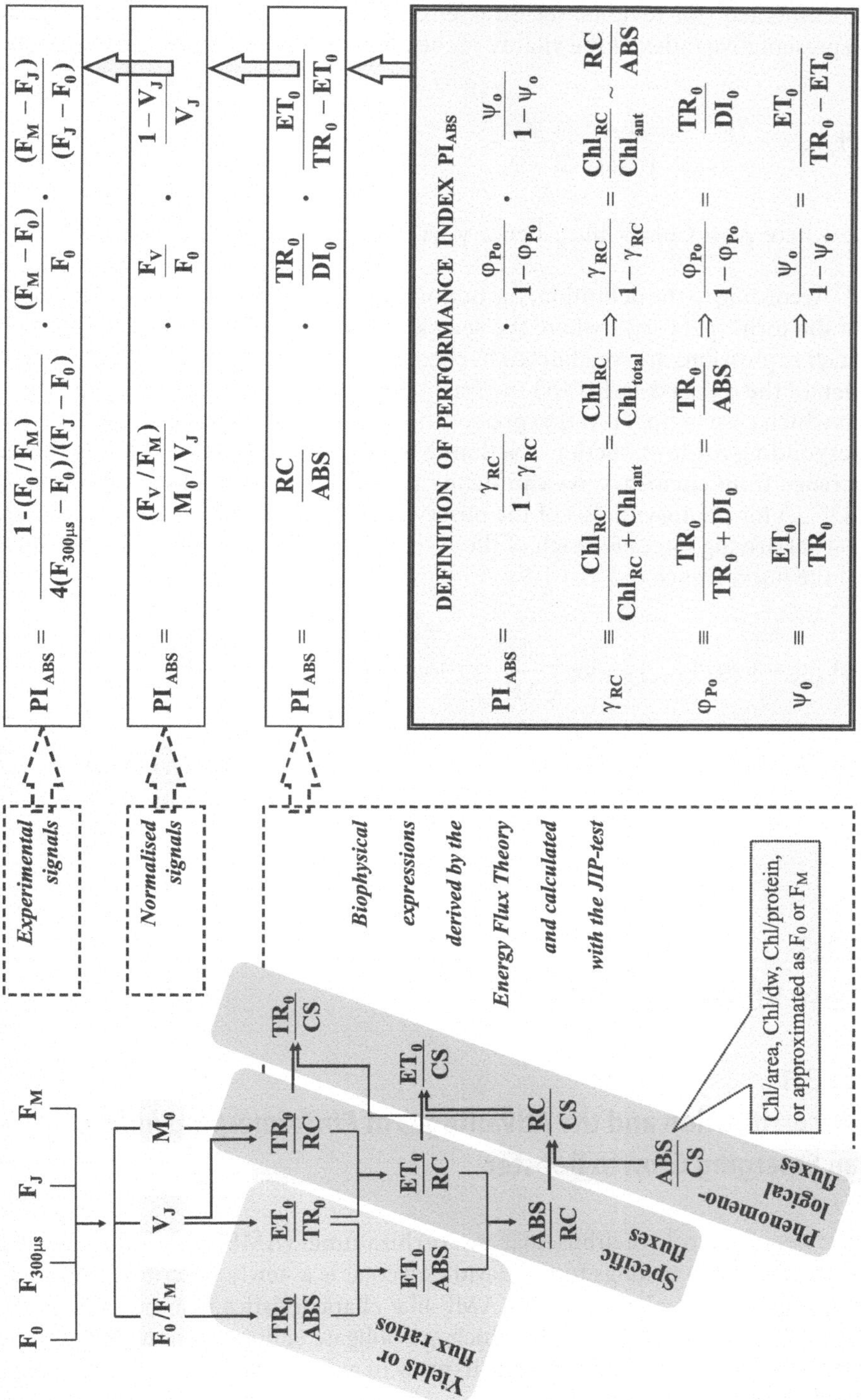

Experimental signals
Normalised signals
Biophysical expressions derived by the Energy Flux Theory and calculated with the JIP-test
DEFINITION OF PERFORMANCE INDEX PI_{ABS}
$PI_{ABS} = \frac{1-(F_0/F_M)}{4(F_{300\mu s}-F_0)/(F_J-F_0)} \cdot \frac{(F_M-F_0)}{F_0} \cdot \frac{(F_M-F_J)}{(F_J-F_0)}$
$PI_{ABS} = \frac{(F_V/F_M)}{M_0/V_J} \cdot \frac{F_V}{F_0} \cdot \frac{1-V_J}{V_J}$
$PI_{ABS} = \frac{RC}{ABS} \cdot \frac{TR_0}{DI_0} \cdot \frac{ET_0}{TR_0-ET_0}$
$PI_{ABS} = \frac{\gamma_{RC}}{1-\gamma_{RC}} \cdot \frac{\varphi_{Po}}{1-\varphi_{Po}} \cdot \frac{\psi_o}{1-\psi_o}$
$\gamma_{RC} = \frac{Chl_{RC}}{Chl_{RC}+Chl_{ant}} = \frac{Chl_{RC}}{Chl_{total}} \Rightarrow \frac{\gamma_{RC}}{1-\gamma_{RC}} = \frac{Chl_{RC}}{Chl_{ant}} \sim \frac{RC}{ABS}$
$\varphi_{Po} = \frac{TR_0}{TR_0+DI_0} = \frac{TR_0}{ABS} \Rightarrow \frac{\varphi_{Po}}{1-\varphi_{Po}} = \frac{TR_0}{DI_0}$
$\psi_0 = \frac{ET_0}{TR_0} \Rightarrow \frac{\psi_o}{1-\psi_o} = \frac{ET_0}{TR_0-ET_0}$
F_0 $F_{300\mu s}$ F_J F_M
F_0/F_M V_J M_0
TR0/ABS ET0/TR0 TR0/RC TR0/CS
ET0/ABS ET0/RC ET0/CS
ABS/RC RC/CS ABS/CS
Yields or flux ratios
Specific fluxes
Phenomeno-logical fluxes
Chl/area, Chl/dw, Chl/protein, or approximated as F_0 or F_M

documented (for reviews, see Strasser et al. 2000, 2004), is a very appropriate representative index of the vitality of the photosynthetic system – macrostate:

$$PI_{ABS} = \frac{\gamma_{RC}}{1-\gamma_{RC}} \cdot \frac{\varphi_{Po}}{1-\varphi_{Po}} \cdot \frac{\psi_o}{1-\psi_o}$$

where $\gamma_{RC} = Chl_{RC}/Chl_{total}$, hence $\gamma_{RC}/(1-\gamma_{RC}) = Chl_{RC}/Chl_{total} \sim RC/ABS$.

According to the definition, the performance index is a product of expressions of the form $[p_i/(1-p_i)]$, where the several p_i stand for probabilities or fractions. Such expressions are well known in chemistry, with p_i representing e.g. the fraction of the reduced and $(1-p_i)$ the fraction of the oxidised form of a compound, in which case $\log[p_i/(1-p_i)]$ expresses the potential or driving force for the corresponding oxido-reduction reaction (Nernst's equation). Extrapolating this inference from chemistry, we can define the $\log(PI_{ABS})$ as the total driving force (DF_{ABS}) for photosynthesis of the observed system, created by summing up the partial driving forces for each of the several energy bifurcations (all at the onset of the fluorescence rise O-J-I-P).

$$DF_{ABS} = \log(PI_{ABS}) = \log\left(\frac{RC}{ABS}\right) + \log\left(\frac{\varphi_{Po}}{1-\varphi_{Po}}\right) + \log\left(\frac{\psi_o}{1-\psi_o}\right)$$

By presenting a clear distinction between experimental signals, normalised signals and biophysical parameters, Fig. 21.7 depicts also how PI_{ABS} can be directly calculated using any of these sets.

21.3 Case Study

21.3.1 Mycorrhization and the Advantages of *Piriformospora indica*, an Emerging Growth Booster

The beneficial role of arbuscular mycorrhiza fungi (AMF) is well documented. *P. indica*, which belongs to the Basidiomycota, is a newly described root endophyte (Verma et al. 1998) with AMF-like characteristics (Varma et al. 2001). Moreover, in contrast to AMF which are obligate endosymbionts, *P. indica* has the added advantage of being able to grow in axenic cultures – it is cultivable in vitro (Varma et al. 1999).

P. indica has growth- and yield-promoting effects on a broad range of plants, including medicinal plants: shoot and root length, biomass, basal stem, leaf area, overall size, number of inflorescences and flowers and seed production are all enhanced in the presence of fungi (Rai et al. 2001). Inoculation with the fungus and application of fungal culture filtrate also increase tolerance to temperature and drought, as well as to heavy metals. For example, concerning cadmium, which exerts toxic effects on plants, *P. indica* provides alleviation of the causative stress (tolerance up to 300 μg Cd per gramme of air-dried soil). Moreover, *P. indica* has the properties of biofertilizer, bioregulator, phytoremediator, immunomodulator and antioxidants/drugs enhancer (Varma, personal communication). It also provides biocontrol against insects and pathogens (Pham et al. 2004a, b, c).

All these impressive traits make *P. indica* very valuable, both for basic research, as an excellent model organism for the study and understanding of the beneficial plant-microbe interactions and for applied research, as a powerful new candidate tool for improving plant production systems in agroforestry and flori-horticulture applications for sustainable agriculture.

We here apply our approach for a comparative study of the beneficial role of typical arbuscular mycorrhiza fungi (*G. mosseae*, *G. caledonium*) and *P. indica*, on chick peas (*Cicer arietinum* L. Chafa variety) exposed to cadmium stress.

21.3.2 Phenomics of the O-J-I-P Fluorescence Transient for the Study of Cadmium Stress on Chick Peas (*Cicer arietinum* L. Chafa variety) With and Without Symbiosis With *Glomus mosseae*, *G. caledonium* and *Piriformospora indica*

As analysed above (Fig. 21.6), normalisations and differences of the fluorescence transients reveal a deconvolution of the typical O-J-I-P shape into additional bands, carrying useful information about microstates of the photosynthetic system.

Figure 21.8 is an application of this approach to the presented case study. O-J-I-P Chl *a* fluorescence transients are plotted as kinetics of the relative variable fluorescence $W_t = V_t/V_J = (F_t - F_0)/(F_J - F_0)$ on a logarithmic time-scale (like in the lower panel of Fig. 21.4). The presented transients were exhibited by dark-adapted leaves of chickpea (*C. arietinum* L. Chafa variety) measured with the Handy-PEA fluorimeter at the 42nd day after inoculation with *G. mosseae* (black diamonds), *G. caledonium* (black squares) and *P. indica* (black circles). All inoculated plants were under cadmium stress (added on the 21st day). The transients from non-inoculated plants of the same age in the absence (control, grey circles) or presence of Cd (black triangles) are also depicted. We observe that these transients *(open symbols)* show minor differences (similarly to the

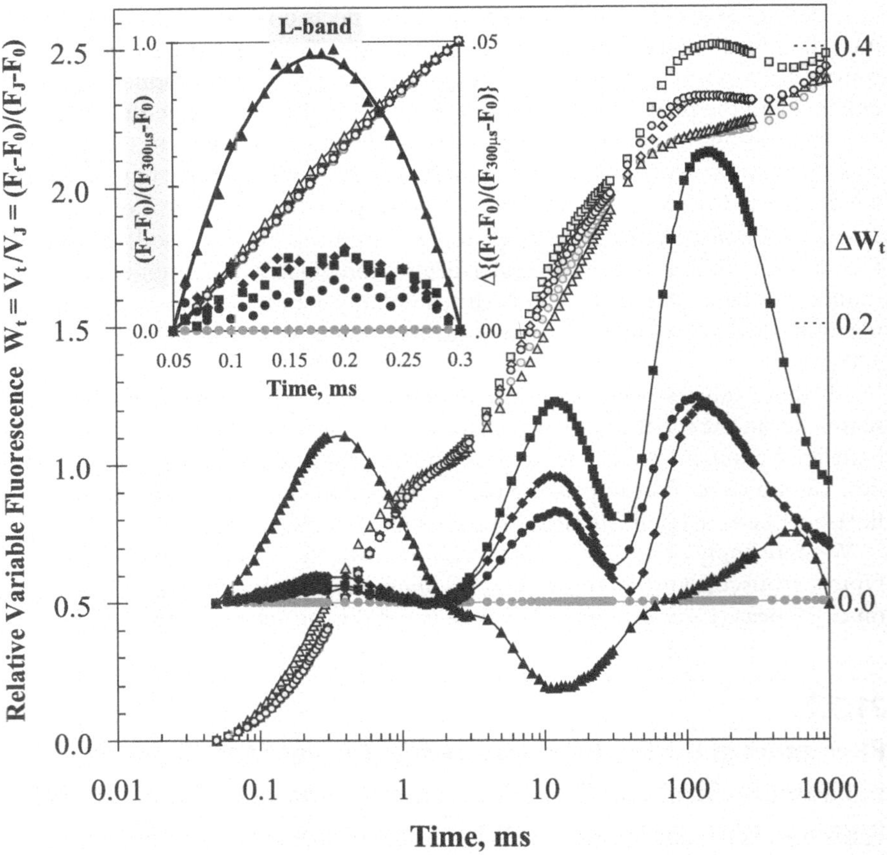

Fig. 21.8 Chl *a* fluorescence transients (each presenting average kinetics of raw fluorescence data from 12 samples) of dark-adapted leaves of chick peas (*Cicer arietinum* L. Chafa variety) measured with the Handy-PEA fluorimeter at the 42nd day after inoculation with *G. mosseae* (*black diamonds*), *G. caledonium* (*black squares*) and *P. indica* (*black circles*). All inoculated plants were under cadmium stress (added on the 21st day). The transients from non-inoculated plants of the same age in the absence of Cd (control, *grey circles*) or presence of Cd (*black triangles*) are also depicted. The transients are presented as kinetics of the relative variable fluorescence $W_t = V_t/V_J = (F_t-F_0)/(F_J-F_0)$, *open symbols, left axis*, and as ΔW_t (treated minus control; *closed symbols, right axis*). The insert depicts, on a linear time-scale from 50 μs to 300 μs, the transients normalised as $(F_t-F_0)/(F_{300\mu s}-F_0)$, as well as their differences $\Delta[(F_t-F_0)/(F_{300\mu s}-F_0)]$ from the control, which reveal the L-band

cases presented in Figs. 21.4, 21.6). However, when plotted as difference kinetics ΔW_t (treated minus control; closed symbols), they reveal major differences concerning the amplitudes of the bands. The difference kinetics demonstrate that: (a) the trend of the impact of all three symbionts is the same and (b) this impact is the almost complete elimination (range 50 μs to 2 ms) or even the overcompensation (2 ms to 1 s) of the major effects of Cd stress on the transients. Similar information is derived from the insert of Fig. 21.8, where the transients are de-

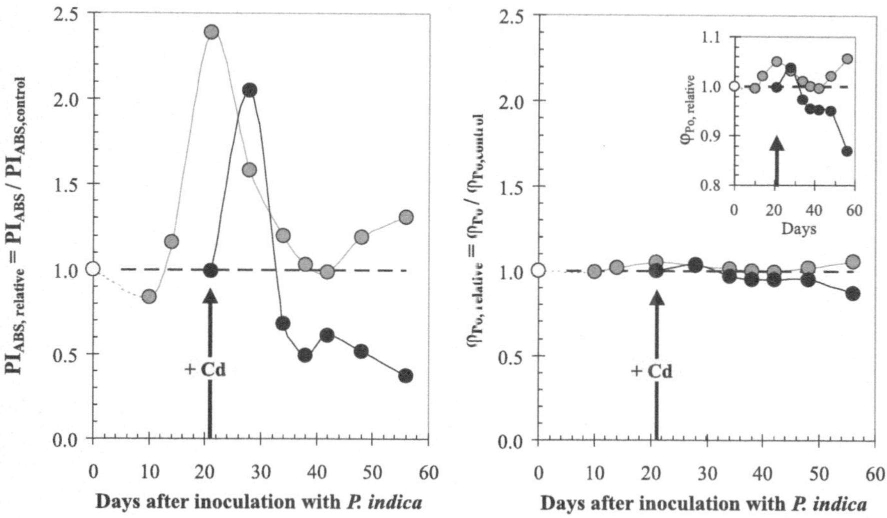

Fig. 21.9 The relative performance index $PI_{ABS}/PI_{ABS,control}$ (*left panel*) and the relative maximum quantum yield of primary photochemistry $\varphi_{Po}/\varphi_{Po,control}$ (*right panel*), from day 10 to day 56 after inoculation of chick peas (*C. arietinum* L. Chafa variety) plants with *P. indica*. Inoculated (*grey circles*) and non-inoculated (*black circles*) plants were put under cadmium stress on the 21st day. The value of the parameter from non-inoculated, and without addition of cadmium, plants of the same age was used as the control value of each parameter (subscript "*control*")

picted, on a linear time-scale from 50 μs to 300 μs, as $(F_t - F_0)/(F_{300\mu s} - F_0)$, along with their difference $\Delta[(F_t - F_0)/(F_{300\mu s} - F_0)]$ from the control. It is worth noting that, when this normalisation is used, the difference transients reveal the L-band (not appearing in ΔW_t; see legend of Fig. 21.6).

Let us now follow the impact of Cd stress with and without symbiosis on parameters derived by the JIP-test.

We first demonstrate a comparison of the impact of *P. indica* on the performance index PI_{ABS} (for definition and formulae, see Fig. 21.7) and the commonly used maximum quantum yield of primary photochemistry φ_{Po} (for definition and formulae see Table 21.1) by depicting in Fig. 21.9 their stress kinetics from day 10 to day 56 after inoculation with *P. indica*. Inoculated (grey circles) and non-inoculated plants (black circles) were put under cadmium stress on the 21st day. The parameters are presented as $PI_{ABS}/PI_{ABS,control}$ (left panel) and $\varphi_{Po}/\varphi_{Po,control}$ (right panel), where $PI_{ABS,control}$ and $\varphi_{Po,control}$ refer to non-inoculated plants of the same age that were not put under cadmium stress (control plants).

We observe that the $PI_{ABS}/PI_{ABS,control}$ undergoes wide changes during the course of the stress. Though Cd addition is shown to affect both non-inoculated and inoculated plants, the beneficial role of the symbiont concerning the tolerance to Cd stress is clearly revealed. $\varphi_{Po}/\varphi_{Po,control}$ appears much less sensitive than $PI_{ABS}/PI_{ABS,control}$ and thus much less appropriate in detecting vitality

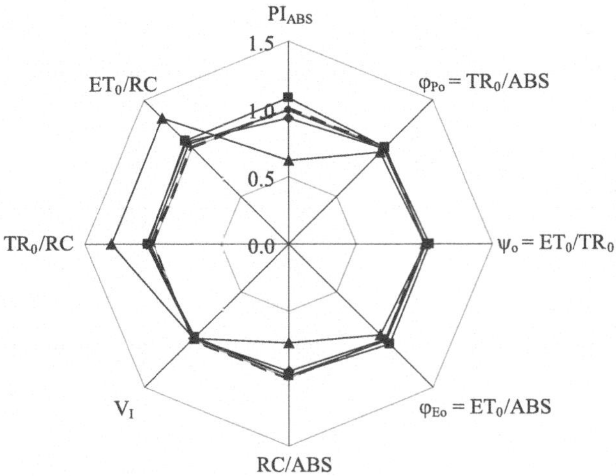

Fig. 21.10 The impact of cadmium stress on different parameters derived by the JIP-test from the fluorescence transients (for terms, definitions and formulae, see Section 21.2.4.4, also Fig. 21.5 and Table 21.1). The figure refers to the 42nd day after inoculation with *G. mosseae* (*diamonds*), *G. caledonium* (*squares*) and *P. indica* (*circles*). All inoculated plants were under cadmium stress (added on the 21st day). The case of non-inoculated plants of the same age under cadmium stress is also depicted (*triangles*). Each case is represented by an octagon, where the value of each parameter is normalised on that of the control case (i.e. non-inoculated, and without addition of cadmium, plants of the same age), which is thus depicted by the regular octagon (*dashed thick line*)

changes. However, magnification of the changes it undergoes (as shown in the insert) reveals that it exhibits a trend similar to that of $PI_{ABS}/PI_{ABS,control}$ for both non-inoculated and inoculated plants.

The spider-plot of Fig. 21.10 presents the impact of cadmium stress on different parameters derived by the JIP-test from the fluorescence transients, for non-inoculated and inoculated (with the three symbionts) plants. The parameters are: PI_{ABS}, φ_{Po}, ψ_o, φ_{Eo}, RC/ABS, V_I, TR_0/RC and ET_0/RC (see text in Section 21.2.4.4, also Fig. 21.5 and Table 21.1). The figure refers to the 42nd day after inoculation with *G. mosseae* (diamonds), *G. caledonium* (squares) and *P. indica* (circles). All inoculated plants were under cadmium stress (added on the 21st day). Non-inoculated plants of the same age under cadmium stress are also depicted (triangles). Each case is represented by an octagon, where the value of each parameter is normalised on that of the control case (i.e. non-inoculated, and without addition of cadmium, plants of the same age), which is thus depicted by the regular octagon (dashed thick line).

Further than depicting in a comparative way the quantitative impact of stress on the individual parameters, each of which is linked to microstates, the presentation of the results with a spider-plot has the advantage of providing an easy

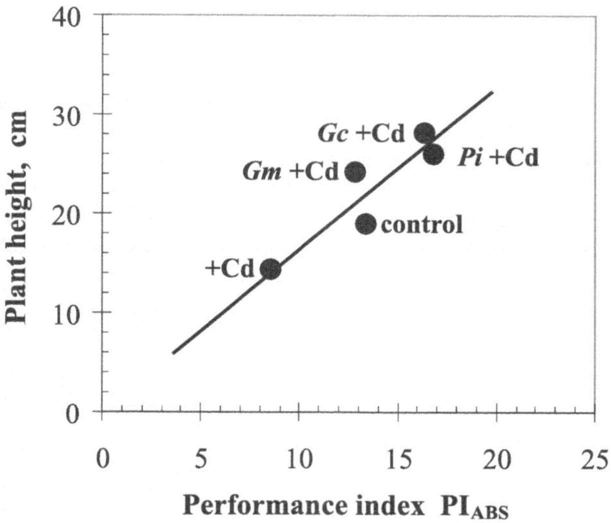

Fig. 21.11 Correlation of the height of the plant (physiological parameter) and the performance index PI_{ABS} (biophysical parameter) derived by the JIP-test, for non-inoculated chick peas (*C. arietinum* L. Chafa variety) plants in the absence (*control*) or presence (+*Cd*) of cadmium, or inoculated with *G. mosseae*, *G. caledonium* and *P. indica* and exposed to cadmium stress (*Gm*+Cd, *Gc*+Cd and *Pi*+Cd, respectively)

recognition of stress effects. We can immediately see the distortion from the regular octagon (control) caused by Cd (triangles), which can be registered as the characteristic pattern of Cd stress; and we can also see that, for plants in symbiosis with any of the three symbionts, almost no distortion from the control pattern (regular octagon) occurs.

21.3.3 Correlation of Physiological with Biophysical Parameters

Further than proving the high sensitivity of the performance index, we also checked whether and how it is related with physiological parameters, commonly used for the evaluation of the impact of symbiosis on the vitality of plants. Figure 21.11 shows indeed a striking correlation between the height of the plant and the performance index PI_{ABS}. The data presented come from non-inoculated chick peas (*C. arietinum* L. Chafa variety) plants in the absence *(control)* and presence (+Cd) of cadmium, as well as from plants inoculated with *G. mosseae*, *G. caledonium* and *P. indica* and exposed to cadmium stress (*Gm*+ Cd, *Gc*+ Cd and *Pi*+ Cd, respectively).

21.4 Conclusions

Biophysical phenomics, as we term our approach, here applied for the evaluation of the effectiveness of mycorrhization, is shown to be powerful for the description of an in vivo vitality analysis (behaviour/performance) of PSII, i.e. for the description of a biophysical phenotype (macrostate), as well as for the recognition and evaluation of stress impacts on microstates (the functional building blocks into which the macrostate is deconvoluted). With this approach we demonstrate the beneficial role of typical AMF and of the equally effective *P. indica*, concerning tolerance to Cd stress.

Our techniques are thus shown to be very suitable for studying the effectiveness of soil microbial activity. The advantages of these techniques can be summarised as follows:

- They provide an early diagnosis of vitality changes (primary stress effects, hence stress tolerance).
- They can be used to screen not only leaves but any green part of the plant.
- They are rapid – only a few seconds are needed for each measurement.
- They can be applied in vivo.
- They can be carried out anywhere – in the field, in the greenhouse or even in tissue cultures – and even on samples as small as 2 mm^2.
- They are not invasive.
- They are inexpensive.

References

Briantais JM, Vernotte C, Krause GH, Weiss E (1986) Chlorophyll *a* fluorescence of higher plants: chloroplasts and leaves. In: Govindjee, Amesz J, Fork DC (eds) Light emission by plants and bacteria. Academic, New York pp 539–583

Clark AJ, Landolt W, Bucher J, Strasser RJ (1998) The response of *Fagus sylvatica* to elevated CO2 and ozone probed by the JIP-test based on the chlorophyll fluorescence rise OJIP. In: De Kok LJ, Stulen I (eds) Responses of plant metabolism to air pollution and global change. Backhuys, Leiden, pp 283–286

Clark AJ, Landolt W, Bucher J, Strasser RJ (2000) Beech (*Fagus sylvatica*) response to ozone exposure assessed with a chlorophyll *a* fluorescence performance index. Environ Pollut 109:501–507

Dau H (1994) Molecular mechanisms and quantitative models of variable photosystem II fluorescence. Photochem Photobiol 60:1–23

Epitalawage N, Eggenberg P, Strasser RJ (2003) Use of fast chlorophyll a fluorescence technique in detecting drought and salinity tolerant chickpea (*Cicer arietinum* L.) varieties. Arch Sci Genève 56:79–93

Govindjee (1995) Sixty-three years since Kautsky: chlorophyll *a* fluorescence. Aust J Plant Physiol 22:131–160

Govindjee, Amesz J, Fork DC (1986) Light emission by plants and bacteria. Academic, New York

Kautsky H, Hirsch A (1931) Neue Versuche zur Kohlensäureassimilation. Naturwissenschaften 19:964

Krause GH, Weiss E (1991) Chlorophyll fluorescence and photosynthesis: the basics. Annu Rev Plant Physiol Plant Mol Biol 42:313–349

Krüger GHJ, Tsimilli-Michael M, Strasser RJ (1997) Light stress provokes plastic and elastic modifications in structure and function of photosystem II in camellia leaves. Physiol Plant 101:265–277

Ouzounidou G, Moustakas M, Strasser RJ (1997) Sites of action of copper in the photosynthetic apparatus of maize leaves: kinetic analysis of chlorophyll fluorescence, oxygen evolution, absorption changes and thermal dissipation as monitored by photoacoustic signals. Aust J Plant Physiol 24:81–90

Papageorgiou G (1975) Chlorophyll fluorescence: an intrinsic probe of photosynthesis. In: Govindjee (ed) Bioenergetics of photosynthesis. Academic, New York, pp 319–371

Pham GH, Singh A, Malla R, Kumari R, Prasad R, Sachdev M, Luis P, Kaldorf M, Tatjana P, Harrmann S, Hehl S, Declerck S, Buscot F, Oelmuller R, Rexer KH, Kost G, Varma A (2004a) Interaction of *P. indica* with other microorganisms and plants. In: Varma A, Abbott L, Werner D, Hampp R (eds) Plant surface microbiology. Springer, Berlin Heidelberg New York, pp 237–265

Pham GH, Kumari R, Singh A, Sachdev M, Prasad R, Kaldorf M, Buscot F, Oelmuller R, Tatjana P, Weiβ M, Hampp R, Varma A (2004b) Axenic cultures of *Piriformospora indica*. In: Varma A, Abbott L, Werner D, Hampp R (eds) Plant surface microbiology. Springer, Berlin Heidelberg New York, pp 593–616

Pham GH, Srivastava A, Saxena AK, Pareek A, Varma A (2004c) Protocol to understand the interaction between rhizobacteria and symbiotic fungus: *Piriformospora indica*. In: Podila G, Varma A (eds) Basic research and applications: Mycorrhizae. (Microbiology series, vol 1) IK International, New York, pp 425–450

Prakash JSS, Srivastava A, Strasser RJ, Mohanty P (2003) Senescence-induced alterations in the photosystem II functions of *Cucumis sativus* cotyledons: probing of senescence driven alterations of photosystem II by chlorophyll *a* fluorescence induction O-J-I-P transients. Indian J Biochem Biophys 40:160–168

Rai M, Acharya D, Singh A, Varma A. (2001) Positive growth responses of the medicinal plants, *Spilanthus calva* and *Withania somnifera* to inoculation by *Piriformospora indica* in field trials. Mycorrhiza 11:123–128

Schmitz P, Maldonado-Rodriguez R, Strasser RJ (2001) Evaluation of the nodulated status of *Vigna unguiculata* probed by the JIP-test based on the chlorophyll *a* fluorescence rise. In: CSIRO (ed) Proceedings of the 12th international congress on photosynthesis. CSIRO, Collingwood, S36-012

Srivastava A, Strasser RJ (1995) How do land plants respond to stress temperature and stress light? Arch Sci Genève 48:135–145

Srivastava A, Strasser RJ (1996) Stress and stress management of land plants during a regular day. J Plant Physiol 148:445–455

Srivastava A, Guisse B, Greppin H, Strasser RJ (1997) Regulation of antenna structure and electron transport in PS II of *Pisum sativum* under elevated temperature probed by the fast polyphasic chlorophyll *a* fluorescence transient OKJIP. Biochim Biophys Acta 1320:95–106

Strasser RJ (1978) The grouping model of plant photosynthesis. In: Akoyunoglou G (ed) Chloroplast development. Elsevier, Dordrecht, pp 513–524

Strasser RJ (1981) The grouping model of plant photosynthesis: heterogeneity of photosynthetic units in thylakoids. In: Akoyunoglou G (ed) Photosynthesis III. Structure and molecular organisation of the photosynthetic apparatus. Balaban International Science Services, Philadelphia, pp 727–737

Strasser RJ (1985) Dissipative Strukturen als Thermodynamischer Regelkreis des Photosyntheseapparates. Ber Deutsche Bot Ges Bd 98:53–72

Strasser RJ, Govindjee (1992) The F0 and the O-J-I-P fluorescence rise in higher plants and algae. In: Argyroudi-Akoyunoglou JH (ed) Regulation of chloroplast biogenesis. Plenum, New York pp 423–426

Strasser RJ, Tsimilli-Michael M (1998) Activity and heterogeneity of PS II probed in vivo by the chlorophyll *a* fluorescence rise O-(K)-J-I-P. In: Garab G (ed) Photosynthesis: mechanisms and effects, vol V. Kluwer Academic, Rotterdam, pp 4321–4324

Strasser RJ, Tsimilli-Michael M (2001) Stress in plants, from daily rhythm to global changes, detected and quantified by the JIP-test. Chim Nouv 75:3321–3326

Strasser RJ, Tsimilli-Michael M (2005) State-changes realising adaptation to stress as the result of an optimized redistribution of functional microstates. In: van Est A, Bruce D (eds) Photosynthesis: fundamental aspects to global perspectives. Allen, Montreal, pp 537–540

Strasser RJ, Srivastava A, Govindjee (1995) Polyphasic chlorophyll *a* fluorescence transient in plants and cyanobacteria. Photochem Photobiol 61:32–42

Strasser RJ, Srivastava A, Tsimilli-Michael M (2000) The fluorescence transient as a tool to characterize and screen photosynthetic samples. In: Yunus M, Pathre U, Mohanty P (eds) Probing photosynthesis: mechanism, regulation and adaptation. Taylor and Francis, London, pp 443–480

Strasser RJ, Tsimilli-Michael M, Srivastava A (2004) Analysis of the chlorophyll *a* fluorescence transient. In: Papageorgiou GC, Govindjee (eds) Chlorophyll fluorescence: a signature of photosynthesis. (Advances in photosynthesis and respiration series, vol 19) Kluwer Academic, Rotterdam, pp 321–362

Tóth SZ, Schansker G, Strasser RJ (2005) In intact leaves, the maximum fluorescence level (FM) is independent of the redox state of the plastoquinone pool: a DCMU-inhibition study. Biochim Biophys Acta 1708:275–282

Tsimilli-Michael M, Strasser RJ (2001) Fingerprints for climate changes on the behaviour of the photosynthetic apparatus, monitored by the JIP-test. A case study on light and heat stress adaptation of the symbionts of temperate and coral reef foraminifers in hospite. In: Walther G-R, Burga CA, Edwards PJ (eds) Fingerprints of climate changes – adapted behaviour and shifting species ranges. Kluwer, New York, pp 229–247

Tsimilli-Michael M, Strasser RJ (2002) Mycorrhization as a stress adaptation procedure. In: Gianinazzi S, Haselwandter K, Schüepp H, Barea JM (eds) Mycorrhiza technology in agriculture: from genes to bioproducts. Birkhauser, Basel, pp 199–209

Tsimilli-Michael M, Krüger GHJ, Strasser RJ (1995) Suboptimality as driving force for adaptation: a study about the correlation of excitation light intensity and the dynamics of fluorescence emission in plants. In: Mathis P (ed) Photosynthesis: from light to biosphere, vol V. Kluwer Academic, Rotterdam, pp 981–984

Tsimilli-Michael M, Krüger GHJ, Strasser RJ (1996) About the perpetual state changes in plants approaching harmony with their environment. Arch Sci Genève 49:173–203

Tsimilli-Michael M, Pêcheux M, Strasser RJ (1998) Vitality and stress adaptation of the symbionts of coral reef and temperate foraminifers probed in hospite by the fluorescence kinetics O-J-I-P. Arch Sci Genève 51:1–36

Tsimilli-Michael M, Pêcheux M, Strasser RJ (1999) Light and heat stress adaptation of the symbionts of temperate and coral reef foraminifers probed in hospite by the chlorophyll a fluorescence kinetics O-J-I-P. Z Naturforsch 54C:671–680

Tsimilli-Michael M, Eggenberg P, Biro B, Köves-Pechy K, Vörös I, Strasser RJ (2000) Synergistic and antagonistic effects of arbuscular mycorrhizal fungi and *Azospirillum* and *Rhizobium* nitrogen-fixers on the photosynthetic activity of alfalfa, probed by the chlorophyll *a* polyphasic fluorescence transient O-J-I-P. Appl Soil Ecol 15:169–182

Van Rensburg L, Krüger GHJ, Eggenberg P, Strasser RJ (1996) Can screening criteria for drought resistance in *Nicotiana tabacum* L. be derived from the polyphasic rise of the chlorophyll *a* fluorescence transient (OJIP)? S Afr J Bot 62:337–341

Van Heerden PDR, Tsimilli-Michael M, Krüger GHJ, Strasser RJ (2003) Dark chilling effects on soybean genotypes during vegetative development: parallel studies of CO2 assimilation, chlorophyll *a* fluorescence kinetics O-J-I-P and nitrogen fixation. Physiol Plant 117:476–491

Varma A, Verma S, Sudha, Sahay N, Britta B, Franken P (1999) *Piriformospora indica* – a cultivable plant growth promoting root endophyte with similarities to arbuscular mycorrhizal fungi. Appl Environ Microbiol 65:2741–2744

Varma A, Singh A, Sudha, Sahay N, Sharma J, Roy A, Kumari M, Rana D, Thakran S, Deka D, Bharti K, Franken P, Hurek T, Blechert O, Rexer K-H, Kost G, Hahn A, Hock B, Maier W, Walter M, Strack D, Kranner I (2001) *Piriformospora indica*: a cultivable mycorrhiza-like endosymbiotic fungus. In: Varma A, Hock B (eds) Mycota IX. Springer, Berlin Heidelberg New York, pp 123–150

Verma S, Varma A, Rexer K-H, Hassel A, Kost G, Sarbhoy A, Bisen P, Buetehorn P, Franken P (1998) *Piriformospora indica* gen. nov., a new root-colonizing fungus. Mycologia 90:895–909

22 Analysis of the Plant Protective Potential of the Root Endophytic Fungus *Piriformospora indica* in Cereals

F. Waller, B. Achatz, and K.-H. Kogel

22.1 Introduction

Piriformospora indica is a recently discovered basidiomycete that infests roots of a large variety of mono- and dicotyledonous plants (Verma et al. 1998; Pham et al. 2004). Endophytic growth of this fungus in roots leads to enhanced plant growth (Varma et al. 1999), reminiscent of the beneficial effects of arbuscular mycorrhiza in host plants. We have recently shown that *P. indica* – upon successful establishment in the roots – reprogrammes barley to salt stress tolerance, resistance to diseases and higher yield. Successful powdery mildew infections in barley leaves are reduced by this root endophyte, due to a yet unknown mechanism of induced resistance (IR) (Waller et al. 2005). As *P. indica* can easily be cultured without a host plant (Varma et al. 1999), it is suitable both as a model system for research and for future applications in agriculture. Here, we present approaches and methods to study the mechanisms behind the observed pathogen resistance induced by *P. indica*. These methods should provide valuable tools for studying the effect of root-interacting fungi on IR in cereals.

Frank Waller, Beate Achatz, Karl-Heinz Kogel: Institute of Phytopathology and Applied Zoology (IPAZ), Justus-Liebig-Universität Gießen, Heinrich-Buff-Ring 26-32, D-35392 Gießen, Germany, email: Frank.Waller@agrar.uni-giessen.de

Beate Achatz: Institute for Vegetables and Ornamental Crops, D-14979 Großbeeren, Germany

Soil Biology, Volume 11
Advanced Techniques in Soil Microbiology
A. Varma, R. Oelmüller (Eds.)

22.2 Plant Responses and Resistance to Pathogens

22.2.1 Local Reactions

Plants are continuously defending themselves against a plethora of attacking viruses, bacteria, fungi and invertebrates. Each plant cell has both preformed and inducible defence capabilities. Among the preformed defences are physical barriers, such as the cell wall or, for example, secondary metabolites with antimicrobial properties. After recognition of the pathogen, induced defence responses may comprise of local cell wall fortifications, the production or activation of secondary metabolites, the localized release of antimicrobial compounds at sites of attack or a hypersensitive reaction (HR), leading to localized cell death which limits the spread of the pathogen in the plant. Local defence responses are accompanied by an enhanced expression of defence-related genes: pathogenesis-related genes (PR genes) respond rapidly to challenges by pathogens and have been widely used as markers for defence reactions in plants.

22.2.2 Systemic Reactions and Resistance in Cereals

A prior activation of plant defence that leads to resistance against pathogens is termed induced resistance (IR; Sticher et al. 1997). IR has been studied extensively in the case of salicylic acid (SA)-mediated *systemic acquired resistance* (SAR) in dicotyledonous plants: Micro-lesions induced by necrotizing pathogens trigger a local accumulation of salicylic acid, with mitogen-activated protein kinases, H_2O_2 and other signals being involved. Whereas the mobile signal leading to systemic responses is still a matter of debate, it is clear that the expression of SAR marker genes, the PR genes, is controlled by the protein NPR1, which is necessary for SAR (Mou et al. 2003; Dong 2004). Another major type of IR is *induced systemic resistance* (ISR) which is triggered by non-pathogenic rhizobacteria. ISR depends on both NPR1 and the jasmonic acid/ethylene pathway, but not on SA (Pieterse et al. 1998). Cereals share many components of resistance pathways with dicotyledonous plants: SA derivatives induce, for example, resistance in cereals (Kogel et al. 1994), and NPR 1 homologues have been shown to be functional in rice (Chern et al. 2005). However, specific IR signalling components in cereals have yet to be characterized in detail (Kogel and Langen 2005).

22.2.3 Beneficial Microbial Endophytes Protecting Cereals from Pathogens

Micro-organisms growing inside of plants are referred to as endophytes. A large number of these are known to protect plants against pathogens. For example, grasses (Poaceae) are frequently associated with fungi of the Clavicipitaceae (Ascomycota), with interactions ranging from mutualism to antagonism (Schardl et al. 2004). In these interactions, the endophyte is strictly confined to upper parts of the plant, grows only intercellularly and has a rather narrow host range. The beneficial effect of these endophytes has been shown to result from a direct antimicrobial and insecticidal activity of alkaloids. Another group of endophytic fungi, the arbuscular mycorrhiza (AM; Glomeromycota; Schüssler et al. 2001) protect plants from various stresses, including root diseases (Dehne 1987; Azcón-Aguilar and Barea 1996; Borowicz 2001; Harrison 2005; Hause and Fester 2005). An improved defence status of mycorrhizal roots is associated with an increased expression of the H_2O_2 scavengers catalase, peroxidase and superoxide dismutase (Blee and Anderson 2000; Pozo et al. 2002). Such an elevated antioxidant activity could protect roots from cell death mediated by necrotrophic root pathogens, which require killing of host cells for a successful infection. Using a split root technique, it has been demonstrated that AM induce systemic protection against root pathogens (Cordier et al. 1998; Pozo et al. 2002). This systemic effect of mycorrhization is restricted to the root and does not protect plants from leaf diseases, but rather increases susceptibility to them (Shaul et al. 1999; Gernns et al. 2001). In addition to this agronomic drawback of the AM symbiosis, AM cannot be cultured axenically, limiting a wide-spread field application. Since a biological approach to protect cereals from pathogens has a significant impact for modern plant production systems, exploiting an axenically cultivatable endophyte with the ability to protect all plant parts from pathogens would be an important step towards a feasible broad-range application of biological measures in agriculture.

22.3 Interaction of *P. indica* with Cereals

P. indica is a basidiomycete fungus from the newly defined order Sebacinales (Hymenomycetes; Verma et al. 1998; Weiß et al. 2004). This endophyte infests roots of a large variety of mono- and dicotyledonous plants and can be axenically cultured (Verma et al. 1998, Pham et al. 2004). It has been shown that hyphae of *P. indica* develop both inter- and intracellular in the root cortex of a number of different plant species, thereby improving plant growth and stress tolerance (Varma et al. 1999, 2000). As the fungus' broad host range and easy cultivation could be valuable for agricultural applications, we tested *P. indica* for the ability to protect barley from abiotic stress and pathogens (Waller et al. 2005).

22.3.1 *P. indica* Colonizes Root Cortical Cells in Barley

We analysed fungal growth in barley roots grown in *P. indica*-inoculated substrate upon staining with 0.01% acid fuchsin lactic acid (Kormanik and McGraw 1982). For microscopy, whole roots as well as longitudinal and cross-sections produced by a cryo-microtome were used. Hyphae develop a dense mesh on the surface of the roots (Fig. 22.1a). Both hyphae and typical pear-shaped chlamydospores were localized intracellularly in the first few cell layers of the root (Fig. 22.1b), but could not be detected in the central root tissues beyond the endodermis.

22.3.2 *P. indica* Enhances Biomass and Yield in Barley

For growth experiments, barley seedlings were planted into pots with *P. indica*-inoculated soil (see Section 22.4.1). After five weeks of cultivation in the greenhouse, the fresh weight of shoots was evaluated. Shoot fresh weight was up to 1.65 times higher than that of control plants grown in soil without *P. indica* (Waller et al. 2005). Tests under field conditions, using Mitscherlich pots with

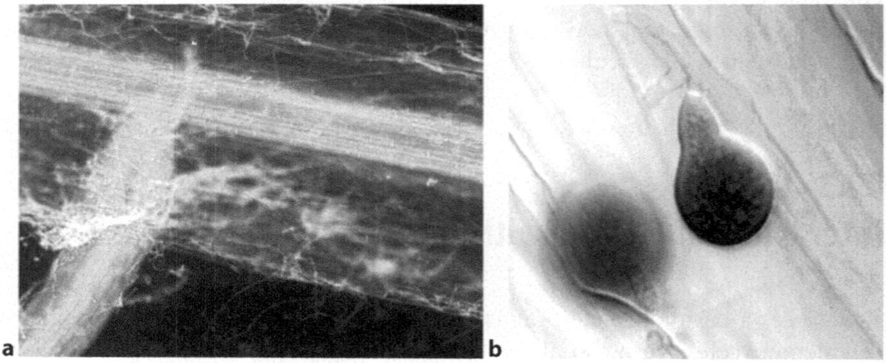

Fig. 22.1 *P. indica* hyphae and spores on a barley root. Two weeks after inoculation of barley roots with *P. indica*, acid fuchsin lactic acid staining reveals a mesh of hyphae surrounding the root (central part and emerging lateral root exhibit autofluorescence; **a**, fluorescence microscopy), as well as typical pear-shaped chlamydospores (**b**, bright-field image)

six plants per pot, revealed that a beneficial effect of *P. indica* on plant growth is present in plants until harvest: Grain yield increased by 11% in the barley elite cultivar *Annabell* as compared with control plants (Waller et al. 2005). This increase was mainly due to a higher number of ears per plant.

22.4 Approaches to Study the Mechanism of *P. indica*-Induced Pathogen Resistance

22.4.1 *P. indica* Induces Disease Resistance Against Root Pathogens

To assess whether *P. indica*-infested plants are more resistant to biotic stress, barley roots were inoculated with macroconidia of the necrotrophic fungal pathogen *Fusarium culmorum* (causing root rot). In the presence of *P. indica*, the devastating effect of *F. culmorum* infection was strongly diminished: Root and shoot fresh weight was reduced only 2-fold in *P. indica*-infested plants as compared with the 12-fold decrease in controls with *F. culmorum* alone. Similar results were obtained when resistance to the root-pathogenic fungus *Cochliobolus sativus* (hemibiotrophic life style) was tested. In axenic culture, *P. indica* did not exhibit antifungal activity to *F. culmorum* nor to *C. sativus*, indicating that the protective potential of the endophytic fungus does probably not rely on antibiosis (Waller et al. 2005).

1. Method for infestation of barley with *P. indica* and cultivation of plants: Barley was grown in pots with a 2:1 mixture of expanded clay (Seramis; Masterfoods, Verden, Germany) and Oil-Dri (Damolin, Mettmann, Germany) in an incubator with a 22 °C/18 °C day/night cycle, a photoperiod of 16 h (240 µmol m^{-2} s^{-1} photon flux) and 60% relative humidity. Plants were fertilized weekly with 20 ml of a 0.1% Wuxal top N solution (N/P/K: 12/4/6; Schering). For inoculation with *P. indica*, 2 g of crushed mycelium were added to 300 g of substrate before sowing. *P. indica* was propagated in liquid *Aspergillus* minimal medium (Peskan-Berghöfer et al. 2004) on a horizontal rotary shaker at 18–22 °C. Mycelium from liquid culture was washed with water to remove remaining traces of medium and crushed using a Waring Blendor (VWR International, Darmstadt, Germany) before adding it to the substrate. For yield evaluations, barley was sown in soil containing *P. indica* mycelium (4 g per 300 g of substrate) and grown for 4 weeks in a growth chamber after which six plantlets were transplanted into 6-l Mitscherlich pots (Stoma, Siegburg, Germany) filled with a mixture of a loam soil and sand (1:2). Soil nutri-

ent additives were 0.25 g of N, 0.4 g of P, 1.6 g of K, and 0.2 g of Mg; N was applied a second time at a rate of 0.25 g per pot, 2 weeks after planting.

2. Method for testing *Fusarium culmorum* in barley:
   To test the effect of *F. culmorum*, barley was grown as described above. Two weeks after planting into *P. indica* containing soil, plants were transferred into pots containing macroconidia of *F. culmorum*. Root and shoot fresh weight was measured two weeks after inoculation with *F. culmorum*.

22.4.2
P. indica Induces Systemic Disease Resistance

We recorded the effect of *P. indica* infestation on leaf infections by the biotrophic barley powdery mildew fungus, *Blumeria graminis* f.sp. *hordei*. A reduction in powdery mildew infection on leaf segments of *P. indica*-infested plants could be observed. Frequencies of mildew colonies decreased by 48% in second youngest leaves and by 58% in youngest leaves of 3-week-old *P. indica* infested plants (Waller et al. 2005).

Beside a reduction in pustule number, we frequently observed a smaller size and a reduced density of pustules. We quantified colonies belonging to three categories "large compact white colonies" (cat. I), "smaller, less dense colonies" (cat. II), and "colonies smaller than 0.3 mm in diameter" (cat. III; Fig. 22.2). In *P. indica*-infested plants, a shift towards smaller colonies was observed. This indicates a resistance mechanism that is limiting the development of the fungus after successful penetration. One possible explanation could be a reduced supply of nutrients to the fungus.

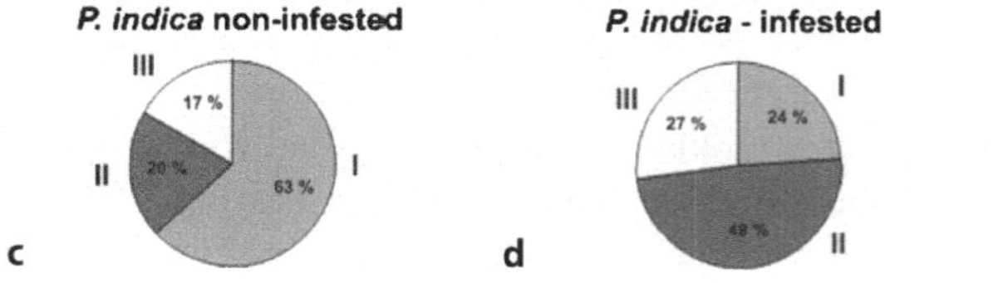

Fig. 22.2 Phenotype of *Blumeria graminis* pustules on barley leaves of *P. indica* infested plants. Shown are pustules on barley leaves 6 days after inoculation with *Blumeria graminis* f.sp. *hordei* (**a**, **b**). We quantified the percentage of colonies belonging to three categories: large, compact white colonies (as can be seen in **a**; cat. I), smaller, less dense colonies (as in **b**; cat. II) and colonies smaller than 0.5 mm in diameter (cat. III). In *P. indica*-infested plants, a shift towards smaller colonies, as compared with *P. indica* non-infested plants, was observed (**c**, **d**)

Microscopic analysis of powdery mildew on barley leaves revealed higher frequencies of HR as well as a cell wall-associated defence visible as cell wall appositions. These observations confirmed that the pathogen is arrested by an active plant response. As *P. indica* grows only in the outer cell layers of the host root and does not infest barley leaves, these data demonstrate a systemic plant response mediated by an endophytic fungus.

1. Method for leaf segment test:
   To assess powdery mildew resistance, leaf segments 7 cm in length were cut about 1 cm distal from the leaf sheath. Leaf segments were placed on 0.7% agar plates containing 40 mg l^{-1} benzimidazol (to inhibit leaf senescence). Inoculation was performed by shaking barley leaves heavily infected with *B. graminis* f.sp. *hordei*, race A6 (Wiberg 1974) in an inoculation tower about 1 m above the plates and manually circulating the air to ensure equal distribution of the spores. Inoculation density was checked by counting the number of spores per square millimetre, using a counting plate of defined size placed beside the plates with the leaf segments and counting the spores in this plate using a microscope. For counting the number of successful interaction sites, an inoculation density of 8–20 spores mm^{-2} was used. Plates were placed in an incubator at 18 °C with a 16 h /8 h light/dark cycle. After 6 days, pustules were visible and could be counted on a defined leaf segment, e.g. 3 cm or 5 cm in length. The severity of powdery mildew infection (disease index) was calculated as colonies produced by *B. graminis* on a defined leaf area. Generally, at least nine leaves were used per experiment and standard deviation as well as significance level calculated (unpaired Student's *t*-test).
2. Method for microscopic classification of interactions with the powdery mildew fungus:
   For cytological analysis, youngest leaves of three-week-old barley plants were inoculated with *B. graminis* f.sp. *hordei* (A6) as described above. Then whole plants were incubated in an incubator at 18 °C with a 16 h/8 h light/dark cycle.
   For H_2O_2 detection, a histochemical staining method using 3,3-diaminobenzidine (DAB) was used (Thordal-Christensen et al. 1997). After inoculation of the whole plant with powdery mildew and incubation for 27–43 h (depending on which stage of infection is visualized), leaves were cut and placed with the cut side in a solution of 1 mg ml^{-1} DAB for approximately 5 h. Subsequently, the leaves were destained [0.15% trichloroacetic acid (w/v) in ethylalcohol/chloroform (4:1 (v/v)]. The solution was changed once during the next 48 h of incubation. Leaf segments were stored in 50% glycerol.
   Staining of fungal structures and microscopy was done as described by Hückelhoven and Kogel (1998): To stain fungal structures for bright-field microscopy, leaves were incubated in 10% blue ink (v/v, Pelikan 4001; Pelikan, Hannover, Germany) in 25% acetic acid for 1 min followed by a washing step to remove excess ink. Autofluorescence was observed by fluorescence microscopy (excitation wavelength 485 nm). For cytological studies, an Axioplan microscope (Zeiss, Jena, Germany) was used. For quantification of interaction types, one hundred or more attacked short cells (cell type A and B of

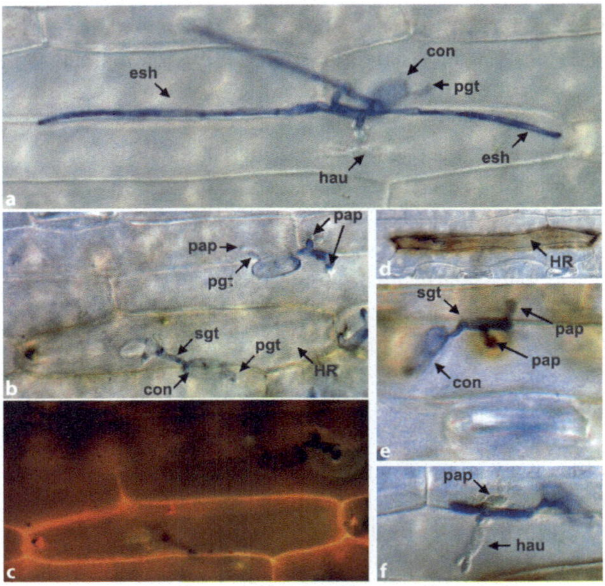

Fig. 22.3 Interaction of the powdery mildew fungus *B. graminis* f. sp. *hordei* with its host plant *Hordeum vulgare*. Shown are interaction sites at 32 h (**d**, **e**, **f**), 48 h (**b**, **c**) and 72 h (**a**) after inoculation with the pathogen *B. graminis* f.sp. *hordei*. After formation of a primary germtube (*pgt*) on the surface of the leaf, conidia (*con*) of the pathogen form a secondary germtube (*sgt*) that penetrates the epidermal leaf cell (**a**, **b**). **a** Overview of a successful penetration, with the fungus developing its nutrition organ, the haustorium (*hau*) and elongated secondary hyphae (*esh*) spreading on the leaf surface; **b** and **c** show the same cell as bright-field (**b**) and fluorescence (**c**) images. Active responses of the plant can stop the biotrophic pathogen from spreading through the plant, either by local cell death, resulting from a hypersensitive reaction (HR) of the penetrated cell (**b**, **c**, **d**) or by local fortifications of the cell wall at the site of attempted penetration (papilla = *pap*; **b**, **e**). Sites of H_2O_2 accumulation are detected by staining with 3,3-diaminobenzidine (DAB), as can be seen in **b** and **d** as the brown staining of the attacked cell and in **e** as the brown stain surrounding the papillae. Autofluorescence is visible at sites of HR (**c**), as phenolic cell wall components accumulate

the epidermis, according to Koga et al. 1990) were scored per leaf. Cellular responses to powdery mildew attack were categorized by counting cells showing (1) an active defence response, (HR, visible as whole cell autofluorescence, DAB staining), (2) a local defence stopping a penetration attempt (non-penetrated cell, visible as the formation of cell wall appositions), or (3) a successful penetration (formation of a haustorium; Fig. 22.3).

22.4.3 Assessment of the Antioxidant Capacity of *P. indica*-Infested Roots

The protective activity by *P. indica* against root pathogens with necrotrophic nourishment strategies prompted us to analyse the antioxidant status of infested roots. Ascorbate levels were consistently higher at one, two and three weeks after root infestation with *P. indica*, while levels of dehydroascorbate (DHA) were reduced. At the same time, activity of ascorbate recycling dehydroascorbate reductase (DHAR) increased. Concomitantly, slightly enhanced total glutathione concentrations and glutathione reductase activities were observed (Waller et al. 2005). It can be reasoned that higher antioxidant levels protect roots from cell death provoked by the root pathogens *F. culmorum* and *C. sativus*. Because production of reactive oxygen species and host cell killing is a prerequisite for successful fungal development and pathogenesis of necrotrophic fungi (Govrin and Levine 2000), we hypothesize that higher antioxidant capacity, such as elevated ascorbate levels, could cause the observed reduction of necrotrophic pathogens in the barley root.

22.4.4 Gene Expression Induced by *P. indica* in Barley Leaves

To gain information on the nature of *P. indica*-induced systemic protection of leaves against powdery mildew infection, we analysed the expression of "marker genes", indicative of specific resistance pathways. Interestingly, a number of genes typically associated with IR are not induced in the interaction with *P. indica*. Genes tested include *pathogenesis-related protein* 1 (*PR 1*), *pathogenesis-related protein* 5 (*PR 5*), *barley chemical induced protein* 1 (*BCI 1*; Beßer et al. 2000), and *jasmonate induced protein* 23 (*JIP 23*; Hause et al. 1996). As barley leaves do not show a constitutive up-regulation of typical marker genes for SA and JA, it is possible that other signalling pathways are involved in inducing systemic resistance after *P. indica* infestation of the roots (Waller et al. 2005). To elucidate the *P. indica* mediated IR mechanism, future strategies include screening of the Affymetrix Barley 1 gene chip (Close et al. 2004; Affymetrix, Santa Clara, Calif., USA) and custom-made microarrays with subtracted cDNA libraries enriched in *P. indica*-induced transcripts.

22.5 Conclusions

Cereals provide the staple crops for feeding a growing world population. Different approaches have to be taken to provide a stable harvest of these crops. Along with high yields, resistance against abiotic stress and pathogens is a prime goal. Identification of *P. indica*, an axenically cultivatable endophyte with the ability to protect the plant systemically from pathogens is an important step towards a broad-range application of more efficient biological measures in agriculture. Understanding the molecular mechanism mediated by *P. indica* will enable us to envisage new approaches to ensure healthy plants producing stable harvests. The methods presented in this chapter provide the means to analyse these mechanisms in all interactions of cereals with beneficial microbes.

Acknowledgements

The authors thank Ralph Hückelhoven for providing Fig. 22.1b. Work in our laboratory was supported by the Deutsche Forschungsgemeinschaft (grant FOR 343).

References

Azcón-Aguilar C, Barea JM (1996) Arbuscular mycorrhizas and biological control of soil-borne plant pathogens – an overview of the mechanisms involved. Mycorrhiza 6:457–464

Beßer K, Jarosch B, Langen G, Kogel K-H (2000) Expression analysis of genes induced in barley after chemical activation reveals distinct disease resistance pathways. Mol Plant Pathol 1:277–286

Blee KA, Anderson AJ (2000) Defence responses in plants to arbuscular mycorrhizal fungi. In: Podila GK, Douds DD (eds) Current advances in mycorrhizae research. APS, St. Paul, pp 27–44

Borowicz VA (2001) Do arbuscular mycorrhizal fungi alter plant–pathogen relations? Ecology 82:3057–3068

Chern MS, Fitzgerald HA, Canlas PE, Navarre DA, Ronald PC (2005) Over-expression of a rice NPR1 homologue leads to constitutive activation of defense response and hypersensitivity to light. Mol Plant Microbe Interact 18:511–520

Close TJ, Wanamaker SI, Caldo RA, Turner SM, Ashlock DA, Dickerson JA, Wing RA, Muehlbauer GJ, Kleinhofs A, Wise RP (2004) A new resource for cereal genomics: 22K barley genechip comes of age. Plant Physiol 134:960–968

Cordier C, Pozo MJ, Barea JM, Gianinazzi S, Gianinazzi Pearson V (1998) Cell defense responses associated with localized and systemic resistance to *Phytophthora parasitica* induced in tomato by an arbuscular mycorrhizal fungus. Mol Plant Microbe Interact 11:1017–1028

Dehne H-W (1987) Zur Nutzung der VA Mykorrhiza als Antistressfaktor. Angew Bot 61:135–143

Dong X (2004) NPR1, all things considered. Curr Opin Plant Biol 7:547–552

Gernns H, Alten HV, Poehling HM (2001) Arbuscular mycorrhiza increased the activity of a biotrophic leaf pathogen – is a compensation possible? Mycorrhiza 11:237–243

Govrin EM, Levine A (2000) The hypersensitive response facilitates plant infection by the necrotrophic pathogen *Botrytis cinerea*. Curr Biol 10:751–757

Harrison MJ (2005) Signaling in the arbuscular mycorrhizal symbiosis. Annu Rev Microbiol 59:19–42

Hause B, Fester T (2005) Molecular and cell biology of arbuscular mycorrhizal symbiosis. Planta 221:184–196

Hause B, Demus U, Teichmann C, Parthier B, Wasternack C (1996) Developmental and tissue-specific expression of JIP-23, a jasmonate-inducible protein of barley. Plant Cell Physiol 37:641–649

Hückelhoven R, Kogel K-H (1998) Tissue-specific superoxide generation at interaction sites in resistant and susceptible near-isogenic barley lines attacked by the powdery mildew fungus (*Erysiphe graminis* f. sp. *hordei*). Mol Plant Microbe Interact 11:292–300

Koga H, Bushnell WR, Zeyen RJ (1990) Specificity of cell type and timing of events associated with papilla formation and the hypersensitive reaction in leaves of *Hordeum vulgare* attacked by *Erysiphe graminis* f.sp. *hordei*. Can J Bot 68:2344–2352

Kogel K-H, Langen G (2005) Induced resistance and gene expression in cereals. Cell Microbiol 7:1555–1564

Kogel K-H, Beckhove U, Dreschers J, Münch S, Rommé Y (1994) Acquired resistance in barley. Plant Physiol 106:1269–1277

Kormanik PP, McGraw A-C (1982) Quantification of vesicular–arbuscular mycorrhizae in plant roots. In: Schenck NC (ed) Methods and principles of mycorrhizal research. American Phytopathological Society, St. Paul, pp 37–45

Mou Z, Fan W, Dong X (2003) Inducers of plant systemic acquired resistance regulate npr1 function through redox changes. Cell 113:935–944

Peskan-Berghöfer T, Markert C, Varma A, Oelmüller R (2004) Association of *Piriformospora indica* with *Arabidopsis thaliana* roots represents a novel system to study beneficial plant–microbe interactions and involves early plant protein modifications in the endoplasmic reticulum and at the plasma membrane. Physiol Plant 122:465–477

Pham GH, Singh A, Malla R, Kumari R, Prasad R, Sachdev M, Rexer K-H, Kost G, Luis P, Kaldorf M, Buscot F, Herrmann S, Peskan T, Oelmüller R, Saxena AK, Declerck S, Mittag M, Stabentheimer E, Hehl S, Varma A (2004) Interaction of *Piriformospora indica* with diverse microorganisms and plants. In: Varma A, Abbott LK, Werner D, Hampp R (eds) Plant surface microbiology. Springer, Berlin Heidelberg New York, pp 237–265

Pieterse CMJ, Van Wees SCM, Van Pelt JA, Knoester M, Laan R, Gerrits H, Weisbeek PJ, Van Loon LC (1998) A novel signaling pathway controlling induced systemic resistance in *Arabidopsis*. Plant Cell 10:1571–1580

Pozo MJ, Cordier C, Dumas-Gaudot E, Gianinazzi S, Barea JM, Azcon-Aguilar C (2002) Localized versus systemic effect of arbuscular mycorrhizal fungi on defence responses to *Phytophthora* infection in tomato plants. J Exp Bot 53:525–534

Schardl CL, Leuchtmann A, Spiering MJ (2004) Symbioses of grasses with seedborne fungal endophytes. Annu Rev Plant Biol 55:315–340

Schüssler A, Schwarzott D, Walker C (2001) A new fungal phylum, the Glomeromycota: phylogeny and evolution. Mycol Res 105:1413–1421

Shaul O, Galili S, Volpin H, Ginzberg I, Elad Y, Chet I, Kapulnik Y (1999) Mycorrhiza-induced changes in disease severity and PR protein expression in tobacco leaves. Mol Plant Microbe Interact 12:1000–1007

Sticher L, Mauch-Mani B, Métraux JP (1997) Systemic acquired resistance. Annu Rev Phytopathol 35:235–270

Thordal-Christensen H, Zhang ZG, Wei YD, Collinge DB (1997) Subcellular localization of H2O2 in plants – H_2O_2 accumulation in papillae and hypersensitive response during the barley–powdery mildew interaction. Plant J 11:1187–1194

Varma A, Verma S, Sudha, Sahay N, Bütehorn B, Franken P (1999) *Piriformospora indica*, a cultivable plant-growth-promoting root endophyte. Appl Environ Microbiol 65:2741–2744

Varma A, Singh A, Sudha, Sahay NS, Sharma J, Roy A, Kumari M, Rana D, Thakran S, Deka D, Bharti K, Franken P, Hurek T, Blechert O, Rexer K-H, Kost G, Hahn A, Maier W, Walter M, Strack D, Kranner I (2000) *Piriformospora indica*: an axenically cultureable mycorrhiza-like endosymbiotic fungus. In: Hock B (ed) The Mycota IX. Springer, Berlin Heidelberg New York, pp 125–150

Verma S, Varma A, Rexer K-H, Hassel A, Kost G, Sarbhoy A, Bisen P, Bütehorn B, Franken P (1998) *Piriformospora indica*, gen. et sp. nov., a new root-colonizing fungus. Mycologia 90:896–903

Waller F, Achatz B, Baltruschat H, Fodor J, Becker K, Fischer M, Heier T, Hückelhoven R, Neumann C, von Wettstein D, Franken P, Kogel K-H (2005) The endophytic fungus *Piriformospora indica* reprograms barley to salt stress tolerance, disease resistance and higher yield. Proc Natl Acad Sci USA 102:13386–13391

Weiß M, Selosse M-A, Rexer K-H, Urban A, Oberwinkler F (2004) Sebacinales: a hitherto overlooked cosm of heterobasidiomycetes with a broad mycorrhizal potential. Mycol Res 108:1003–1010

Wiberg A (1974) Genetical studies of spontaneous sources of resistance to powdery mildew in barley. Hereditas 77:89–148

23 Members of Sebacinales Confer Resistance Against Heavy Metal Stress in Plants

N. Hahn, A. Varma, R. Oelmüller, and I. Sherameti

23.1 Introduction

We study the effect of endophytic fungi on the protection of plants against heavy metal stress. Co-cultivation of several members of the Sebacinales with *Loliom perenne*, *Festuca rubra rubra*, barley and *Arabidopsis* confers resistance to high concentrations of Cd^{2+}. We are using molecular tools to understand the basis of this resistance, using *Arabidopsis* as a model system. Here, we describe protocols which allow the identification of genes and proteins which are involved in conferring Cd^{2+} resistance in *Arabidopsis*. Genes which are differentially expressed in response to Cd^{2+} treatments in *Arabidopsis* roots in the presence and absence of endophytic fungi can be identified by microarray or differential display technics. Further, the separation of protein extracts from differentially treated tissues on two-dimensional gels, and the use of mass spectrometry for the identification of protein spots which differ in their intensity under the different conditions, allow the identification of proteins which are involved in this scenario.

23.2 Scientific Background

Cd^{2+}, a non-essential heavy metal pollutant of the environment, derives from various agricultural, mining or industrial activities as well as car exhaust gases (Foy

Nadin Hahn, R. Oelmüller, I. Sherameti: Institute of General Botany and Plant Physiology, University of Jena, Dornburger Strasse 159, D-07743 Jena, Germany,
email: sherameti@yahoo.com

A. Varma: Amity Institute of Microbial Sciences, Amity University, Sector 125, Noida 201303, India, email: ajitvarma@aihmr.amiity.edu

Soil Biology, Volume 11
Advanced Techniques in Soil Microbiology
A. Varma, R. Oelmüller (Eds.)

et al. 1978). Because Cd^{2+} is highly soluble in water and thus rapidly distributed in aquatic ecosystems (Lockwood 1976), it exerts an enormous toxicity. Plants acquire Cd^{2+} mainly from contaminated water through the root system. Above a certain Cd^{2+} level, toxic effects become visible. Chlorosis, for instance, may be caused by iron deficiency or uptake of high Cd^{2+} levels, because both uptake and distribution of heavy metals, in particular Cd^{2+}, in the plants interact with the iron metabolism (Haghiri 1973; Root et al. 1975; Siedlecka 1999). Furthermore, Cd^{2+} appears to cause phosphorus deficiencies and manganese transport problems (Godbold and Hüttermann 1985; Guerinot and Eiche 1999) and interferes with the uptake, transport and cellular availability of several other elements, such as Ca^{2+}, Mg^{2+}, PO_4^{3-} or K^+. Cd^{2+} also inhibits enzyme activities (Lockwood 1976; Marschner 1995), causes chromosomal abberations (Avanzi 1999) and blocks cell division and proliferation (Rosas et al. 1984). Many studies also contribute to the understanding of the cellular and subcellular localization of Cd^{2+} and its distribution throughout the plant (c.f. Küpper et al. 2000; Ager et al. 2002).

The uptake is only poorly understood. Studies with different species suggest that Cd^{2+} can be transported together with Zn^{2+} and Fe^{2+} (Korshunova et al. 1999; Moreau et al. 2002). There is also evidence that Cd^{2+} uptake is mediated by a transporter system for Mn^{2+} (Himeno et al. 2002). The increase in Cd^{2+} in the external medium causes an increase in Mn^{2+} uptake and translocation to the shoots, further evidence that Cd^{2+} and Mn^{2+} are co-transported (Ramos et al. 2002). Most of the Cd^{2+} accumulates in leaves in the cell wall fraction and this accumulation is fairly independent of the Cd^{2+} level in the nutrient solution (Ramos et al. 2002). Within the cell, the lowest Cd^{2+} concentration is found in the chloroplasts. In *Arabidopsis*, Cd^{2+} is preferentially sequestered within the trichome of the leaf surface (Ager et al. 2002).

Detoxification of Cd^{2+} within the cell occurs mainly by phytochelatins (Grill et al. 1985, 1987; Cobbett et al. 1998; Cobbett 2001). Phytochelatins are synthesized from glutathione. Phytochelatin-deficient mutants of *Arabidopsis* have confirmed the important role of glutathione as a substrate for phytochelatin biosynthesis and its role in Cd^{2+} detoxification. The *A. thaliana* CAD1 (AtPCS1) gene encodes a phytochelatin synthase and cad1 mutants are Cd^{2+} hypersensitive (Cazale and Clemens 2001). Two copies of this gene are present in the *Arabidopsis* genome and both are expressed (Cazale and Clemens 2001). There are several reports demonstrating that elevated levels of Cd^{2+} stimulate antioxidant enzyme activities, such as glutathione reductase and superoxide dismutases (c.f. Fornazier et al. 2002) or inhibitors of antioxidative enzymes such as superoxide dismutases or peroxidases (Gallego et al. 1999; Mascher et al. 2002).

More recently, substantial progress has been made in understanding heavy metal homeostasis in plants. Heavy metal P-type ATPase transporters (HMA; Williams and Mills 2005) belong to an ancient family of metal pumps with diverse functions in plants. They play an essential role in zinc homeostasis in *Arabidopsis* (Hussain et al. 2004). Three of these transporters (HMA2, 3, 4) are closely related to each other. HMA2 and HMA4 expression occurs predominantly in the vascular tissue of roots, stems and leaves, and they play a role in zinc translocation. *Hma2* and *hma3* mutations confer increased sensitivity to Cd^{2+} (Hussain et al. 2004). HMA4 was able to complement an *Escherichia coli* mutant impaired in Zn^{2+}, but not in Cu^{2+} homeostasis. Heterologous expression of HMA4 in *Saccharomyces* made the yeast more resistant to Cd^{2+} (Mills et al. 2003). A null mutant of HMA4 in *Arabidopsis* exhibited a lower translocation of Zn^{2+} and Cd^{2+} from the root to the shoot, while an overexpressor displayed an increase in the Zn^{2+} and Cd^{2+} content (Verret et al. 2005). Bernard et al. (2004) could show that the *Thlaspi caerulescens* homolog of HMA4 is highly expressed in a Cd^{2+} hyperaccumulator.

23.3 Differential Display to Understand Cd^{2+} Resistance Mediated by Endophytic Fungi

Differential display technology is described in Chapter 20 in this book. Using this technique we have identified several Cd^{2+}-regulated genes in *Arabidopsis* roots. One of these genes (accession number AF412407) codes for HMA4 (Fig. 23.1). Interestingly, the expression level of this gene in *Arabidopsis* roots co-cultivated with the endophytic fungus *Piriformospora indica* is two times lower than in control plants, although these plants were grown without Cd^{2+} (Fig. 23.2). This might explain why an endophyte can confer heavy metal resistance to plants.

23.4 Studies on Protein Level

Two-dimensional gel electrophoresis is used to identify proteins in *Arabidopsis* roots which differ in their amounts after different treatments [e.g. in the presence or absence of *P. indica* and/or Cd^{2+} (100–200 μmol)]. For better analysis we separate soluble and membrane-associated proteins. Soluble protein extracts are obtained after homogenation of roots in a buffer containing 100 mM Tris, pH 7.0, 10 mM $MgCl_2$, 2,2% SDS 1 mM β-mercapto-ethanol. The slurry is first centrifuged at 40 000 g for 20 min, before high-speed centrifugation at 100 000 *g* for 10 min. After determination of the protein concentration, the supernatant is used for two-dimensional gel electrophoresis.

The pellet of the last centrifugation is used for the separation of membrane proteins. The membranes are resuspended in 100 mM Tris, pH 7.0, 10 mM $MgCl_2$, 10 mM mercapto-ethanol and kept at 75 °C for 20 min.

1. To precipitate the membrane proteins, 40–60 μg protein in 100 μl buffer is used.

```
                        10        20        30        40        50        60
                ....*....|....*....|....*....|....*....|....*....|....*....|
consensus     1 MMKTTLKVEGMTCGHCVKTVEKALEEVDG-----------VASVDVDLEKGT-------- 41
query        16 LQKSYFDVLGICCTSEVPIIENILKSLDG-----------VKEYSVIVPSRT-------- 56
1JK9 B       5 DTYEATYAIPMHCENCVNDIKACLKNVPGinslnfdieqqIMSVESSVAPSTiintlrnc 64
gi 19552399   7 LKQTTLRSDEFSCPSCVSKIENKLNGLDG-----------VDNAEVKFSSGR-------- 47
gi 2498247    1 -MKATFQVPSITCNHCVDKIEKFVGEIEG-----------VSFIDVSVEKKS-------- 40
gi 20090199   1 MGETTLVVDDMICKYCRDTVTRLITSING-----------VSRVSVNVPERT-------- 41
gi 15643089   1 -MRYVLYVPDISCNHCKMRISKALEELG------------VKNYEVSVEE---------- 37
gi 13541083   1 MKTIKMRIYGMTCNDCVATVEKGLKSVDG-----------VLWVSVSLPDGS-------- 41
gi 16082329   1 MKKVEMKIYGMTCDDCAVTVKNGLESVDG-----------VLSAEVSLPEKR-------- 41

                        70        80        90       100       110       120
                ....*....|....*....|....*....|....*....|....*....|....*....|
consensus    42 ---------------------------ATVTFDSNKVDIEAIIEA-------------IE 61
query        57 ---------------------------VIVVHDSLLISPFQIAKA-------------LN 76
1JK9 B      65 gkdaiirgagkpnssavailetfqkytIDQKKDTAVRGLARIVQVgenktlfditvngVP 124
gi 19552399  48 ---------------------------ILVDHDPSKVSIKDLVAA-------------VA 67
gi 2498247   41 ---------------------------VVVEFDAP-ATQDLIKEA-------------LL 59
gi 20090199  42 ---------------------------VNVTYDSRITDSHVIRMT-------------LL 61
gi 15643089  38 ---------------------------KKVVVET--ENLDSVLKK-------------LE 55
gi 13541083  42 ---------------------------AVVKVDDS-VDPEKLEDAe-----------vFK 62
gi 16082329  42 ---------------------------AEVVIDESKISPEKLEDAr-----------vFK 63

                       130
                ....*....|
consensus    62 DAGYKVEEIK 71
query        77 EARLEANVRv 86
1JK9 B     125 EAGNYHASIH 134
gi 19552399  68 EVGYTAKPSA 77
gi 2498247   60 DAGQEVV--- 66
gi 20090199  62 EAGYKNIWET 71
gi 15643089  56 EIDYPVESYQ 65
gi 13541083  63 KTRYRGEVRD 72
gi 16082329  64 VTRYRGEVRK 73
```

Fig. 23.1 The "Conserved Domain Database" at the NCBI server recognizes a copper chaperone domain. The figure shows an alignment of a consensus sequence, the gene identified in this paper and several typical proteins from various organisms which share the conserved region. The localization of conserved residues is visualized by *bold* letters. *Lower case* letters indicate gaps to optimize the alignment. Numbers on the *right* and *left* of the column indicates the position of the amino acid as deposited in the Databank. *gi 19552399* Copper chaperone from *Corynebacterium glutamicum*. *gi 2498247* Copper ion binding protein from *Helicobacter pylori*. *gi 20090199* Heavy metal-associated protein from *Methanosarcina acetivorans*. *gi 15643089* Heavy metal binding protein from *Thermotoga maritima*. *gi 13541083* Copper chaperone from *Thermoplasma volcanium*. *gi 16082329* Mercuric resistance operon regulatory protein merP related protein from *Thermoplasma acidophilum*

2. 400 µl methanol is added, vortexed and– after centrifugation– the pellet is recovered.
3. 100 µl chloroform is added, vortexed, an additional 200 µl water was added, vortexed again and– after centrifugation– the upper phase is carefully removed.
4. 300 µl methanol is added to the rest, vortexed and centrifuged again. The pellet contains the membrane-associated proteins.
5. The pellet is washed twice with methanol (500 µl), dried in a Speed-Vac and the proteins resuspended in the appropriate sample buffer for 2D gel electrophoresis.

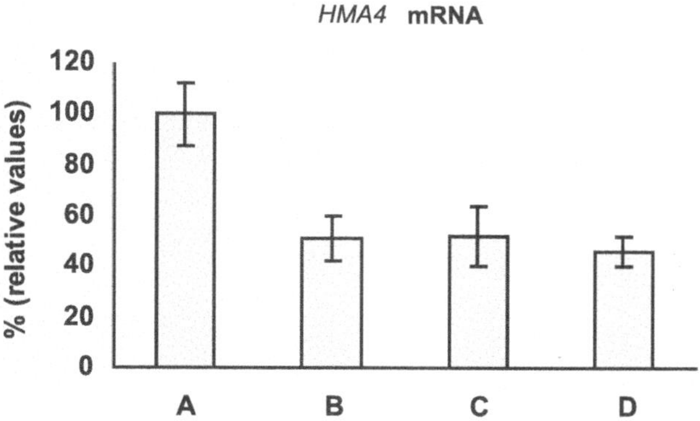

Fig. 23.2 mRNA levels of HMA4 in 14-day-old *Arabidopsis* roots. *A* Control, seedlings without treatment. *B* Seedlings co-cultivated with *Piriformospora indica*. *C* Seedlings cultivated on 200 μM Cd^{2+} from day 9 to day 14. *D* Seedlings cultivated on 200 μM Cd^{2+} from day 9 to day 14 in the presence of *Piriformospora indica*. The mRNA levels for the control seedlings (*A*) were taken as 100. Based on four independent microarrays

23.4.1
Two-Dimensional Gel Electrophoresis, Preparation of Proteins

1. 180 μg protein in 100 μl extraction buffer is precipitated with methanol, dried and resuspended in 380 μl of sample buffer [8 M urea, 2 M thiourea, 30 mM dithioereitrol, 4% (w/v) CHAPS, 20 mM Tris base, 0.5% bromophenol blue, 0.5% IPE buffer (pH 3–10, Amersham Pharmacia), 0.05% dodecyl-β-D-maltoside].
2. 350 μl of the supernatant is added to 1.75 ml of 0.5% (v/v) IPE buffer for isoelectric focusing (Amersham Pharmacia, Freiburg, Germany).
3. For the second dimension the gel system of Schäger and von Jagow (1987) is used.
4. Gels are stained with silver (Fig. 23.3).

23.4.2
Mass Spectrometry, Preparation of Samples by Tryptic Digestion

Silver-stained gel spots are excised and the proteins extracted into 500 μl of 50 mM ammonium bicarbonate, supplemented with 60 ng/μl trypsin. After lyophilization, the pellet was resuspended in 5 μl of water/acetonitrile/formic acid (95:5:0.1) prior to LC-MS analysis. Peptide analyses, analyte sampling, chromatography and acquisition of data were performed on a LC (Famos-Ultimate; LC-Packings) coupled with an LCQ Deca XP ITMS according to the manufacturer's instructions. Using these techniques we can identify several proteins which are up- or down-regulated by the endophytic fungus *P. indica* in the absence or

Fig. 23.3 Two-dimensional gels from root plasma membrane of seedlings grown in the absence (*left*) or presence (*right*) of 100 μm Cd^{2+}. Protein spots which differ in the two preparations are marked

presence of Cd^{2+} and which might be involved in conferring heavy metal resistance in plants. Analysis of null mutants (cf. Chapter 20) and over-expressers in the genes for these proteins demonstrates whether they play a role in *P. indica*-mediated heavy metal resistance.

References

Ager FJ, Ynsa MD, Dominguez-Solis JR, Gotor C, Respaldiza MA, Romero LC (2002) Cadmium localization and quantification in the plant *Arabidopsis thaliana* using micro-PIXE. Nucl Instr Methods Phys Res B Beam Interact Mater Atoms 189:494–498

Avanzi M (1999) Observazioni sull'attivita citologica di alcuni composti chimic. Caryologica 3:234–248

Bernard C, Roosens N, Czernic P, Lebrun M, Verbruggen N (2004) A novel CPx-ATPase from the cadmium hyperaccumulator *Thlaspi caerulescens*. FEBS Lett 569:140–148

Cazale AC, Clemens S (2001) *Arabidospis thaliana* expresses a second functional phytochelatin synthase. FEBS Lett 507:215–219

Cobbett CS (2001) Heavy metal detoxification in plants. Phytochelatin biosynthesis and function, IUBMB Life 51:183–188

Cobbett CS, May MJ, Howden R, Rolls B (1998) The glutathione-deficient cadmium-sensitive mutant, cad2-1, of *Arabidopsis thaliana* is deficient in β-glutamylcysteine synthase. Plant J 16:73–78

Fornazier RF, Ferreira RR, Vitoria AP, Molina SMG, Lea PJ, Azevedo RA (2002) Effect of cadmium on antioxidant enzyme activities in sugar cane. Biol Plant 45:91–97

Foy C, Chaney AL, White MC (1978) The physiology of metal toxicity in plants. Annu Rev Plant Physiol 29:511–566

Gallego SM, Benavides MP, Tomaro MC (1999) Effect of cadmium ions on antioxidant defense system in sunflower cotyledons. Biol Plant 42:49–55

Godbold DL, Hüttermann A (1985) Effect of Zn, Cd and Hg on root elongation of *Picea abies* (Karst.) seedlings and the significance of these metals to forest die-back. Environ Pollut 38:295–303

Grill E, Winnacher EL, Zenk MH (1985) Phytochelatins: the principle heavy-metal complexing peptides of higher plants. Science 230:674–676

Grill E, Winnacker EL, Zenk MH (1987) Phytochelatins, a class of heavy-metal-binding peptides from plants, are functionally analogous to metallothionines. Proc Natl Acad Sci USA 84:439–443

Guerinot ML, Eiche D (1999) Zeroing in on zink uptake in yeast and plants. Curr Opin Plant Biol 2:244–249

Haghiri H (1973) Cadmium uptake by plants. J Environ Qual 2:93–96

Himeno S, Yanagiya T, Enomoto S, Kondo Y, Imura N (2002) Cellular cadmium uptake mediated by the transport system of manganese. Tohoku J Exp Med 196:43–50

Hussain D, Haydon MJ, Wang Y, Wong E, Sherson SM, Young J, Camakaris J, Harper JF, Cobbett CS (2004) P-type ATPase heavy metal transporters with roles in essential zinc homeostasis in *Arabidopsis*. Plant Cell 16:1327–1329

Korshunova YO, Eide D, Clark WG, Guerinot ML, Pakrasi HB (1999) The IRT1 protein from *Arabidopsis thaliana* is a metal transporter with a broad substrate range. Plant Mol Biol 40:137–144

Küpper H, Lombi E, Zhao-Fang J, McGrath S-P (2000) Cellular compartimentation of cadmium and zinc in relation to other elements in the hyperaccumulator *Arabidopsis halleri*. Planta 212:75–84

Lockwood MP (1976) Effects of pollutants on aquatic organisms. Cambridge University Press, New York

Marschner H (1995) Mineral nutrition of higher plants. Academic, London, pp 189–179

Mascher R, Lippmann B, Holzinger S, Bergmann H (2002) Arsenate toxicity: effects on oxidative stress response molecules and enzymes in red clover plants. Plant Sci 21:1–10

Moreau S, Thomson TM, Kaiser BN, Trevaskis B, Guerinot ML, Udvardi MK, Puppo A, Day DA (2002) GmZIP1 encodes a symbiosis-specific zinc finger transporter in soybean. J Biol Chem 277:4738–4746

Mills RF, Krijger GC, Baccarini PJ, Hall JL, Williams LE (2003) Functional expression of AtHMA4, a P1B-type ATPase of the Zn/Co/Cd/Pb subclass. Plant J 35:164–176

Ramos I, Esteban E, Lucena JJ, Garate A (2002) Cadmium uptake and subcellular distribution in plants of *Lactuca* sp.– Mn interaction. Plant Sci 162:761–767

Root RA, Miller RJ, Koeppe DE (1975) Uptake of cadmium– its toxicity and effect on the iron-to-zinc ratio in hydroponically grown corn. J Environ Qual 4:473–476

Rosas M, Carbajal ME, Gomez-Arroyo S, Belmont R, Vilalogos-Pietrini R (1984) Cytogenetic effects on cadmium accumulation on water hyacinth (*Eichoreia cratsiper*) Environ Res 33:386–395

Schägger H, Jagow G von (1987) Tricine-sodium dodecyl sulphate polyacrylamide gel electrophoresis for the separation of proteins in the rage of 1 to 100 kDa. Anal Biochem 166:368–379

Siedlecka A, Krupa Z (1999) Cd/Fe-interaction in higher plants– its consequences for the photosynthetic apparatus. Photosynthetica 36:321–331

Verret F, Gravot A, Auroy P, Preveral S, Forestier C, Vavasseur A, Richaud P (2005) Heavy metal transport by AtHMA4 involves the N-terminal degenerated metal binding domain and the C-terminal His11 stretch. FEBS Lett 579:1515–1522

Williams LE, Mills RF (2005) (1B)-ATPases– an ancient family of transition metal pumps with diverse functions in plants. Trends Plant Sci 10:491–502

24 Screening of Plant Growth-Promoting Rhizobacteria

C.S. Nautiyal and S.M. DasGupta

24.1 Introduction

Microorganisms are essential for the maintenance of sustainable ecosystems and microbial biodiversity. Microorganisms include bacteria, actinomycetes, and fungi, occupy an important niche in every ecosystem, and are important in recycling the elements in nature and in the decomposition of organic matter. Of the microorganisms, bacteria are the most common type, possibly because they grow rapidly and have the ability to utilize a wide range of substances as either carbon or nitrogen sources. The most prominent group of bacteria present in the rhizosphere are the non-sporulating gram-negative rods. Fungi and actinomycetes are also present but their populations are smaller than the bacteria.

Colonization of the plant root system is the very first step in nearly all interactions between plants and soil-borne microbes. The region of contact between root and soil where soil is affected by roots was designated as the "rhizosphere" by Hiltner in 1904. He believed that the rhizosphere microorganisms play an important role in plant development. Sorensen in 1997 defined the rhizosphere as the volume of soil surrounding the plant roots in which bacterial growth is stimulated.The rhizosphere has attracted much interest, since it is a habitat in which several biologically important process and interactions take place.

The rhizosphere refers in general to the portion of soil adjacent to the roots of living plants. It supports a diverse and densely populated microbial community, and is subjected to chemical transformations caused by the effect of root exudates and metabolites of microbial degradation. The bacterial communities as-

C. Shekhar Nautiyal: Plant–Microbe Interaction Laboratory, National Botanical Research Institute, Rana Pratap Marg, Lucknow, 226001, email: csnnbri@yahoo.com

Sangeeta Mehta DasGupta: Amity Institute of Microbial Sciences, Amity University, Uttar Pradesh, Sector 125, Noida, 201303

Soil Biology, Volume 11
Advanced Techniques in Soil Microbiology
A. Varma, R. Oelmüller (Eds.)

sociated with this microzone are thought to be determined by the quantity and composition of root exudates that serve as substrates for microbial growth. Root exudates can also selectively affect the growth of bacteria and fungi that colonize the rhizosphere by serving as selective growth substrates for soil microorganisms. These microbial associations may result in endophytic, symbiotic, associative, or parasitic relationships within the plant, depending on the type of microorganisms, soil nutrient status, and soil environment. The best known groups are symbiotic members of the family Rhizobiaceae, mycorrhizal fungi and plant growth-promoting rhizobacteria (PGPR).

24.2 Candidature for Being a Rhizobacteria

Plant growth-promoting rhizobacteria (PGPR) are free-living bacteria that have a beneficial effect on plants, as they enhance emergence, colonize roots, and stimulate growth. In addition to bacteria present on the root surface (rhizoplane) and in the rhizosphere, there are significant numbers of bacteria present in the root interior that are also beneficial for plant growth.

Rhizobacteria essentially possess the property of rhizosphere competence, meaning that they are able to compete for the habitat and nutrition in the rhizosphere and have the ability to outnumber the resident microflora of the rhizosphere. Microorganisms that can grow in the rhizosphere are ideal for use as bioinoculants, since the rhizosphere provides the front-line defense for roots against attack by pathogens. Bacterial root colonization comprises a series of steps: migration towards plant roots, attachment, distribution along the root, and finally growth and establishment of the population.

The relative rates of utilization of root exudates may determine different degrees of competence. Alternatively, dissimilarities in rhizosphere competence may be attributable to the differences in the extent of bacterial attachment to the root surface, especially since adhering cells presumably are more likely to be transported with the extending root or are closer to the source of exudates than are bacteria at some distance from the root. Adherence of a variety of bacterial species to roots has indeed been the subject of considerable study (Dazzo 1980). However, the extensive colonization of roots by some bacteria may simply reject their ability to survive in large numbers in the absence of the plant (Acea et al. 1988). In view of the potential benefits that can be gained by successful establishment of rhizobia, free-living N_2-fixing bacteria, species capable of controlling plant pathogens, and plant growth-promoting bacteria in the rhizosphere, and the inconsistent results obtained to date following inoculation with these microorganisms, it is important to establish the properties of bacteria that are responsible for rhizosphere competence. Findings suggest the importance of initial cell density in determining the extent of colonization of the rhizosphere. They also show that growth rate and attachment, at least alone, are not major factors in root colonization, although they may add to the rhizosphere competence of bacteria that can survive in large numbers in soil. Because of the potential value of inoculants added to soil or seeds in consistently increasing N_2 fixation, stimulating plant growth or resulting in the control of plant pathogens, additional study is warranted to establish more completely the attributes of bacteria that contribute to rhizosphere competence. Therefore, a study to evaluate several hypotheses to explain why bacteria differ in their capacity to colonize the root zone is important. The colonizing ability of a strain can at present be evaluated only in vivo. Sufficient information on the factors involved in this complex process is not yet available to enable an assessment to be made by investigating the biochemical or genetic makeup of a strain. However, through the use of genetic means, some factors which play a role in the colonization of root surfaces are being recognized.

24.3 Screening Methods

24.3.1 Criteria for Screening

The isolation and development of plant-beneficial bacterial strains applicable to a variety of crops, soils, and locations will depend upon the development of improved detection and screening procedures that more rapidly screen and identify beneficial strains (Nautiyal 1997a, b).

24.3.2 Selection of Screening Methods

Screening methods for microorganisms can be selected on the basis of their abundance and cultivability. For example those in abundance can directly be screened from rhizospheric samples, whereas those in lesser amounts can be labeled in situ with fluorescent dyes. Also, using random or specific primers, those difficult to culture or non-culturable can be screened. Thus it is important to select a screening method that gives the best yield in terms of bacterial count.

24.3.3 Classic Methods

24.3.3.1 Direct Screening from Soil

Bacteria which are in abundance can directly be screened from the soil adhering to the root by using the following technique. Soil adhering to the roots should be gently shaken onto a sterile paper of known weight and weighed again.Serial dilution of the soil samples (10% soil in 0.85% saline MilliQ water; MQW) should be plated on nutrient agar.Bacteria representative of the predominant morphologically distinct colonies present on the plates are selected at random and purified on minimal medium based on AT salts.

24.3.3.2 Screening from Rhizosphere

Roots should be thoroughly washed with tap water for 2 min to remove all the loosely adhering soil particles, followed by washing with sterile 0.85% (w/v) saline MQW. The roots are then macerated in 0.85% saline MQW with a mortar and pestle. Serial dilution of the root homogenate and soil samples (10% soil in 0.85% saline MQW) should then be plated on nutrient agar. Bacteria representative of the predominant morphologically distinct colonies present on the plates are selected at random and purified on minimal medium based on AT salts.

24.3.3.3 Elective Culture Methods

If specific chemical compounds are added to soil in the field or in the laboratory and incubated under specific conditions, the organisms capable of growing under those conditions multiply and come to comprise a greater percentage of the total microbiota. Alternatively, if specific inhibitory chemicals or incubation conditions are used, specific parts of the microbiota are decreased in overall percentage. This very simple concept is the basis for a large fraction of the industrial uses for soil microbiology, such as isolating hydrocarbon-degrading bacteria, isolating bacteria that degrade various pollutants such as pesticides, PCBs, organochlorines, isolating bacteria (including actinomycetes) or fungi that produce new antibiotics, and many other examples. Because of the diversity of the microbiota in soil, this technique can often be applied to isolate microorganisms with any desired property.

24.3.3.4
Selective Culture Media

All media used in microbiological laboratories are selective to some degree or another; there are no truly non-selective media. It is possible to make media and incubation conditions very selective by chemical and physical modifications. This is often very useful to isolate and count particular groups of bacteria or other organisms from soil samples.

There are routine ways to produce selective media:

1. Add compounds used by an organism as a nutrient source.
2. Omit compounds required by most other organisms (omit nitrates and other fixed nitrogen sources to isolate nitrogen fixing bacteria).
3. Add selectively biocidal compounds (penicillin to inhibit gram-positive bacteria, neomycin and streptomycin as general bacterial inhibitors, actidone and nystatin as general fungal inhibitors).
4. Change physical properties [pH, pE (redox potential), etc.].
5. Alter incubation conditions (temperature, water content, osmotic pressure, light, etc.).

Using combinations of these techniques, it is possible to design very selective media; *Pseudomonas* isolation medium is very specific for its named bacterial group. In theory, almost any physiological group of organisms can be cultured selectively. A single-stage isolation from a soil sample can be changed to a multi-stage isolation process by replica plating. Individual colonies are transferred in their original orientation on the plate by pressing a pad of sterile velvet cloth onto the surface of the plate, removing a small sample of each colony and pressing it onto the surface of a fresh plate. This can be a different growth medium so that only a part of the original population is able to grow on the new medium. In this way, progressive selective media can be used to isolate bacteria with combinations of properties. To enhance the detection of the particular group of microorganism under study, it is also possible to improve the diagnostic precision of the media by using some properties of the organisms (pigment, fluorescence under ultraviolet, biochemical reactions with extra, added substrates, etc.).

24.3.3.5
Non-Selective Media

So-called "non-selective" media are only media and incubation conditions designed to isolate as large a part of the microbiota in soil as is possible. Truly non-selective media do not exist. The least selective media today may isolate 1–10% of the total soil bacteria and maybe 5–15% of the fungal population of soils. The media used for bacteria and fungi are different.

24.3.4 Modern Methods

24.3.4.1 Fluorescence Methods

Many older methods using direct microscopic examination of soil samples are still in use today because of their simplicity. They are especially useful when examining smaller soil samples, such as pieces of organic materials or mineral grains. There are two main types of methods used to visualize the microorganisms in these samples: classic stains such as phenol aniline blue and fluorescent stains such as fluorescein isothiocyanate. The first can be examined after staining with any bright-field, white-light microscope, assuming that light can be transmitted through the object being examined. The second uses a stain that emits light at a visible wavelength when illuminated with far-violet or ultraviolet light. This can be incident illumination that does not have to pass through the object. The mercury arc lamp is a strong source of ultraviolet light that is filtered though an excitation filter (to exclude all but ultraviolet light) and passed to a dichroic mirror. This mirror is coated with a very thin metal film that reflects ultraviolet light and does not allow it to pass through. It does allow visible light to go through and not be reflected. The ultraviolet light is passed down through the objective lens and focussed onto the object from the top (which is why the object can be opaque). If the fluorescent dyes staining the microorganisms fluoresce in the visible spectrum, the emitted light is collected by the objective lens and passes through the dichroic mirror to the eyepiece lens and the eye of the observer. The eyepieces always contain a barrier filter (usually yellow) that prevents any of the ultraviolet light from reaching the eyes of the observer.

The most common fluorescent stains are acridine orange, fluorescein isothiocyanate (FITC), and rhodamine (fluoresces red). They react with parts of protein molecules – the sulfhydryl groups – and attach strongly to the protein molecules. Another example is calcofluor; it reacts with cellulose, chitin, and similar polysaccharides and is useful for staining fungi and Actinomycetes. It is also relatively non-toxic and can be used as a vital stain to examine living cells. Other fluorescent stains include europium chelate [europium (iii) thenoyltrifluoroacetonate] that stains nucleic acids and 4'-6'-diamidino-2-phenyl-indole (DAPI), ethidium bromide, and bisbenzimide (Hoechst 33258) that all stain DNA.

Another group of fluorescent stains are the fluorescent probes. They differ from FITC and rhodamine in that they are not fluorescent until they come into contact with the correct environment. Typically this is the lipid within microbial cells. Only then do they fluoresce and emit visible light. Examples of this group are DANSYL chloride and the 8-anilino-1-naphthalene sulfonic acid salts (Mg-ANS, Na-ANS). Their major advantage is that they can be applied to soil samples and immediately examined without removing excess, unreacted stain. FITC and rhodamine need extensive washing to remove unreacted stain.

24.3.4.2
Fluorescent Antibody and Related Methods (Immunofluorescence Methods)

The fluorescent antibody technique is the only one that can simultaneously locate and identify microorganisms in intact soil samples or sections.

The antibodies to microbial cells are produced by injecting the cells under study into a suitable animal (guinea pigs or rabbits are commonly used). After incubation, the animals produce antibodies to the microbial cells that can be isolated from serum samples of the animals. The antibodies are proteins and so can be reacted with FITC to produce FITC-antibody conjugates. These FITC-antibodies only adhere to the correct microbial cells if applied to a soil sample. When excess FITC-antibody conjugate has been removed by washing, only those microbial cells fluoresce and they can be simultaneously located and identified by epifluorescence microscopy (as for FITC staining).

It has been used in soil microbiology to identify nitrogen-fixing *Rhizobium* spp, *Bacillus* spp, various fungal genera such as *Aspergillus*, and a few *Actinomycetes.*

One problem is the relatively non-specific nature of many antibody preparations. Many bacteria in the same general taxonomic group have similar chemical structures on their cells and produce a complex of antibodies from those structures that overlap with the complex produced by the other similar bacteria. Thus if the antibody complex is used to form the conjugate with FITC, that FITC-antibody will cross-react with the other bacteria as well as with the target species. Usually this reaction will be weaker but still significant. One way to "purify" the complex is to remove the cross-reacting antibodies by reacting them with the actual bacterial cells from the unwanted cross-reacting species. Any common antibodies (the cross-reacting group) will be adsorbed onto the surfaces of the added cells and removed from the complex. The remaining antibodies are then much more specific to the target species.

A more recent modification of the method uses monoclonal or polyclonal antibodies produced in other microbial cells to obtain larger quantities of antibody for conjugation with FITC. Many of these antibodies are now available commercially from suppliers and some are available already conjugated and therefore labelled with FITC and/or rhodamine.

24.3.4.3
Enzyme-linked Immunosorbant Assays

Enzyme-linked immunosorbant assays (ELISAs) have found some use in soils when the population sought exceeds 10 000 cells/ml. The technique has been applied mainly to *Rhizobia* in soil and roots of legumes. The major difficulty is removal of the microbial cells from the substrates; both direct lysis in situ and removal of cells followed by lysis have been used.

24.3.5 Molecular Methods

24.3.5.1 Gene Probe and Nucleic Acid Hybridization

These techniques rely on detecting specific sequences of nucleic acids in the organisms under study. If the sequences used are carefully chosen to be diagnostic, this technique can find specific organisms in soils and other environmental samples. The gene probe is a short segment of nucleotides that binds specifically with the homologous sequence in the target microorganism. If the segment is labelled with radioactive ^{32}P, any binding to the target nucleotides can be detected by the presence of the radioactivity after reaction.

24.3.5.2 Polymerase Chain Reaction

The polymerase chain reaction (PCR) has recently been applied to microbial ecology. In this technique, extracted DNA is melted to form single strands, annealed with primers, and the DNA is extended from the primers by nucleotide addition using DNA polymerase enzyme. The primers are chosen to link to regions of DNA of interest (close to a diagnostic target sequence).

24.3.5.3 Bioluminescence Marker Genes

A bioluminescence marker gene is typically the *lux* gene of *Vibrio fischeri*. This gene causes photoluminescence in bacteria (emits light). If the gene can be inserted into the target organisms, they become photoluminescent and this property can be used to detect them and track their fate in soil and water samples. This technique has been used with *Escherichia coli* and *Pseuodomonas* target organisms. It has been extended by fusing other genes with the *lux* gene and inserting both into cells. Naphthalene degradation is promoted by a *nah* gene which has been combined with the *lux* gene to make a diagnostic pair. This can track both the bacteria and their activity in soil or rhizospheric samples.

24.4 Metagenomics

Metagenomics, the genomic analysis of a population of microorganisms that provides an access to the physiology and genetics of uncultured organisms, has emerged as a powerful technique in recent times. Metagenomic analysis involves isolating DNA from an environmental sample, cloning the DNA into a suitable vector, transforming the clones into a host bacterium, and screening the resultant transformants. Metagenomic analysis has several advantages over culture or PCR-based methods. For example, it: (a) provides access to uncultured microorganisms, (b) does not require prior knowledge of gene sequences, and (c) recovers complete genes.

Two methods are generally adopted for isolation of DNA: the direct lysis method and the cell extraction method. However, the direct analysis method has been reported to yield at least 10-fold more DNA than the cell lysis method. This method is based on the initial extraction of extracellular DNA with alkaline buffer, followed by the direct lysis of the cells in the soil by chemical and mechanical means and then quantitative extraction of released DNA. For purification, multiple electrophoresis runs are commonly employed.

Following purification, the metagenomic DNA is partially digested with restriction enzymes. Restriction digestion requires a starting DNA minimally three times longer than the desired insert. Then a library of the DNA based on the size required is prepared and a suitable vector is choosen for its insertion into a bacterial cell. There is a limited choice of vectors for cloning metagenomic DNA. The most commonly used vectors are: plasmids, fosmids, cosmids and bacterial artificial chromosomes (BACs). Plasmids are ideal for small-insert libraries (less than 15 kb). Fosmids and cosmids are suitable for libraries with moderate size inserts (38–52 kb). These vectors have much greater cloning efficiencies than BACs. Nonetheless, BAC is the choice for cloning for various reasons. BACs based on the *E. coli* factor can carry large inserts (>300 kb). Once a gene of interest is identified, phylogenetic anchors such as 16S rRNA gene and the archaeal DNA repair gene *rad*A can be searched in the flanking DNA to provide a link of phylogeny with the functional gene.

The applications of metagenomics are immense. The heart of this approach is that it provides us with access to the genome heterogeneity of both culture-dependent and culture-independent microorganisms. Metagenomics has already opened new avenues of research by enabling unprecedented analysis of genome and evolution in environmental contexts and providing access to far more microbial diversity. The early screening campaigns of metagenomic libraries centered around the cloning of genes encoding phylogenetically conserved molecular traits to explore microbial diversity.

24.5 Tracking of GEMs

In order to follow the fate of a genetically engineered microorganism (GEM) in the environment, it is necessary to detect and quantify it in time and space. Both the organism and the genetic information that constitutes the modification must be tracked simultaneously and independently so that both loss of the new information from the GEM and its possible lateral transfer to indigenous microorganisms can be assessed.

Traditional methods for the detection and enumeration of specific microbes generally involve sample dilution and plating for single colonies on solidified selective media. The medium may be selective for a natural property of the organism or for a newly acquired property (e.g. lactose utilization, resistance to nalidixic acid) that has been introduced specifically for the purpose of tracking the GEM and which differs from the introduced property that constitutes the crucial new functional aspect of the GEM. Exceptionally, this latter property may also serve as a basis for specific selection or detection of the GEM on solid media, thereby enabling both the organism and the newly acquired genetic information to be independently tracked by plating techniques. Where the newly acquired information does not itself confer a phenotypic property detectable by plating procedures, it can be directly linked to a marker which does have this property (e.g. genes encoding lactose utilization, catechol 2,3-dioxygenase, etc.).

24.6 Conclusions

Realistically, only those microorganisms which can grow in the rhizosphere are suitable for use as biocontrol agents, as the rhizosphere provides the first line of defense for the roots of a plant against attack by soil-borne pathogens. Thus it is necessary to screen the rhizospheric bacteria having plant growth-promoting ability. Also the introduction of bioinoculants having plant growth-promoting and biocontrol activity will be successful if they have rhizosphere competence to exert the desired effect to the plants.

Depending on that availability and culturability of microorganisms there are several classic, modern, and molecular methods to screen them. Looking at the biodiversity of microorganisms, we are presently able to culture only 1–2% of them. This is where metagenomics plays an important role. By virtue of it, many genes beneficial for plant health can be isolated from the rhizosphere and can then be cloned in a bacterium for its expression.

These screening methods would by large help us to formulate better bioinoculants to replace the chemical fertilizers, which continuously challenge the soil and human health.

References

Acea MJ, Moore CR, Alexander M (1988) Survival and growth of bacteria introduced into soil. Soil Biol Biochem 20:509–515

Akopyants NS, Fradkov A, Diatchenko L, Hill JE, Siebert PD, Lukyanov SA, Sverdlov ED, Berg DE (1998) PCR-based subtractive hybridization and differences in gene content among strains of *Helicobacter pylori*. Proc Natl Acad Sci USA 95:13108–13113

Blomberg GV, Lugtenberg BJJ (2001) Molecular basis of plant growth promotion and biocontrol by rhizobacteria. Curr Opin Plant Biol 4:343–350

Brady SF, Chao CJ, Handelsman J, Clardy J (2001) Cloning and heterologous expression of a natural product biosynthetic gene cluster from eDNA. Org Lett 3:1981–1984

Britto de Oliveira RG, Wolters AC, Elsas JD van (1995) Effects of antibiotics in soil on the population dynamics of transposon Tn*5* carrying *Pseudomonas fluorescens*. Plant Soil 175:323–333

Dandurand LMC, Knudsen GR (2002) Sampling microbes from the rhizosphere and phyllosphere. In: Hurst CJ, Crawford RL, Knudsen GR, McInerney MJ, Stetzenbach LD (eds). Manual of environmental microbiology, 2nd edn. ASM, Washington, D.C., pp 516–526

Dazzo FB (1980) Adsorption of microorganisms to roots and other plant surfaces. In: Bitton G, Marshall KC (eds) Adsorption of microorganisms to surfaces. Wiley, New York, 253–316

Deacon JW, Berry LA (1993) Biocontrol of soil-borne plant pathogens: concepts and their application. Pest Sci 37:417–426

Devliegher W, Arif MAS, Verstaete W (1995) Survival and plant growth promotion of detergent-adapted *Pseudomonas fluorescens* ANP15 and *Pseudomonas fluorescens* 7NSK2. Appl Environ Microbiol 61:3865–3871

Dewar K, Sabbagh L, Cardinal G, Veilleux F, Sanschagrin F, Birren B, Levesque RC (1998) *Pseudomonas aeruginosa* PAO1 bacterial artificial chromosomes: strategies for mapping, screening, and sequencing 100 kb loci of the 5.9 Mb genome. Microb Comp Genomics 2:105–117

Preston GM, Bertrand N, Paul BR (2001) Type III secretion in plant growth-promoting *Pseudomonas fluorescens* SBW25. Mol Microbiol 41 999–1014

Gal M, Preston GM, Massey RC, Spiers AJ, Rainey PB (2003) Genes encoding a cellulosic polymer contribute toward the ecological success of *Pseudomonas fluorescens* SBW25 on plant surfaces. Mol Ecol 12:3109–3121

Gillespie DE, Rondon MR, Handelsman J (2004) Metagenomic libraries from uncultured microorganisms. In: Osborn AM (ed) Molecular microbial ecology. BIOS Scientific, Oxford (in press)

Guyon P, Petit A, Tempé J, Dessaux Y (1993) Transformed plants producing opines specifically promote growth of opine-degrading agrobacteria. Mol Plant Microbe Interact 6:92–98

Kim U, Birren B, Slepak T, Mancino V, Boysen C, Kang H, Simon M, Shizuya H (1996) Construction and characterization of a human bacterial artificial chromosome library. Genomics 34:213–218

Kluepfel DA (1993) The behaviour and tracking of bacteria in the rhizosphere. Annu Rev Phytopathol 31:441–472

Lam ST, Torkewitz NR, Nautiyal CS, Dion P (1997) Microorganisms with mannopine catabolizing ability. US Patent 5 610 044

Lan R, Reeves PR (2000) Intraspecies variation in bacterial genomes: the need for a species genome concept. Trends Microbiol 8:396–401

Landa BB, Mavrodi DM, Thomashow LS, Weller DM (2003) Interactions between strains of 2,4-diacetylphloroglucinol-producing *Pseudomonas fluorescens* in the rhizosphere of wheat. Phytopathology 93:982–994

Li DM, Alexander M (1990) Factors affecting co-inoculation with antibiotic-producing bacteria to enhance rhizobial colonisation and nodulation. Plant Soil 129:195–201

Lugtenberg BJJ, Dekkers L, Bloemberg GV (2001) Molecular determinants of rhizosphere colonization by *Pseudomonas*. Annu Rev Phytopathol 39:461–490

Lynch JM (1990) The rhizosphere. Wiley, Chichester

Albareda M, Dardanelli MS, Sousa C, Megias M, Tempranol F, Rodriguez-Navarro D (2006) Factors affecting the attachment of rhizospheric bacteria to bean and soybean roots. FEMS Microbiol Lett 259:67–73

Mathre DE, Cook RJ, Callan NW (1999) From discovery to use traversing the world of commercializing biocontrol agents for plant disease control. Plant Dis 83:972–983

Mavrodi DV, Mavrodi OV, McSpadden BB, Gardener BB, Landa DM, Weller W, Thomashow LS (2002) Identification of differences in genome content among *phlD*-positive *Pseudomonas fluorescens* strains by using PCR-based subtractive hybridization. Appl Environ Microbiol 68:5170–5176

Mehta S, Nautiyal CS (2001) An efficient method for qualitative screening of phosphate solubilizing bacteria. Curr Microbiol 43:51–56

Nairn JD, Chanway CP (1999) Recovery of rhizosphere-colonizing GEM from inside wheat roots. Can J Microbiol 45:612–615

Nautiyal CS (1997a) A method for selection and characterisation of rhizosphere competent bacteria of chickpea. Curr Microbiol 34:12–17

Nautiyal CS (1997b) Selection of chickpea rhizosphere competent *Pseudomonas fluorescens* NBRI1303 antagonistic to *Fusarium oxysporum* f. sp. *ciceri*, *Rhizoctonia bataticola* and *Pythium* sp. Curr Microbiol 35:52–58

Nautiyal CS (2002) A biologically pure culture of bacteria which suppresses diseases caused by pathogens in chickpea crops and a culture of bacteria comprising a strain of *Pseudomonas fluorescens*. US Patent 6495362

Nautiyal CS (2006) Biological control of plant diseases by natural and genetically engineered fluorescent *Pseudomonas* spp. In: Ray RC, Owen P, Ward W (eds) Microbial biotechnology in horticulture. Science, Enfield, pp 125–162

Nautiyal CS, Dion P (1990) Characterization of the opine-utilizing microflora associated with samples of soil and plants. Appl Environ Microbiol 56:2576–2579

Nautiyal CS, Mehta S, Rehman A, Singh HB (2004) Bioinoculants for sustainable agriculture. In: Singh SP, Singh HB (eds) Eco-agriculture with bioaugmentation: an emerging concept. Rohitashwa, Lucknow, pp 148–162

Nautiyal CS, Mehta S, Singh HB (2006) Biological control and plant growth-promoting *Bacillus* strains from milk. J Microbiol Biotechnol 16:184–192

Nishimura M, Nakamura S, Hayashi N, Asakawa S, Shimizu N, Kaku H, Hasebe A, Kawasaki S (1998) Construction of a BAC library of the rice blast fungus *Magnaporthe grisea* and finding specific genome regions in which its transposons tend to cluster. Biosci Biotechnol Biochem 62: 1515–1521

Ochiai H, Inoue Y, Hasebe A, Kaku H (2001) Construction and characterization of a *Xanthomonas oryzae* pv. *oryzae* bacterial artificial chromosome library. FEMS Microbiol Lett 200:59–65

O'Connell KP, Goodman RM, Handelsman J (1996) Engineering the rhizosphere expressing a bias. Trends Biotechnol 14:83–88

Quaiser A, Ochsenreiter T, Klenk HP, Kletzin A, Treusch AH, Meurer G, Eck J, Sensen CW, Schleper C (2002) First insight into the genome of an uncultivated crenarchaeote from soil. Environ Microbiol 4:603

Randon MR, August PR, Bettermann AD, Brady SF, Grossman TH, Liles MR, Loiacono KA, Lynch BA, MacNeil IA, Minor C, Tiong CL, GilmanM, Osburne MS, Clardy J, Handelsman J, Goodman RM (2000) Cloning the soil metagenome: a strategy for accessing the genetic and functional diversity of uncultured microorganisms. Appl Environ Microbiol 66:2541–2547

Steffan RJ, Goksoyr J, Bej AK, Atlas RM (1988) Recovery of DNA from soils and sediments. Appl Environ Microbioll 54:2908

Stein JL, Marsh TL, Wu KY, Shizuya H, De Long EF (1996) Characterization of uncultivated prokaryotes: isolation and analysis of a 40 kilobase-pair genome fragment from a planktonic marine archaeon. J Bacteriol 178:591–599

25 Research Methods in Arbuscular Mycorrhizal Fungi

A. Gaur and A. Varma

25.1 Introduction

Mycorrhizas are widespread associations between plant and fungi and are characterized by a bi-directional transfer of nutrients, where plants provide sugar to the fungi and these help the plants in the acquisition of mineral nutrients from the soil (Smith and Barker 2002). Additionally, mycorrhizal fungi also aid in soil-water extraction, increasing the drought tolerance of the host (Subramanian et al. 1997; Mathur and Vyas 2000). These associations are also reported to improve the plant's ability to tolerate heavy metal toxicity (Khan 2001), as well as attacks by pathogens (Calvet et al. 1993; Filion et al. 1999; Fusconi et al. 1999) and herbivores (Gehring and Whitham 1991; Borowicz 1997). At the single plant level, these benefits result in increased mass production and plant competitive ability.

Arbuscular mycorrhizal (AM) fungi are soil microorganisms that establish a mutual symbiosis with the majority of higher plants, providing a direct physical link between soil and plant roots (Strullu 1991). These fungi are the most ancient (Redecker et al. 2000) and widespread mycorrhizal associations. AM fungi were recently placed in a new monophyletic phylum, the Glomeromycota, encompassing three orders (Schussler et al. 2001) and five families (Morton and Redecker 2001). Associations with these fungi are widespread among tropical trees, shrubs and herbs (Harley and Smith 1983), including members of the Araucariaceae (Smith and Read 1997). About 95% of the world's plant species belong to characteristically mycorrhizal families (Smith and Read 1997) and potentially benefit from AM fungus-mediated mineral nutrition (Jeffries and Barea 1994) due to the fundamental role played by these glomalean fungi

---

Amity Institute of Microbial Sciences, Amity University, Uttar Pradesh, Sector-125,
Noida 201301, UP, India, email: ajitvarma@aihmr.amity.edu
Noida 201303, UP, India, email: agaura@amity.edu

Soil Biology, Volume 11
Advanced Techniques in Soil Microbiology
A. Varma, R. Oelmüller (Eds.)

in biogeochemical element cycling. AM symbioses occur in almost all habitats and climates, including disturbed soils (Enkhtuya et al. 2002) and those derived from mine activities (Bi et al. 2003). Today, there is increased research interest in understanding the basic physiology and ecology of mycorrhiza in vivo, in vitro, and in situ. Only through such extensive efforts can we hope to reliably manage mycorrhiza in forestry and agriculture.

The purpose of this chapter is to highlight the basic methods in mycorrhizal research. A wide range of methods are currently being used by various research groups; and each method needs to be assessed for the particular application required.

25.2 Assessment of AM Fungal Propagules in Soil

The spores of AM fungi are larger than those of most fungi (ranging from 10 μm to 1000 μm in diameter) and can easily be observed under a dissecting microscope. However, spore counts often underestimate the numbers of AM fungi since colonized roots and hyphae can also serve as propagules. Therefore, various assays have been used to estimate total propagule number (Sylvia 1994).

25.2.1 Soil Sampling

Soil samples are collected from the rhizosphere of a plant. One kilogram of soil in each sample is collected at a depth of 30–50 cm. Six composite samples are collected to represent the soil sample of each site and kept in plastic bag at temperature around 2–5 °C. The wet soil samples are air-dried at room temperature before storage. These soil samples are then extracted to separate spores.

25.2.2 Spore Extraction

AM fungal spores are extracted from soil by wet sieving and decanting, as described by Gerdemann and Nicolson (1963), and by sucrose centrifugation, as described by Smith and Skipper (1979). The soil sample (100 g) is suspended in 1 l water by gentle stirring. A dispersant such as sodium hexametaphosphate can be used with clay soils. Heavier particles are then allowed to settle for a few seconds and the liquid is decanted through a 450 μm sieve to remove large particles of organic matter and allow the spores pass through. Next, the suspension is passed through a 100 μm sieve and then through a 63 μm sieve. The spores and small amount of debris that remain on the 63 μm sieve are poured into a centrifuge tube containing water and centrifuged at 1800 rpm. The upper solution is poured off, 40% sucrose is added to the debris at the bottom and the mixture is then centrifuged for 2 min at 1800 rpm. The upper solution is separated for examination under the stereoscopic microscope. The spores which are collected under the microscope are stored in Ringer's solution (Daniels and Graham 1976) for identification.

25.2.3 Quantification of Spore Numbers

The entire sieving after the wet sieving and decanting method is examined in nematode-counting dishes (Doncaster 1962) under a dissecting microscope at 60× magnification. These counting dishes enable a complete sample to be counted more rapidly and facilitate the separation of viable spores (which mostly sink) from non-viable ones (which mostly float). The non-viable floating spores, mostly concentrated in the scum on the meniscus at the edge of the dish, are empty or gas-filled shells. Viable spores, viz. those with cytoplasm and oil globules (readily checked by piercing with a needle), rarely float and those that do are mostly very small spores or large spores which have dried out and become difficult to re-wet. The extracted spores can also be counted on filter paper after filtering the sieving through a Whatman filter paper (Gaur and Adholeya 1994). However, there are certain limitations to direct counting of spores. First, some AM fungi produce spores too small to be extracted or counted using these techniques [e.g. *Glomus tenuis* (Greenall) Hall (1977)]. Second, it has been claimed that some AM fungi may not produce spores (Baylis 1969). And third, even the spores that are large enough to be extracted and counted may not all be recovered (Clarke and Mosse 1978). The most probable number (MPN) technique (or method of ultimate dilution) for enumerating viable microorganisms has been proposed as a possible solution to the problems faced when using the usual methods of counting AM fungal propagules (Sylvia 1994).

25.2.4 Infectivity Assays

A more straightforward approach for comparing AM populations among soils is an infectivity assay. The drawback is that actual propagule numbers are not estimated. The MPN of infective propagules can be estimated as described by Porter (1979). A 10-fold dilution series is first made by mixing 10 g soil for each replicate of treatment with 90 g diluent sterilized sand. The second dilution is made by mixing 10 g soil/sand mixture from the first dilution with a further 90 g diluent sterilized sand. This dilution is done up to 10^{-5} with five replicates. Then, 10 g aliquots of 10^{-3}, 10^{-4} and 10^{-5} dilutions are placed in plastic pots, each containing 350 g sterilized sand. Three sorghum seeds are sown in one pot and, a week later, the seedlings are thinned to one. The plants are harvested after 6 weeks under glasshouse conditions, where they are maintained by periodically supplying nutrient solution devoid of phosphorus. The sorghum roots are washed free of soil, autoclaved in 10% KOH and stained in trypan blue-locto-glycerol. Infection is observed under 40× magnification and the MPN is calculated as described by Alexander (1982).

In an another procedure described by Gaur et al. (1998), the inoculum is mixed with 20, 40, 60 and 80% of the sterilized soil and transferred to 5× 9-cm plastic pots after homogenization. Five replicates are prepared for each level of dilution. Eight germinated seeds of *Zea mays* are planted per pot and cultivated for 12 days in a greenhouse before the roots are washed, stained and processed for estimation of primary infection points. The number of primary entry points is counted on a whole-root system under a stereoscopic microscope (40×).

25.2.5 Identification of AM Fungi

The ability to make a good, semi-permanent, diagnostic slide is critical in making a species determination for a specimen of AM fungi (Schenck and Perez 1990). Semi-permanent mountants, such as polyvinyl alcohol–lactic acid–glycerol (PVLG; Table 25.1; Koske and Testier 1983) or Hoyer's (Cunningham 1972), allow slides to remain useable for years.

Pick out 40–100 typical, clean spores with a pipette or other device and place them in a watch glass containing distilled water. A drop of PVLG is then placed on a clean and dry microscope slide. Add 10–25 spores with a minimum amount of water to the mountant. Gently mix the spores and mountant together with a

Table 25.1 Composition of important reagents used in mycorrhizal techniques: 1. PVLG mountant

Polyvinyl alcohol	8.33 g
Distilled water	50 ml
Lactic acid	50 ml
Glycerine	5 ml
Polyvinyl alcohol (24–32 centripose viscosity) can be used and dissolved in water by heating (90 °C) overnight	

needle or other object to slightly disperse the spores. Allow the mountant to set for 4–5 min to become more viscous and then add a coverslip gently onto the mountant without the formation of any air bubble. Do not apply pressure to the coverslip in this process. Let the mountant with spores dry overnight in a flat position. Clean off any excess mountant with a muslin cloth moistened with a solvent such as ethanol. Seal the edges of the coverslip with a clear fingernail polish or other sealant and allow it to dry. Repeat the steps with a second drop of PVLG on another slide. Gently break the spore walls under coverslip of this second slide by applying light pressure on the coverslip with the back of the needle. It is very important that the break of the spore wall be adequate. The spores should not be crushed during the process. The spores can also be stained in a drop of Melzer's reagent (Morton 1991; Table 25.2) by mixing with PVLG in a ratio of 1:1. The AM species can be identified using spore color, size wall structure and other morphological structures (Schenck and Perez 1990).

25.2.6 Use of Fatty Acids for Identification of AM Fungi

Fatty acids derived from abundant phospholipids of AM fungi (located in membrane structures) and neutral lipids (located in storage structures) are potentially useful for estimating the biomass of infective AM propagules (Olsson et al. 1995). In addition some fatty acids have potential as specific markers for AM fungi. For example, fatty acids 16:1ω5c, 18:1ω7c, 20:3, 20:4 and 20:5 have been detected either exclusively or in higher amounts in AM spores and the roots of plants colonized by AM fungi, prompting attempts at identification, characterization or differentiation of AM fungi on the basis of fatty acid profiles.

Madan et al. (2002) performed fatty acid methyl ester (FAME) analysis on the spores of four AM fungi (*Glomus coronatum*, *G. mosseae*, *Gigaspora margarita*, *Scutellospora calospora*) and showed 16:1ω5c to be the dominant fatty acid present. In addition, spores of *G. margarita* contained large quantities of 18:1ω9c and three 20-C fatty acids (20:1ω9c, 20:2ω6c, 22:1ω9c) that were not present in the spores of the other two species. The results of this study confirmed the use of 16:1ω5c as a marker fatty acid for AM fungi in controlled environments and suggested that 18:1ω9c, 20:1ω9c, 20:2ω6c and 22:1ω9c could be used as possible markers for the detection of *G. margarita*.

Table 25.2 Composition of important reagents used in mycorrhizal techniques: 2. Melzer's reagent – mixed 1:1 (v/v) with PVLG

Iodine	1.5 g
Potassium iodide	5 g
Distilled water	100 ml

25.3
Quantification of AM Fungal Root Colonization in Root

Arbuscular mycorrhizal fungal structure in roots is usually not observed without appropriate staining because internal structures are obscured by the natural pigments and cell contents within roots. Clearing procedures, which use chemical agents to remove cell contents and cell wall pigments, are a valuable method for viewing internal features in plant tissues (Gardner 1975).

The preparation of plant roots for quantification of the extent of AM colonization is probably the most frequently performed procedure in AM research. Biological stains have been selected which bind to the fungal structures without much background staining of the plant tissue. Over the years, several techniques have been published which document methods for clearing and subsequent staining of roots to reveal the mycorrhiza. The first of these, described by Phillips and Hayman (1970), used trypan blue (TB) and this method has been widely adopted. A modification was described by Koske and Gemma (1989), in which some of the toxic reagents were eliminated from the process, although TB remained.

25.3.1
Clearing and Staining Roots

Clearing and staining procedures require that root samples should be washed free of soil. It is important that clearing with KOH (Kormanik and McGraw 1982; Brundrett et al. 1984) and staining solution volumes are sufficient for the amount of roots being processed and that roots are not tightly clumped together for uniform contact with solutions. To ensure uniform staining, the roots should be chopped into small (1–2 cm) segments.

1. Wash root specimens stored in capsules under running tap water thoroughly. Place them in beaker containing 5–10% KOH solution for about 15–30 min.
2. Pour off the KOH solution and rinse the capsules well in a beaker using at least three complete changes of tapwater or until no brown color appears in the rinse water.
3. Cover the capsules in the beaker with alkaline H_2O_2 at room temperature for 10 min or until the roots are bleached.
4. Rinse the capsules in the beaker thoroughly, using at least three complete changes of tap water to remove the H_2O_2.
5. Cover the capsules in the beaker with 1% HCl and soak for 3–4 min. Then pour off the solution. Do not rinse after this step because the specimens must be acidified for proper staining.
6. Cover the capsules in the beaker with staining solution (0.01% acid fuchsin in lactoglycerol or 0.05% trypan blue in lactophenol) and keep them overnight for staining.

7. After removing from the capsules, place the root specimens in a glass Petri plate or multiwell plate for destaining. The destaining solution (50% glycerol) is the standard used in Step 6 but, of course, without the stain. Semi-permanent slides of stained roots can be made with PVLG mountant. For temporary slides, the stained roots can be observed in plain lactoglycerol.

25.3.2 Modifications of Staining Procedure

Modifications to standard clearing and staining procedures have been proposed for safety reasons. Vierheilig et al. (1998) demonstrated adequate staining of mycorrhizal roots with ink–vinegar solutions as a safe alternative to the hazardous, toxic and potentially carcinogenic stains. Other authors have proposed the use of acid fuchsin (AF; Kormanik and McGraw 1982; Merryweather and Fitter 1991) or chlorazol black E (Brundrett et al. 1984; Table 25.3). However, a major problem of most stains is that they are known or suspected carcinogens (Coombes and Haveland-Smith 1982) and, in order to solve this, Grace and Stribley (1991) proposed replacing TB with aniline blue or methyl blue, although Brundrett et al. (1996) suggest that there is insufficient evidence that these latter two stains are not also toxic. In addition, effective clearing of roots involves the use of KOH, which is caustic.

Vital staining procedures that measure succinate dehydrogenase activity can be used to confirm that the mycorrhizal fungus hyphae being enumerated are metabolically active (Schaffer and Peterson 1993; Tisserant et al. 1993; Vivas et al. 2003). Abdel-Fattah (2001) reported histochemical staining of succinate dehydrogenase and alkaline phosphatase (vital stain) activities as enzyme markers in various AM fungal structures. Grace and Stribley (1991) reviewed the use of stains in the literature for 1989 and 1990 and found that 68% of authors used TB, 18% CBE, 9% AF and 5% some other procedure. A further technique, auto-fluorescence (fluorescence microscopy) was first described by Ames et al.

Table 25.3 Composition of important reagents used in mycorrhizal techniques: 3. Staining solution

0.01% acid fuschin	0.01 g acid fuschin in 100 ml lactoglycerol
0.05% trypan blue	0.05 g trypan blue in 100 ml lactoglycerol
0.03% chlorozol black E (CBE)	0.03 g CBE in lactoglycerol (dissolve CBE in water before adding equal volumes of lactic acid and glycerol)
Lactoglycerol	
Lactic acid	876 ml
Glycerine	64 ml
Distilled water	60 ml

(1982) and involves subjecting roots to ultraviolet illumination, under which the arbuscules auto-fluoresce. The method was found to work well and no significant differences were found between the extents of colonization detected by this method and by that of Phillips and Hayman (1970). Subsequent workers also reported the auto-fluorescence of fungal structures other than arbuscules (Jabaji-Hare et al. 1984).

25.3.3 Measurement of Root Colonization by AM Fungi

Measurement of mycorrhizal root colonization is done after clearing and staining the roots. Root length can be measured simultaneously with mycorrhizal colonization by a gridline intersection procedure (Giovannetti and Mosse 1980) or separately by making slides and viewing them under a compound microscope (McGonigle et al. 1990). Here we describe the assessment of mycorrhizal colonization using the Biermann and Linderman (1981) method (frequency distribution method) in which the colonization is assessed (using a compound microscope) as a proportion of the root length colonized by mycorrhizal fungi.

1. Spread randomly selected, stained root segments (1 cm in length) in lactoglycerol within a Petri dish marked with a 1-cm grid to facilitate scanning and view them under a stereomicroscope at 12× to 50×.
2. Calibrate the ocular micrometer with the stage micrometer by placing it on the eyepiece of a compound microscope.
3. Mount 5–10 root pieces on each glass slide and calibrate the ocular micrometer with the stage micrometer at the particular magnification of the compound microscope and observe the root pieces.
4. Estimate the proportions of each root segment consisting of vesicles, arbuscules and hyphae, to the nearest 10%.
5. Record the frequency distributions from samples containing 25, 50, 100 root segments. The percent root length with mycorrhizal fungi in the sample is calculated from the frequency distribution.

25.4 Extraction and Quantification of Extra-Radical Mycelium of AM Fungi in Soils

One of the important advances in the past decade of mycorrhizal research was the increased emphasis on the structure, organization and function of the extraradical mycelium (ERM).

The first step in ERM assessment, for example determining lengths and metabolic activity, is extracting the ERM from the soil or growth substrate. One set of ERM extraction techniques is based on vacuum filtration of a soil suspension through a membrane filter. This membrane filtration technique (MFT) was introduced by Hanssen et al. (1974). The technique was modified recently by several authors and is now widely used for the assessment of ERM lengths of AM fungi (Boddington et al. 1999). Vilariño et al. (1993) developed another extraction technique using a rotating wire frame to retrieve the ERM fragments from an agitated soil suspension. A third set of extraction techniques is based on sucrose flotation and centrifugation (Schubert et al. 1987). All these techniques are suitable for quantification of the lengths of ERM with respect to the spatial distribution of the hyphae in soil (Dodd 1994). Their disadvantage, however, is that they severely disturb the ERM network during sample processing. Thus, the identification of structures such as branched absorbing structures (BAS; Bago et al. 1998a, b) or the measurement of morphological parameters such as hyphal branching or anastomoses formation is often difficult. A simple "inserted membrane technique" (IMT) for sampling mycorrhizal extraradical mycelium (ERM) was developed as an alternative to the commonly used MFT (Baláz and Vosátka (2001). The ERM was extracted by insertion of cellulose nitrate or cellulose acetate membrane filters (0.45–0.6 μm pore size) into the mycorrhizosphere of host plants. The membranes with adhered ERM were removed at harvest and stained: (a) with trypan blue for estimation of total hyphal length and (b) with enzyme stains to indicate the viability of the ERM.

25.5 Assessment of Growth Response of Effective Isolates

Any measure of the benefit provided by mycorrhizas depends on the relative contribution of root and mycorrhiza-mediated nutrient uptake to plants (Janos 1980). Mycorrhizal dependency has often been quantified by calculating the yield ratio between mycorrhizal plants and uninoculated control plants grown in a particular soil at a single soil P level (Koide et al. 1988; Manjunath and Habte 1991). However, it is better to analyze mycorrhizal benefits across a range of soil P levels, by producing nutrient response curves.

Mycorrhizal dependency is a plant property which refers to the degree of its responsiveness to mycorrhizal colonization. It can be measured by quantifying the growth improvement owing to the mycorrhizal performance, such as the relative non-mycorrhizal contribution compared with mycorrhiza-mediated nutrient uptake (Plenchette et al. 1983; Brundrett 1991). Mycorrhizal dependency is the result of morphological and physiological plant traits which are modulated by both the nutrient availability of the soil, particularly P, and the effectiveness of the mycorrhizal fungus involved (Khalil et al. 1994). It can vary greatly from one plant species to another and even between cultivars or ecotypes within a single species (Azcon and Ocampo 1981). Some plant species can be obligatorily mycorrhizal for P uptake (Janos 1980; Merryweather and Fitter 1996).

25.6 Inoculum Production of AM Fungi

Since AM fungi are obligate symbionts, they are always produced on roots. The method of culture and inoculum production of AM fungi vary from the pot culture techniques of Brundrett and Juniper (1995) to the currently used techniques, such as on-farm production (Sieverding and Barea 1991; Douds et al. 2005a, b, 2006), nutrient film technique (Mosse and Thompson 1984), aeroponics (Jarstfer and Sylvia 1995) and axenic culture (Fortin et al. 2002). Apart from the host plant (Sreenivasa and Bagyaraj 1988), many factors such as temperature (Furlan and Fortin 1973), light (Ferguson and Menge 1982), pot size and soil fertility (Menge et al. 1978) and the particle size of the growth substrate (Gaur and Adholeya 2000) are known to affect inoculum production of AM fungi.

25.6.1 On-Farm Production of AM Fungi

On-farm inoculum production is a promising technique for large-scale AM fungal inoculum production where the inoculum is produced on-farm, directly on the site of its application, using local resources. The mycorrhizal inocula can be prepared by harvesting the roots of growing plants and applied in the rest of the field over a period of time. The soil left in the nursery after removing the roots also contains large amounts of AM fungal propagules; and it serves as the source for further and continued production of inocula for in-house use for the farmer. Gaur and Adholeya (2002) conducted experiments in marginal soil for enhancing crop production along with producing a higher number of AM fungal propagules. The procedure is described in detail by Sieverding and Barea (1991) and can produce 5000 l of soil inocula from a 25 m^2 plot. Gaur and Adholeya (2002) reported production of five fodder crops (*Zea mays*, *Medicago sativa*, *Trifolium alexandrinum*, *Avena sativa*, *Sorghum vulgare*) in marginal soil along with producing a high number of indigenous AM propagules.

25.6.1.1 Procedure for On-Farm Inoculum Production of AM

Large-scale production of AM fungi begins with a starter culture. The starter culture can be procured either by isolating or by ordering it from various laboratories that maintain pure cultures of specific interest.

Soil in nursery beds should be sterilized either with methyl bromide or formalin by drenching it up to at least 45 cm depth with either of the solutions. After treating with chemicals the soil should be covered for 3 days and then kept open for at least 8 days before commencing any operation. Sunlight for sterilization involves covering the soil with transparent polythene sheets for a minimum of 20 days.

1. Preparation of nursery bed:
   The nursery soil should be raised up to 30 cm. Making surrounding furrows of a similar depth can do this. Soil should be thoroughly mixed and preferably sieved. If the soil is compact, sand may be mixed for good mycorrhizal development, in a ratio of 2:1.
2. Sowing and inoculation:
   Furrows of 6 cm depth are made in the nursery beds and AM propagules, mixed with any suitable carrier, are placed in the furrows. The inoculum should contain at least 30–40 spores per gram of substrate. The inoculum should be covered with a thin layer of soil on which host seeds (preferably monocots) should be sown.
3. Maintenance and monitoring:
   The beds should be watered when required and should be kept free from weeds. After 3 months, the extent of colonization and spore production could be assessed.

25.6.2 Traditional Culture Methods

The most frequently used technique for increasing propagule number is the propagation of AM fungi on a suitable host in disinfested soil using pot cultures. Examples of the plants that have been used successfully include alfalfa, maize, onion and sudan grass. Hosts can be propagated from seeds that may be disinfested with sodium hypochloride or hydrogen peroxide. Hepper (1984) reviewed procedures for disinfestations and for germinating spores. All the components of the culture system are disinfected before the initiation of pot culture. The most commonly used method is heat pasteurization, where large batches of soil may be treated by heating to 85 °C for two 8-h periods with 48 h between treatments in a commercial soil pasteurizer. Conducive environmental conditions for culturing AM fungi are a balance of light intensity, adequate moisture and moderate temperature without detrimental addition of fertilizers or pesticides (Jarstfer and Sylvia 1992). Good light quality and high photosynthetic flux density are necessary for high root colonization and spore production (Whitbeck 2001). Soil moisture affects AM fungal development directly or indirectly (Al-Karaki et al. 1998). Amendments with fertilizers and chemicals can have both beneficial and detrimental effects on the development of colonized root systems and sporulation. Responses to P and N fertilization may be strain-dependent (Douds and Schenck 1990) and are affected by relative amounts of N and P.

To initiate pot cultures, a layer of inoculum is placed 1–2 cm below the seed or cuttings. Initial isolates are obtained by trapping the infested soil collected from the field. However, these mixed cultures should rapidly be processed for purification and single-species cultures initiated. Detailed methods for pot culturing and extensive discussion on these methods are provided by Jarstfer and Sylvia (1992). Cultures reaching a high propagule density (10 spores/g) after a number of multiplication cycles can be stored using suitable methods (Staddon and Fitter 2001) after air-drying. Furthermore, AM fungi have been cultured with plant host in different substrates such as sand, peat, expanded clay, perlite, vermiculite, soilrite (Mallesha et al. 1992), rockwool (Heinzemann and Weritz 1990) and glass beads (Redecker et al. 1995).

25.6.2.1 Method

1. Rhizosphere soil is collected, with shoots of trap plant cut at the crown, and roots are finely chopped and mixed with the soils using a sharp chopper.
2. The chopped roots and soil are mixed 1:1 (v/v) with autoclaved coarse sand in a mechanical mixer, or massaged well in a durable plastic bag.
3. The soil mix is then transferred to a 15-cm plastic pot.
4. Seeds of suitable trap plant are planted in the pot.
5. The pot cultures are maintained in a greenhouse for at least 3 months and sporulation is checked from time to time. Sanitary tests may also be carried out to ensure no contamination from parasitic fungi occurs.
6. Fertilizer application is kept to a minimum, to encourage AMF proliferation.
7. Trap culture pots are later left to dry under shade for up to 2 weeks.
8. The spores are harvested using the sieving and decanting technique or the density-gradient centrifugation technique.
9. The monospecific spores are now ready for inoculation onto seedlings of the desired crops.

25.6.3 AM Fungal Culture Using Aeroponic and Hydroponic Culture

The major benefit of aeroponic and hydroponic culture systems is that colonized roots and spores are produced free of any substrate, permitting more efficient production and distribution of inocula. Here, plants are inoculated with AM fungi and grown in sand or vermiculite for 4–5 weeks under conditions conducive for rapid colonization, after which they are washed and non-destructively checked for colonization and then they are transferred into the system.

In aeroponics, the plant's root system is exposed constantly to an aerated mist of dilute nutrient solution. Unlike soil culture, hydroponics or other traditional growth cultures, aeroponics shows good root hair development due to the highly aerated environment surrounding the root system. Aeroponic culture allows control of root zone temperature, nutrition, moisture and gaseous phase. An aeroponic system for the production of AM fungi was first used by Sylvia and Hubbell (1986). Mohammad et al. (2000) reported the production of *Glomus intraradices* in an aeroponic system where they compared the conventional atomizing disc with the ultrasonic nebulizer technology as misting sources. Laurent et al. (1999) used this culture method to produce *Acacia mangium* saplings associated with AM fungi.

The hydroponics or nutrient film technique was adapted for AM fungus inoculum production by Mosse and Thompson (1984). Culture host plants are placed on an inclined tray over which flows a layer of nutrient solution. As in the aeroponic culture, seedlings must be precolonized in another media. Dugassa et al. (1995) presented a hydroponic system for culturing and maintaining the vesicular–arbuscular mycorrhizal fungus.

25.6.4 Monoaxenic Culture of AM Fungi

Recently, Ri T-DNA-transformed roots were used to obtain colonized root cultures. Bécard and Fortin (1988) presented a detail evaluation of the root organ culture technique and reported basic improvements necessary for AM fungus colonization of roots. Cultures are initiated by transfer of pregerminated, surface-sterilized spores or surface-sterilized, colonized root pieces into Petri plates of minimal media (Bécard and Fortin 1988) or modified Strullu-Romand medium (Declerck et al. 1996).

The establishment of in vitro root-organ cultures has greatly influenced our understanding of the AM symbiosis. Root-organ cultures were first developed by White and co-workers (White 1943; Butcher and Street 1964; Butcher 1980). These authors used excised roots on synthetic mineral media supplemented with vitamins and a carbohydrate source. Pioneering work by Mosse and Hepper (1975) used root cultures obtained from *Lycopersicum esculentum* Mill. (tomato) and *Trifolium pratense* L. (red clover) to establish in vitro mycorrhiza with *Glomus mosseae* Nicolson & Gerd. The authors demonstrated for the first time that spores of an AM fungus could be successfully used to colonize excised roots growing on a mineral-based medium. Later, Strullu and Romand (1986, 1987) showed that it was also possible to re-establish mycorrhiza on excised roots of Fragaria × Ananassa Duchesne (strawberry), *Allium cepa* L. (onion) and tomato, using the intraradical phase (i.e., vesicles or entire mycorrhizal root pieces) of several species of *Glomus* as inoculum. A natural genetic transformation of plants by the ubiquitous soil bacterium *Agrobacterium rhizogenes* Conn.

(Riker et al. 1930) produces a condition known as hairy roots. This stable transformation (Tepfer 1989) produces Ri T-DNA transformed plant tissues that are morphogenetically programmed to develop as roots. Their modified hormonal balance makes them particularly vigorous and allows profuse growth on artificial media (Tepfer 1989). The first in vitro sporulation of an AM fungus was obtained by Bécard and Fortin (1988) using carrot hairy roots colonized by *Glomus intraradices* Schenck & Smith. Plenchette et al. (1996) reported *Glomus versiforme* associated in vitro with Ri T-DNA transformed carrot root and after 4 months of cultivation, numerous axenic AM propagules were obtained.

25.6.5 Storage of AM Fungal Inoculum

Spore of AM fungi are generally stored at 4 °C in dried po-culture soil (Ferguson et al. 1982). Cryopreservation of spores at –60 °C to –70 °C has also been reported (Douds and Schenck 1990). Cultures of AM fungi should be dried slowly with the host plant and frozen in situ.

25.7 Conclusions

AM fungi, the most widespread symbionts on earth, are receiving attention because of the increasing range of their application in sustainable agriculture and ecosystem management. Since AM fungi are obligate symbionts, most studies have been conducted on a host plant grown in a sterilized medium using pot culture methods. Procedures such as single-spore culture isolates of AM fungi have been a valuable resource, not only for plant growth experiments, but also for taxonomic and biochemical studies. Several techniques for establishing single-spore isolate have used germinated and ungerminated spores. The current chapter has covered basic techniques in AM fungal research such as the isolation of AM fungal spores from soil, their identification, the establishment of pot cultures in the greenhouse, methods for isolating extra-radical mycelium from soil, vital staining of mycorrhizal roots and methods for AM fungal inoculum production. New approaches to the study of the biology of AM fungi have also been developed, involving growing these fungi in Ri T-DNA transformed root cultures in which some AM fungus species develop profusely and form viable spores.

References

Abdel-Fattah Gamal M (2001) Measurement of the viability of arbuscular–mycorrhizal fungi using three different stains; relation to growth and metabolic activities of soybean plants. Microbiol Res 156:359–367

Alexander M (1982) Most probable number method for microbial populations. In: Page AL, Miller RH, Keene DR (eds) Methods of soil analysis. Part 2. Chemical and microbiological properties. Soil Science Society of America, Madison, pp 815–820

Al-Karaki GN, Al-Raddad A, Clark RB (1998) Water stress and mycorrhizal isolate effects on growth and nutrient acquisition of wheat. J Plant Nutr 21:891–902

Allen MF, Allen EB, Friese CF (1989) Responses of the nonmycotrophic plant *Salsola kali* to invasion by vesicular ± arbuscular mycorrhizal fungi. New Phytol 111:45–49

Ames RN, Ingham ER, Reid CPP (1982) Ultraviolet-induced autofluorescence of arbuscular mycorrhizal root infections: an alternative to cleaning and staining methods for assessing infections. Can J Microbiol 28:351–355

Azcon R, Ocampo JA (1981) Factors affecting the vesicular–arbuscular infection and mycorrhizal dependency of thirteen wheat cultivars. New Phytol 87:677–685

Bago B, Azcón-Aguilar C, Goulet A, Piché Y (1998a) Branched absorbing structures (BAS): a feature of the extraradical mycelium of symbiotic arbuscular mycorrhizal fungi. New Phytol 139:375–338

Bago B, Azcón-Aguilar C, Piché Y (1998b) Architecture and developmental dynamics of the external mycelium of the arbuscular mycorrhizal fungus *Glomus* intraradices grown undermonoxenic conditions. Mycologia 90:52–62

Baláz M, Vosátka M (2001) A novel inserted membrane technique for studies of mycorrhizal extraradical mycelium. Mycorrhiza 11:291–296

Baylis GTS (1969) Host treatment and spore production by endogone. NZ J Bot 7:173–174

Bécard G, Fortin JA (1988) Early events of vesicular–arbuscular mycorrhiza formation on Ri T-DNA transformed roots. New Phytol 108:211–218

Bi YL, Li XL, Christie P, Hu ZQ, Wong MH (2003) Growth and nutrient uptake of arbuscular mycorrhizal maize in different depths of soil overlying coal fly ash. Chemosphere 50:863–869

Biermann B, Linderman RG (1981) Quantifiying vesicular arbuscular mycorrhizae: a proposed method towards standardization. New Phytol 87:63–67

Boddington CL, Bassett EE, Jakobsen I, Dodd JC (1999) Comparison of techniques for extraction and quantification of extraradical mycelium of arbuscular mycorrhizal fungi in soils. Soil Biol Biochem 31:479–482

Borowicz VA (1997) A fungal root symbiont modifies plant resistance to an insect herbivore. Oecologia 112, 534–542

Bowen GD (1984) Future directions in plantation nutrition research. In: Bowen GD, Nambiar EKS (eds) Nutrition of plantation forest. Academic, London, pp 489–505

Brundrett MC, Juniper S (1995) Non-destructive assessment of germination and single-spore isolation of VAM fungi and production of pot cultures from single spores. Soil Biol Biochem 27:85–91

Brundrett MC, Bougher N, Dell B, Grove T, Malajczuk N (1996) Working with mycorrhizas in forestry and agriculture. (ACIAR monograph 32) ACIAR, Canberra

Brundrett MC, Piché Y, Peterson RL (1984) A new method for observing the morphology of vesicular–arbuscular mycorrhizae. Can J Bot 62:2128–2134

Brundrett MC (1991) Mycorrhizas in natural ecosystems. In: Macfayden A, Begon M, Fitter AH (eds) Advances in ecological research, vol 26. Academic, London, pp 171–313

Butcher DN (1980) The culture of isolated roots. In: Ingram DS, Helgelson JP (eds) Tissue culture methods for plant pathologists. Blackwell Scientific, Oxford, pp 13–17

Butcher DN, Street HE (1964) Excised root culture. Bot Rev 30:513–586

Calvet C, Pera J, Barea MJ (1993) Growth response of marigold (*Tagetes erecta* L.) to inoculation with *Glomus mosseae*, *Trichoderma aureoviride* and *Pythium ultimum* in a peat-perlite mixture. Plant Soil 148:1–6

Clarke CA, Mosse B (1978) Recovery of VA mycorrhizal spores after germination. Rothamsted Exp Stn Rep 1978:239

Coombes RD, Haveland-Smith RB (1982) A review of the genotoxicity of food, drug and cosmetic colours and other azo, triphenylmethane and xanthene dyes. Mut Res 98:101–248

Cunningham JL (1972) A miracle mounting fluid for permanent whole-mounts of microfungi. Mycologia 64:906–911

Daniels BA, Graham SO (1976) Effects of nutrition and soil extracts on germination of *Glomus mosseue* spores. Mycologia 68:108–116

Declerck S, Strullu DG, Plenchette C (1996) In vitro mass production of the arbuscular mycorrhizal fungus, Glomus versiforme, associated with Ri T-DNA transformed carrot roots. Mycol Res 100:1237–1242

Dodd JC (1994) Approaches to the study of extraradical mycelium of arbuscular mycorrhizal fungi. In: Gianinazzi S, Schüepp H (eds) Impact of arbuscular mycorrhizas on sustainable agriculture and natural ecosystems. Birkhäuser, Basel, pp 147–166

Doncaster CC (1962) A counting dish for nematodes. Nematologica 7:334–337

Douds DD, Schenck NC (1990) Crypreservation of spores of vesicular–arbuscular mycorrhizal fungi. New Phytol 115:667–674

Douds DD Jr, Nagahashi G, Pfeffer PE, Reider C, Kayser W (2006) On-farm production of AM fungus inoculum in mixtures of compost and vermiculite. Bioresour Technol 97:809–918

Douds DD, Schenck NC (1990) Increased sporulation of vesicular–arbuscular mycorrhizal fungi by manipulation of nutrient regimes. Appl Environ Microbiol 56:413–418

Douds DD Jr (2002) Increased spore production by *Glomus intraradices* in the split-plate monoxenic culture system by repeated harvest, gel replacement, and resupply of glucose to the mycorrhiza. Mycorrhiza 12:163–167

Douds DD, Nagahashi G, Pfeffer PE, Kayser WM, Reider C (2005a) On-farm production and utilization of mycorrhizal fungus inoculum. Can J Plant Sci 85:15–21

Douds DD, Nagahashi G, Pfeffer PE, Kayser WM, Reider C (2005b) On-farm production of AM fungus inoculum in compost–vermiculite mixtures. Appl Environ Microbiol 71:15–21

Dugassa DG, Grunewaldt-Stocker G, Schönbeck F (1995) Growth of *Glomus intraradices* and its effect on linseed (*Linum usitatissimum* L.) in hydroponic culture. Mycorrhiza 5:279–282

Enkhtuya B, Rydlová J, Vosátka M (2002) Effectiveness of indigenous and non-indigenous isolates of arbuscular mycorrhizal fungi in soils from degraded ecosystems and man-made habitats. Appl Soil Ecol 14:201–211

Ferguson JJ, Menge JA (1982) The influence of light intensity and artificially extended photoperiod upon infection and sporulation of *Glomus fasciculatus* on sudangrass and on root exudation of sudan grass. New Phytol 92:183–191

Ferguson JJ, Woodhead SH, Ross JP, Daniels BA (1982) Production of endomycorrhizal inoculum. In: Schenck NC (ed) Methods and principles of mycorrhizal research. American Phytopathology Society, St. Paul, pp 47–59

Filion M, St-Arnaud M, Fortin JA (1999) Direct interaction between the arbuscular mycorrhizal fungus *Glomus intraradices* and different rhizosphere microorganisms. New Phytol 141:525–533

Fisher RA, Yates F (1963) Statistical tables for biological, agricultural and medical research. Oliver and Boyd, Edinburgh

Fortin JA, Guillaume B, Declerck S, Dalpé Y, St-Arnaud M, Coughlan AP, Piché Y (2002) Arbuscular mycorrhiza on root-organ cultures. Can J Bot 80:1–20

Furlan V, Fortin JA (1973) Formation of endomycorrhizae by endogone calospora on *Allium cepa* under three temperature regimes. Nat Can 100:467–447

Fusconi A, Gnavi E, Trotta A, Berta G (1999) Apical meristems of tomato roots and their modifications induced by arbuscular mycorrhizal and soilborne pathogenic fungi. New Phytol 142:505–516

Gardner RO (1975) An overview of botanical clearing technique. Stain Technol 50:99–105

Gaur A, Adholeya A, Mukerji KG (1998) A comparison of AM fungi inoculants using *Capsicum* and *Polianthes* in marginal soil amended with organic matter. Mycorrhiza 7:307–312

Gaur A, Adholeya A (1994) Estimation of VAM spores in soil: a modified method. Mycorrhiza News 6:10–11

Gaur A, Adholeya A (2000) Effects of the particle size of soil-less substrates upon AM fungus inoculum production. Mycorrhiza 10:43–48

Gaur A, Adholeya A (2002) Arbuscular-mycorrhizal inoculation of five tropical fodder crops and inoculum production in marginal soil amended with organic matter. Biol Fertil Soil 35:214–218

Gehring CA, Whitham TG (1991) Herbivore-driven mycorrhizal mutualism in insect-suscetible pinyon pine. Nature 353:556–557

Gerdemann JW, Nicolson TH (1963) Spores of mycorrhizal endogone extracted from soil by wet seiving and decanting. Trans Br Mycol Soc 46:235–244

Giovannetti M, Mosse B (1980) An evaluation of techniques for measuring vesicular-arbuscular mycorrhizal infection in roots. New Phytol 84:489–500

Grace C, Stribley DP (1991) A safer procedure for routine staining of vesicular arbuscular mycorrhizal fungi. Mycol Res 95:1160–1162

Hall IR (1977) Species and mycorrhizal infections of New Zealand Endogonaceae. Trans Br Mycol Soc 68:341–356

Hanssen JF, Thingstad TF, Gohson J (1974) Evaluation of hyphal lengths and fungal biomass in soil by a membrane filter method. Oikos 25:102–107

Harley JL, Smith SE (1983) Mycorrhizal symbiosis. Academic, London, pp 284

Heinzemann J, Weritz J (1990) Rockwool: a new carrier system for mass multiplication of vesicular-arbuscular mycorrhizal fungi. Angew Bot 64:271–274

Hepper CM (1984) Isolation and culture of VA mycorrhizal (VAM) fungi. In: Powell CL, Bagyaraj DJ (eds) VA mycorrhiza. CRC Press, Boca Raton, pp 95–112

Jabaji-Hare S, Perumalla CJ, Kendrick WB (1984) Auto fluorescence of vesicles, arbuscules, and intercellular hyphae of a vesicular arbuscular fungus in leek (*Allium porrum*) roots. Can J Bot 62:2665–2669

Janos DP (1980) Mycorrhizae influence tropical succession. Trop Succ 12:56–64

Jarstfer AG, Sylvia DM (1992) Inoculum production and inoculation strategies for vesicular-arbuscular mycorrhizal fungi. In: Meting B (ed) Soil microbial ecology; application in agriculture and environmental management. Dekker, New York, pp 349–377

Jarstfer AG, Sylvia DM (1995) Aeroponic culture of VAM fungi. In: Varma A, Hock B (eds) Mycorrhiza structure, function, molecular biology and biotechnology. Springer, Berlin Heidelberg New York, pp 521–559

Jeffries P, Barea JM (1994) Biogeochemical cycling and arbuscular mycorrhizas in the sustainability of plant soil systems. In: Gianinazzi S, Schuepp H (eds) Impact of arbuscular mycorrhizas on sustainable agriculture and natural ecosystems. Birkhauser, Basel, pp 101–115

Khalil S, Loynachan TE, Tabatabai MA (1994) Mycorrhizal dependency and nutrient uptake by improved and unimproved corn and soybean cultivars. Agron J 86:949–958

Khan AG (2001) Relationships between chromium biomagnification ratio, accumulation factor, and mycorrhizae in plants growing on tannery effluent-polluted soil. Environ Int 26:417–423

Koide R, Li M, Lewis J, Irby C (1988) Role of mycorrhizal infection in the growth and reproduction of wild vs cultivated plants. I. Wild vs cultivated oats. Oecologia 77:537–543

Kormanik PP, McGraw AC (1982) Quantification of vesicular±arbuscular mycorrhizae in plant roots. In: Schenk NC (ed) Methods and principles of mycorrhizal research. American Phytopathology Society, St Paul, pp 37–45

Koske RE, Testier B (1983) A convenient permanent slide mounting medium. Mycol Soc Am Newsl 34:59

Koske RE, Gemma JN (1989) A modified procedure for staining roots to detect VA mycorrhizas. Mycol Res 92:486–505

Laurent FM, Leea S, Tham FY, Jie H, Diem HG (1999) Aeroponic production of *Acacia mangium* saplings inoculated with AM fungi for reforestation in the tropics. For Ecol Manag 122:199–207

Madan RC, Pankhurst BH, Smith H (2002) Use of fatty acids for identification of AM fungi and estimation of the biomass of AM spores in soil. Soil Biol Biochem 34:125–128

Mallesha BC, Bagyaraj DJ, Pai G (1992) Perlite–soilrite mix as a carrier for mycorrhiza and rhizobia to inoculate *Leucaena leucocephala*. Leucaena Res Rep 13:32–33

Manjunath A, Habte M (1991) Root morphological characteristics of host species having distinct mycorrhizal dependency. Can J Bot 69:671–676

Mathur N, Vyas A (2000) Influence of arbuscular mycorrhizae on biomass production, nutrient uptake and physiological changes in *Ziziphus mauritana* Lam. under water stress. J Arid Environ 45:191–195

McGonigle TP, Miller MH, Evans DG, Fairchild GL, Swan JA (1990) A new method which gives an objective measure of colonization of roots by vesicular–arbuscular mycorrhizal fungi. New Phytol 115:495–501

Menge JA, Steirle D, Bagyaraj DJ, Johnson ELV, Leonard RT (1978) Phosphorus concentrations in plants responsible for inhibition of mycorrhizal infection. New Phytol 80:575–578

Merryweather JW, Fitter AH (1991) A modified method for elucidating the structure of the fungal partner in a vesicular arbuscular mycorrhiza. Mycol Res 95:1435–1437

Merryweather J, Fitter A (1996) Phosphorus nutrition of an obligately mycorrhizal plant treated with the fungicide benomyl in the field. New Phytology 132:307–311

Mohammad A, Khan AG, Kuek C (2000) Improved aeroponic culture of inocula of arbuscular mycorrhizal fungi. Mycorrhiza 9:337–339

Morton JB (1991) INVAM newsletters, vol 1–5. West Virginia University, Morgantown

Morton JB, Redecker D (2001) Two new families of Glomales, Archaeosporaceae et Paraglomaceae, with two new genera, Archaeospora and Paraglomus, based on concordant molecular and morphological caracters. Mycologia 93:181–195

Mosse B, Thompson JP (1984) Vesicular–arbuscular endomycorrhizal inoculum production. I. Exploratory experiments with beans (*Phaseolus vulgaris*) in nutrient flow culture. Can J Bot 62:1523–1530

Mosse B, Hepper CM (1975) Vesicular–arbuscular infections in root-organ cultures. Physiol Plant Pathol 5:215–223

Olsson PA, Bååth E, Jakobsen I, Söderström B (1995) The use of phospholipids and neutral lipid fatty acids to estimate biomass of arbuscular mycorrhizal fungi in soil. Mycol Res 99:623–625

Phillips JM, Hayman DS (1970) Improved procedures for clearing roots and staining parasitic and vesicular–arbuscular mycorrhizal fungi for rapid assessment of infection. Trans Br Mycol Soc 55:158–161

Plenchette C, Declerck C, Diop TA, Strullu DG (1996) Infectivity of monoaxenic subcultures of the arbuscular mycorrhizal fungus *Glomus versiforme* associated with Ri-T-DNA-transformed carrot root. Appl Microbiol Biotechnol 46:545–548

Plenchette C, Fortin JA, Furlan V (1983) Growth responses of several plant species to mycorrhizae in a soil of moderate P-fertility. Part I. Mycorrhizal dependency under field conditions. Plant Soil 70:199–209

Porter WM (1979) The most probable number method for enumerating infective propagules of vesicular arbuscular mycorrhizal fungi in soil. Aust J Soil Res 17:515–519

Redecker D, Morton JB, Bruns TD (2000) Ancestral lineages of arbuscular mycorrhizal fungi (*Glomales*). Mol Phyl Evol 14:276–284

Redecker D, Thierfelder H, Werner D (1995) A new cultivation system for arbuscular mycorrhizal fungi on glass beads. Angew Bot 69:189–191

Riker AJ, Banfield WM, Wright WH, Keitt GW, Sagen HE (1930) Studies on infectious hairy root of nursery apple trees. J Agric Res 41:507–540

Schaffer GF, Peterson R (1993) Modifications to clearing methods used in combination with vital staining of roots colonized with vesicular–arbuscular mycorrhizal fungi. Mycorrhiza 4:29–35

Schenck NC, Perez Y (1990) Manual for the identification of VA mycorrhizal fungi. Synergistic, Gainesville

Schubert A, Marzachí C, Mazziteli M, Cravero MC, Bonfante- Fasolo P (1987) Development of total and viable extraradical mycelium in the vesicular–arbuscular mycorrhizal fungus *Glomus clarum* Nicol. & Schenck. New Phytol 107:183–190

Schussler A, Schwarzott D, Walker C (2001) A new fungal phylum, the Glomeromycota: phylogeny and evolution. Mycol Res 105:1413–1421

Sieverding E, Barea JM (1991) Perspectives de la inoculation de sistemas de production vegetal con hongos formadores de micorrizas VA. In: Olivares J, Barea JM (eds) Fijacion y movilizacion biologica de nutrients. (Coleccion nuevas tendencias, vol II) CSIC, Madrid, pp 221-245

Smith GW, Skipper HD (1979) Comparison of method to extract spores of vesicular arbuscular mycorrhizal fungi. Soil Sci Soc Am J 43:722–725

Smith SE, Barker SJ (2002) Plant phosphate transporter genes help harness the nutritional benefits of arbuscular mycorrhizal symbiosis. Trends Plant Sci 75:189–190

Smith SE, Read DJ (1997) Mycorrhizal symbiosis. Academic, London

Sreenivasa MN, Bagyaraj DJ (1988) Selection of a suitable host for mass multiplication of *Glomus fasciculatum*. Plant Soil 106:289–290

Staddon PL, Fitter AH (2001) The differential vitality of intraradical mycorrhizal structures and its implications. Soil Biol Biochem 33:129–132

Strullu DG (1991) Les relations entre les plantes et les champignons. In: Strullu DG, Garbaye J, Perrin R, Plenchette C (eds) Les mycorhizes des arbres et plantes cultivees. Lavoisier, Paris, pp 9–49

Strullu DG, Romand C (1986) Méthode d'obtention d'endomycorhizes à vésicules et arbuscules en conditions axéniques. CR Acad Sci Paris Sci Vie 303:245–250

Strullu DG, Romand C (1987) Culture axénique de vésicules isolées à partir d'endomycorhizes et ré-association in vitro à des racines de tomate. CR Acad Sci Paris Sci Vie 305:15–19

Subramanian KS, Charest C, Dwyer LM, Hamilton RI (1997) Effects of arbuscular mycorrhizae on leaf water potencial, sugar content, and P content during drought and recovery of maize. Can J Bot 75:1582–1591

Sylvia DM (1994) Vesicular–arbuscular mycorrhizal (VAM) fungi. In: SSSA (ed) Methods of soil analysis, part 2. Microbiological and biochemical properties. (SSSA book series, no 5) Soil Science Society of America, Madison, pp 351–378

Sylvia DM, Hubbell DH (1986) Growth and sporulation of vesicular–arbuscular mycorrhizal fungi in aeroponic and membrane systems. Symbiosis 1:259–267

Tepfer D (1989) Ri T-DNA from *Agrobacterium rhizogenes*: a source of genes having applications in rhizosphere biology and plant development, ecology and evolution. In: Kosuge T, Nester EW (eds) Plant–microbe interactions, vol 3. McGraw–Hill, New York, pp 294–342

Tisserant B, Gianinazzi-Pearson V, Gianinazzi S, Gollotte A (1993) In planta histochemical staining of fungal alkaline phosphatase activity for analysis of efficient arbuscular endomycorrhizal infections. Mycol Res 97:245–250

Vierheilig H, Coughlan AP, Wyss U, Piché P (1998) Ink and vinegar, a simple staining technique for arbuscular–mycorrhizal fungi. Appl Environ Microbiol 6:5004–5007

Vilariño A, Arines J, Schüepp H (1993) Extraction of vesicular–arbuscular mycorrhizal mycelium from sand samples. Soil Biol Biochem 25:99–100

Vivas A, Marulanda M, Mez G, Barea JM, Azcon R (2003) Physiological characteristics (SDH and ALP activities) of arbuscular mycorrhizal colonization as affected by *Bacillus thuringiensis* inoculation under two phosphorus levels. Soil Biol Biochem 35:987–996

Whitbeck JL (2001) Effects of light environment on vesicular–arbuscular mycorrhiza development in Inga leiocalycina, a tropical wet forest tree. Biotropica 33:303–311

White PR (1943) A handbook of plant tissue culture. Cattel, Lancaster

26 Field Trials of Bioinoculants

I. Ortaş and A. Varma

26.1 Introduction

The most widespread symbiotic association between micro-organisms and higher plants are arbuscular mycorrhizae (AM), which are present in a range of horticultural, agricultural and forestry plants. The term mycorrhiza, which literally means *fungus-root* (*myco*, fungus; *rhiza*, root), was first applied to fungus–tree associations described in 1885 by the German forest pathologist A.B. Frank. Mycorrhiza is a mutalistic symbiosis (non-pathogenic association) between soil-borne fungi and the roots of high plants. The symbiosis between fungi and plant root is a bi-directional movement of nutrients where carbon flows to the fungus and inorganic nutrients move to the plant, thereby providing a critical linkage between the plant root and soil in the rhizosphere. Mycorrhizal fungi usually proliferate both in the root and in the soil. In natural ecosystems, in nutrient-poor or moisture-deficient soils, nutrients taken up by the extrametrical hyphae can lead to improved plant growth and reproduction. Since mycorrhiza-inoculated plants take more nutrients, they are more competitive and better able to tolerate environmental stresses than non-mycorrhizal plants. It has been estimated that 90% of all plant species belong to genera that characteristically form mycorrhizae (Smith and Read 1997). Mycorrhizal infection occurs in 83% of dicotyledonous and 79% of monocotyledonous plants (Peterson et al. 2004).

A mycorrhizal root system seems able to selectively absorb phosphorus (P) from deficient soils (Fig. 26.1). In all these kinds of mycorrhiza, it is usual to find hyphal connections from the infected root into the soil. The hyphae may extend considerable distances (centimeters). The role of hyphae in mycorrhizal

Ibrahim Ortaş: University of Çukurova, Faculty of Agriculture, Department of Soil Science, 01330 Balcali, Adana, Turkey

Ajit Varma: Amity Institute of Microbial Sciences, Amity University Uttar Pradesh, Sector 125, Expressway, Noida 201303, UP, India, email: ajitvarma@aihmr.amiity.edu

Soil Biology, Volume 11
Advanced Techniques in Soil Microbiology
A. Varma, R. Oelmüller (Eds.)

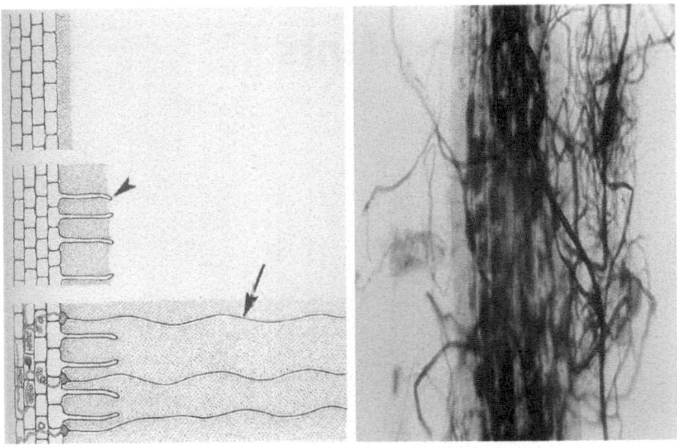

Fig. 26.1 The role of mycorrhizal hyphae on nutrient depletion zone (Peterson et al. 2004)

infection and nutrient uptake has not been studied intensively because a convenient technique was not developed until recently. It has been estimated that the amount of external hyphae is 80 cm/cm root length in onions (Sanders and Tinker 1973). However it is not easy to measure the amount of external hyphae formed in soil by mycorrhizal fungi.

26.2 Effect of Mycorrhizal Infection on Nutrient Uptake

AM infection can increase plant growth and nutrient uptake, especially an increase in P uptake by the host. The contribution of mycorrhizae is considered to be a function of an increase in P uptake due to mycorrhizal infection. Plant roots infected with AM fungi are known to have a higher P absorption ability compared with non-mycorrhizal plants in P-deficient soils (Abbott and Robson 1982).

Arbuscular mycorrhizal fungi (AMF) are generally known to benefit plant nutrient uptake such as P in soil of low fertility (Ortaş 2003). Mycorrhizal inoculation have a positive effect on maize plant P uptake, which was exhibited even when high level of P was applied; but high P fertilization reduced the degree of root colonization and also the quantity of external hyphae (Posta and Fuleky 1997).

In arid and semi-arid areas such as Turkey there are limitations on the amount of water used for plant cultivation and most of the soils are low in nutrient content, such as P, Zn and Fe, which are diffusion-limited in soils. Even if there were no deficiency of nutrient, there are several environmental stress factors such as temperature and, consequently, accumulation of salt in the soil due to evaporation. Thus there is a tendency to use the natural sources such as mycorrhizal fungi to reduce P fertilizer application and hence obtain better plant growth in nutrient-deficient soils. Also mycorrhizal inoculation reduces the quantity of P fertilizer normally required (Charron et al. 2001). Soil chemical and biological factors strongly affect P management. It seems to be very important to manage P in soil, since it is an ecological necessity for the future of soil quality.

For a given dry weight, mycorrhizal plants usually have higher P concentrations in their plant tissue than non-mycorrhizal plants (Stribley et al. 1980). How mycorrhizal plants obtain more P from soil than non-mycorrhizal plants is not yet fully understood. Several mechanisms have been proposed to define the AM effect on improving the absorption of available phosphate and these have been mentioned. Mycorrhizae may induce both quantitative and qualitative changes in plant P utilization (Smith and Read 1997). The amount of acid phosphatase present in AM hyphae (Tarafdar and Marschner 1994) and increased phosphatase activity of root surfaces as a result of infection (Allen et al. 1981) may liberate inorganic P from organic P sources, making P available for uptake. Tinker (1975) suggested that the roots of mycorrhizal plants may alter the rhizosphere chemistry by changing soil pH and may produce exudates such as organic acids which may increase the availability of phosphorus by liberating phosphate ions in the soil. There is still a wide gap in the understanding of the mechanisms involved in increased P availability in the soil by mycorrhiza-infected roots.

In addition to P, AM fungi enhance the acquisition of other nutrients, such as N and K, and immobile micro-nutrient cations, particularly Zn and Cu (Li et al. 1991).

Fries et al. (1998) tested the effect of different levels of P application on mycorrhizal formation and they found that under low P levels, a mycorrhiza-inoculated plant accumulated a greater amount of shoot dry weight, root P concentration and protein concentration than a non-inoculated plant. Medeiros et al. (1994) showed that mycorrhiza-inoculated plants had significantly higher uptake of P, K, Fe, and S than non-mycorrhizal plants. Plants colonized with AM fungi generally have greater growth and acquisition of mineral nutrients and often have a greater ability to withstand drought, compared with non-mycorrhizal plants (Al-Karaki and Clark 1998).

26.3 Effect of Soil Fumigation and Mycorrhizal Inoculation on Plant Growth Under Field Conditions

In the plain of Cukurova there is a serious problem with soil-borne disease such as plant parasitic nematodes, soil-borne plant pathogens, root-rot and some weed pests. Nearly 25% of yield reduction occurs from year to year. In order to prepare a safe seedbed and healthy yield, farmers are using a high amount of chemicals. Due to a combination of soil-borne pathogens, nematodes and weeds, soil fumigation with products such as methyl bromide (MeBr) has been essential for horticultural practice in this area. Since MeBr eliminates both desirable organisms such as arbuscular mycorrhizal fungi (AMF) and undesirable soil organisms, the plant growth and nutrient uptake, especially P and Zn uptake, have significantly declined.

Soil fumigation, as a partial soil sterilization, affects soil chemical properties in addition to the removal of viable mycorrhizal fungi and other micro-organisms. The fertility of sterilized soil may be different than non-sterile soil. Partial soil sterilization generally stimulates subsequent plant growth when compared with non-sterile soils (Ortaş et al. 2004). The main aim of partial soil sterilization in mycorrhizal studies is to eliminate indigenous mycorrhizal spores and pathogenic microbial activity in the soils, but this procedure often alters the chemical and biological properties of the soil.

Soil fumigation may have a dual effect on plant growth, such as increased growth by elimination of soil-borne pathogens, or conversely, stunted growth by exacerbation of existing P deficiency. As a result of reduction of mycorrhizal colonization in low-P soils with soil fumigation, there is a Br accumulation in the soil and plant uptakes in Br are more than ten times greater than control plants (Haas et al. 1987).

Since soil fumigation reduces useful organisms such as mycorrhizae, it is necessary to reinoculate mycorrhizae. Especially mycorrhiza-dependent plants need more mycorrhizal inoculation. It is very important to use mycorrhiza at least for horticulture plants which are transplanted to the soil as a seedling. It may be easy to produce mycorrhiza-inoculated seedlings.

For these reasons, several field experiments were set up to investigate the interaction effect of MeBr, mycorrhizae and P fertilizer on plant yield, growth, nutrient uptake and mycorrhizal formation. The aim of the research is to investigate the effect of MeBr, mycorrhizae and P fertilizer interaction on plant yield, growth, nutrient uptake and mycorrhizal formation. The overall results revealed that yields were lower in sterile (fumigated) plots than in the non-sterile (non-fumigated) ones. Conversely, MeBr application reduced yield compared with the non-fumigated one whether or not the plants were inoculated. As can be seen in Fig. 26.2, under field conditions, the yield of onion plants grown in the MeBr-treated plot was reduced. But mycorrhizal inoculation compensated the yield reduction, compared with the non-inoculated plots.

In this experiment rhizosphere soil was also used as a mycorrhizal source; and it was found that, under field conditions, indigenous mycorrhizal inoculation increased the onion yield in sterile plots. The interpretations of the results show that mycorhizal inoculation may have had some other benefits to the plant, such as protecting it against soil-borne pathogens and environmental stress.

The impact of mycorrhizal fungi is usually assessed by measuring plant growth and P uptake following inoculation of the fungi into sterilized soils (Hetrick et al. 1986). However, growth responses are erratic and sometimes occur when AMF are added to non-sterile soil (Ortaş et al. 1996). In some cases, the root colonization is less in non-sterile soil than in sterilized soil.

Farmers use MeBr before horticultural crops are planted, for the elimination of undesirable soil organisms. At the same time, they kill off all organisms. Since the organisms have a long-term effect on sustainability and quality of soil, it is

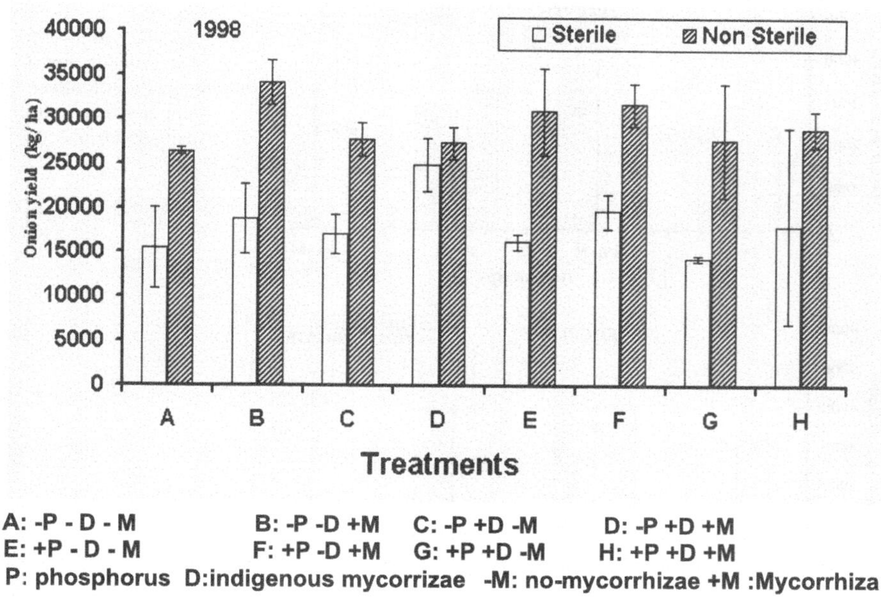

Fig. 26.2 Effect of MeBr on onion yield under field conditions with and without indigenous and selected mycorrhizae

sound to use alternative fumigant sources rather than MeBr. For example, using organic sources and mycorrhiza and their combination are alternative sources.

Ortaş et al. (2003) showed that AMF was very active in plants grown on non-fumigated soil and that AMF activity increased plant growth and nutrient uptake. In non-fumigated plots it seemed still that AMF were active since there was a high mycorrhizal infection.

Ortaş et al. (2003) showed that mycorrhizal inoculation increased plant yield significantly, compared with non-inoculated plants. When zero P was applied, the effect of mycorrhizal inoculation on plant yield was higher than yield increased with additional P application. When zero P was applied, mycorrhizal inoculation increased tomato yield up to 52%, eggplants up to 28% and pepper up to 36%, but with P addition, mycorrhizal inoculation increased yield up to 28%, 14% and 21%, respectively, compared with non-inoculated plants (Fig. 26.3). Mycorrhizal inoculation also increased plant zinc and copper uptake (Ortaş et al. 2003).

In fumigated plots P and Zn content reduced dramatically, which was related to a reduction in AMF colonization (Ortaş et al. 2003). Mycorrhizal inoculation increased the root Mn concentration but not the shoot Mn concentration.

It seems that plant yield supplied by mycorrhizal inoculation cannot be explained only by the effect of mycorrhizal inoculation on nutrient uptake. Olsen et al. (1999) found that mycorrhizal inoculation increased the pepper and tomato growth and they claimed that the growth response of vegetable crops grown within the greenhouse from colonization by an established mycorrhizal

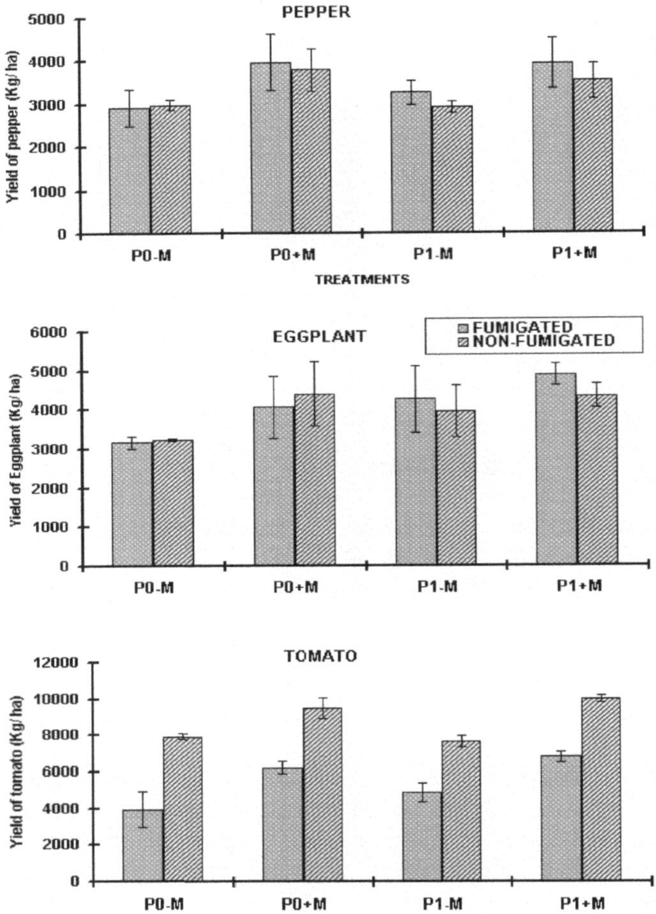

Fig. 26.3 The effect of MeBr and mycorrhizal inoculation on eggplant, tomato and pepper yields under field conditions

mycelium appeared to depend on a critical balance of P and C supply. Since the soil P level was medium and the plant P content had not been affected by mycorrhizal inoculation, it meant the soil P level was enough for both mycorrhizal and non-mycorrhizal plants. In the same field the experiment was repeated with several mycorrhizal inoculum for three years and the conclusion was that the effect of mycorrhizal inoculation on plant growth under field condition depends on year, and mycorrhizal inoculum potential.

It appears that there are some other benefits from mycorrhizae for horticultural plants, such as controlling disease and increasing plant resistance. We conclude that, although mycorrhizal inoculation increases some vegetable yield, this increase is not easily explained through a better nutrient uptake by AMF plants than by un-colonized plants. Mycorhizal inoculation may have some other benefits to plants, such as protection against soil-borne pathogens and environmental stress.

Most of these studies have been performed in the greenhouse under controlled conditions without the influence of complex interactions of other environmental variables. When bringing any mycorrhizal question to the field, one of the more difficult problems is the creation of a suitable non-mycorrhizal control, since a majority of plants is normally mycorrhizal. Fungicides can be useful in distinguishing the mycorrhizal effects on plants in the field from certain other influences.

Eggplant, tomato and pepper are among the most valuable vegetables grown for fresh-market production in Cukurova region, Adana-Turkey. Due to a combination of soil-borne pathogens, nematodes and weeds, the use of soil fumigation such as MeBr has been essential for horticultural practice in this area.

26.4 Effect of Mycorrhizal Inoculation on Plant Growth and Nutrient Uptake under Non-Sterile Field Conditions

Indigenous AM fungi have been found in most non-sterile soils and experimentally it has been shown that introduced mycorrhizal inoculum can infect the host plant under non-sterilized soil conditions (Abbott and Robson 1978). This treatment usually alters soil fertility as the result of an alteration of soil chemical and biological properties (Ortaş and Harris 1996). Although soil fumigations stimulate plant growth through eliminating the soil-borne pathogens and weeds, fumigation usually stunts plant growth due to a reduction in the viable AM population in low-fertility soils (Ellis et al. 1995).

At low-level P applications, sweet corn yield increased as a result of mycorrhizal inoculation (Fig. 26.4). Additionally, mycorrhizal inoculation increased

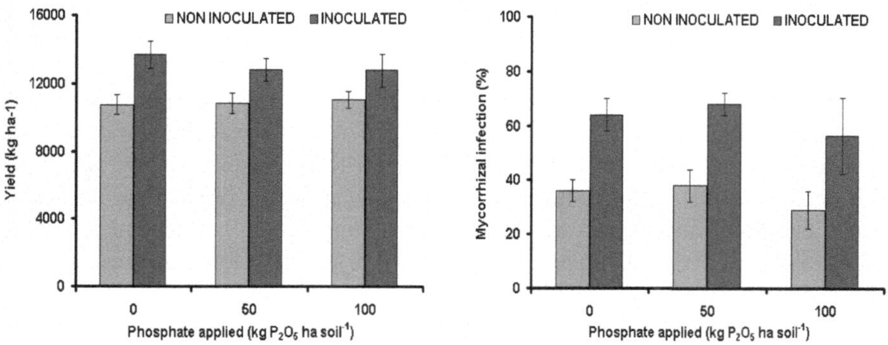

Fig. 26.4 The effect of different rates of P application (0, 50, 100 kg/ha P_2O_5) and mycorrhizal inoculation on sweet corn yield and root inoculation

Table 26.1 The effect of phosphorus application and mycorrhizal inoculation on N, P and K concentration (%) in shoots of sweet corn at silking. ± Standard error

Treatment	N	P	K
–Mycorrhizae			
P0	2.31±0.01	0.16±0.01	0.80±0.28
P1	2.44±0.02	0.20±0.00	0.90±0.14
P2	2.44±0.12	0.22±0.01	0.60±0.03
+Mycorrhizae			
P0	2.60±0.00	0.22±0.01	1.20±0.28
P1	2.84±0.12	0.23±0.01	1.30±0.14
P2	2.58±0.11	0.24±0.03	1.00±0.28

Table 26.2 The effect of phosphorus application and mycorrhizal inoculation on micronutrient content (Zn, Fe, Cu, Mn; mg/kg dry weight) in shoots of sweet corn at silking

Treatment	Zn	Fe	Cu	Mn
–Mycorrhizae				
P0	10.5±1.8	91.6±9.1	3.0±0.3	90.4±13.3
P1	14.5±2.1	103.2±2.0	3.8±0.8	92.9±9.8
P2	15.0±2.8	159.7±29.6	3.5±0.4	102.8±16.7
+Mycorrhizae				
P0	16.2±2.0	97.7±2.1	3.1±1.0	109.7±8.1
P1	15.8±0.8	109.7±12.0	4.0±0.3	100.4±11.0
P2	15.4±1.6	129.9±9.8	4.7±0.4	116.6±1.7

plant N, P, K concentrations significantly (Table 26.1). Furthermore, plant Zn and Mn concentrations increased; however, Fe and Cu concentrations remained the same during the experiment (Table 26.2). In non-inoculated plants, the P concentration of sweet corn shoots increased as the P fertilization increased, but in inoculated plants there was no significant increase. Mycorrhizal inoculation increased significantly root colonization but, with the higher P level addition, the extent of AMF colonization was reduced. It was concluded that, although soils have potential indigenous spores which can effectively infect plant roots, additional mycorrhizal inoculation increases root infection significantly and consequently increases plant nutrient uptake and yield.

As can be seen from Fig. 26.4, the increasing P addition also reduced the root infection, especially with 100 kg/ha P_2O_5 application. Our previous experiments also showed similar results in the same soil.

As can be seen from Table 26.1, mycorrhizal inoculation significantly increased sweet corn plant K content. It seems that the most important K uptake is by mycorrhizal inoculation. So far most work has focused on P uptake; however K is very important element in terms of plant quality (Ortaş and Sari 2003).

Under field conditions without using soil sterilization it is important to manage the indigenous mycorrhizae when the soil nutrients, especially phosphorus, are limited under the field conditions. For sustainable P management soil and crop management can help to get maximum benefit from indigenous mycorrhizae (Ortaş and Sari 2003). The research area was a preserved area for a long time and no pesticide and herbicide were used. So it was expected that the area is rich in soil biological fertility especially in indigenous mycorrhizae.

In order to see the effect of mycorrhizal inoculation on micro-nutrient uptake under field conditions a field experiment was set up in the Research Farm of the University of Cukurova, Faculty of Agriculture, Adana-Turkey. In this experiment onion, garlic, chickpea and horse bean plants were used as test plants. Cocktail mycorrhizae were used as mycorrhizae species.

Since chickpea and horse bean are nitrogen-fixing plants they take more micro-nutrient. Mycorrhiza inoculation significantly increased plant micro-nutrient uptake as well.

The results showed that, in mycorrhizal plots, the yields of onion, garlic, chickpea and horse bean plants was higher than in non-mycorrhizal plants (Table 26.3). Mycorrhizal inoculation increased the shoot Cu and Zn content (Table 26.4).

At the lowest P supply, shoot dry matter production was significantly depressed (Table 26.3). This decreasing effect of low P supply was particularly obvious when soils were sterilized and not inoculated with mycorrhizae. Inoculation of soil with mycorrhizae species significantly increased the plant growth and P uptake of plants, especially under low P supply (Table 26.3). In low P application, plant roots were strongly infected and consequently increased plant growth, but in high P level application there was a slight reduction in root infection. The results show that mycorrhizal inoculation is an effective practice for improving crop production in P-deficient soils.

In another experiment carried out under field conditions mycorrhizal inoculation was successfully applied in non-sterile soil conditions for wheat, which is a strategical plant for the region. During 1999 and 2000 a successive field

Table 26.3 Effect of mycorrhizal inoculation and P application on onion, garlic, chickpea and horsebean yield under field conditions

Treatment	Onion	Garlic	Chickpea	Horsebean
	Yield (kg/ha)			
–P–M	2812±200	4927±526	15 778±120	68 611±2585
+P–M	3229±210	9621±1294	23 500±102	87 389±3064
–P+M	3681±125	7883±431	25 944±236	78 222±2834
+P+M	3768±220	11 050±100	25 667±157	107 056±4878

Table 26.4 Effect of mycorrhizal inoculation and P application on onion, garlic, chickpea and horsebean plant nutrient uptake and root infection under field conditions

Treatment	Fe	Cu	Zn	Mn	P	Infection
	(mg/kg dry weight)				(%)	
Onion						
−P−M	52.8±5.0	7.1±0.1	13.6±1.2	39.6±0.6	0.24±0.01	7±3
+P−M	78.3±6.7	7.5±0.9	14.2±2.7	58.2±11.0	0.28±0.01	10±3
−P+M	53.0±4.5	7.2±0.9	17.6±3.2	56.9±16.1	0.28±0.04	42±25
+P+M	67.6±1.8	6.4±0.8	16.1±3.6	42.6±3.7	0.32±0.05	35±12
Garlic						
−P−M	60.6±9.9	7.7±0.6	13.8±0.7	25.6±0.6	0.25±0.01	13±8
+P−M	100.6±4.2	8.6±4.9	13.9±1.0	26.7±0.5	0.29±0.01	14±12
−P+M	65.9±17.5	11.5±0.1	16.8±2.3	27.0±0.0	0.32±0.03	41±7
+P+M	250.2±237.0	12.5±7.1	15.4±2.2	29.8±2.7	0.30±0.01	32±3
Chickpea						
−P−M	376.7±0.0	14.0±0.0	26.7±0.0	65.4±0.0	0.22±0.00	10±0
+P−M	232.5±24.9	14.9±0.8	26.9±0.6	109.6±27.2	0.24±0.01	8±2
−P+M	235.6±60.4	20.2±4.2	27.3±2.1	94.2±6.3	0.31±0.01	35±7
+P+M	255.6±13.4	21.6±7.3	27.7±2.1	93.1±9.8	0.33±0.01	42±7
Horsebean						
−P−M	590.2±77.14	17.7±0.8	17.0±4.6	114.6±32.9	0.17±0.02	27±0
+P−M	390.1±97.651	18.4±12.0	14.4±4.7	78.9±0.8	0.18±0.03	15±7
−P+M	488.3±190.84	24.0±9.9	21.3±3.9	88.4±17.0	0.22±0.03	38±12
+P+M	449.6±69.437	15.2±3.6	26.7±1.2	103.9±7.8	0.24±0.04	38±2

experiment were set up on Menzilat soil series (typical xerofluvent) which is located in the Research Farm of the University of Cukurova, Faculty of Agriculture, Adana/Turkey. In that experiment 0, 100 and 200 kg/ha P_2O_5 were applied as triple superphosphate. Mycorrhizal inoculum was applied (by hand) 50 mm under the seeds. After two years evaluation it was found that mycorrhizal inoculation under field conditions significantly increased wheat yield (Fig. 26.5). Also, increasing P application increased wheat yield. In the same experiment mycorrhizal inoculation also increased plant P, Zn and Cu content, compared with the control plant.

The results show that mycorrhizal inoculation increased wheat yield, but at the same time indigenous soil mycorrhizal spores significantly inoculated plant roots and consequently the plant got a benefit from indigenous mycorrhizae.

Plant yield increased with increasing P addition in non-inoculated plots. But in the inoculated plot increasing the P addition increased plant yield up to 50 kg/ha P_2O_5. In further addition, P did not increase plant yield (Fig. 26.5). Also, root inoculation reduced with increasing P addition.

Plant species and cultivars are also different in term of nutrient uptake and their colonization by mycorrhizal fungi. Baon et al. (1993) tested eight barley cultivars for P efficiency by comparing their efficiency with *G. etinicatum* inoculum. Responsiveness to mycorrhizae was negatively correlated with agronomic P efficiency and P utilization efficiency. Similarly, Hetrick et al. (1996) used ten wheat cultivars compared at three P regimes and found that mycorrhizal responsiveness declined with increasing P for the six "responsive" cultivars, but four "non-responsive" cultivars were unaffected.

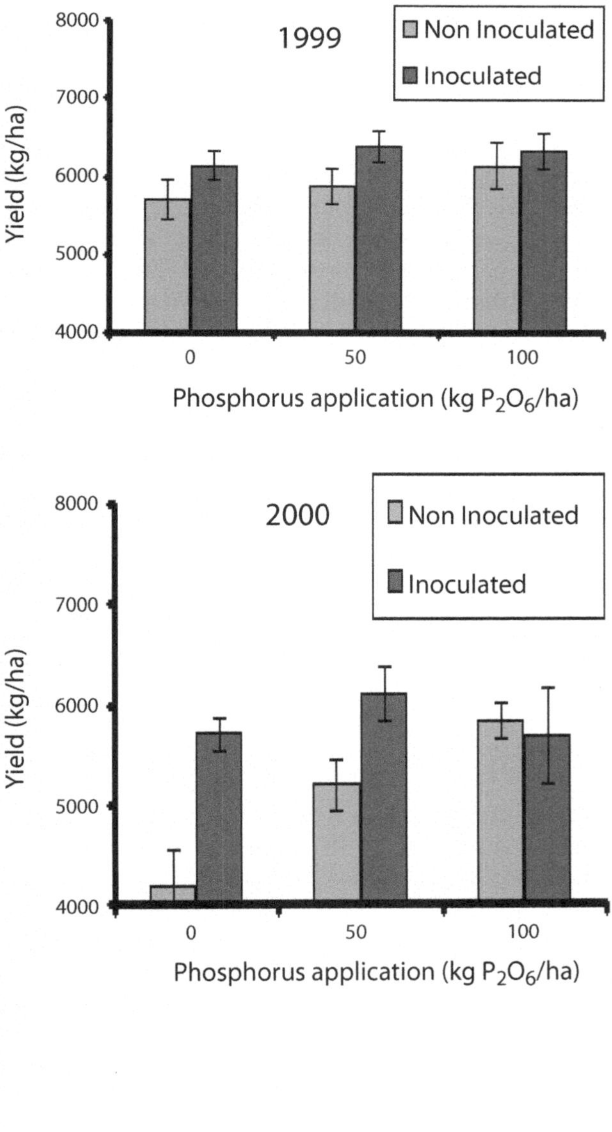

Fig. 26.5 The effect of mycorrhizal inoculation and P application on wheat yield under field condition

26.5 Soil and Crop Management System

Plant responses to mycorrhizal inoculation can be affected by several factors, such as mycorrhizal dependency of the host crop, the nutrient status of the soil and the inoculum potential of the mycorrhizal fungi. Also, soil and crop management practices such as tillage, crop rotation and fallowing may adversely affect populations of mycorrhizal fungi in the field. Soil disturbance significantly reduces the native inoculum potential and consequently reduces nutrient uptake and plant growth (Ortaş et al. 2002). Management practices can influence the types of vesicular–arbuscular mycorrhizal fungi found in agricultural soils. Culturing in soils from degraded ecosystems significantly influences the effectiveness of indigenous AMF isolated from disturbed and undisturbed soils (Enkhtuya et al. 2000). The development of AMF isolates is reduced in soils with more adverse chemical properties, irrespective of the isolate origin (Enkhtuya et al. 2000). Understanding the contributions of soil micro-organisms to soil stabilization at the molecular level will lead to ways to enhance inputs for sustainable agricultural systems.

Plant and soil management should be implemented for mycorrhiza-dependent higher plants, especially less soil distribution, less irrigation and less fertilizer application. In the past 50 years, it has been accepted that taking the maximum yield per unit of land has depredated the soil; and the indigenous mycorrhizal organisms have also been depredated.

The management of mycorrhizal populations in the field is certainly feasible and requires a clear understanding of the ecology of plant communities and different farming systems which affect the populations of mycorrhizal fungi and their diversity and the nutrient uptake and growth of crops. Mycorrhizal inoculation had a positive effect on maize plant P uptake, which was still exhibited even when a high level of P was applied. But high P fertilization reduced the degree of root colonization and also the quantity of external hyphae (Posta and Fuleky 1997). The major benefit of the mycorrhizal symbiosis for crops is improved P uptake; and the management of mycorrhizal fungi will be most critical when soil P is limiting. In temperate zones, P is sometimes applied in excess of crop demand. When the soil P is so high, then root infection percentages are reduced.

Micro-nutrients are very important in terms of human health. Since agriculture is the main source of macro- and micro-nutrients for human food, it is very important to produce a balance of feed plants for the food chain from soil to human. It has been also shown that mycorrhizal inoculation can increase micro-nutrients such as Zn and Cu (Kothari et al. 1990). Liu et al. (2000) reported that the total content of Zn and Cu in maize shoots was higher in mycorrhizal than in non-mycorrhizal plants grown in soils with low P addition. Similarly, Tarkalson et al. (1998) and Ryan and Angus (2003) reported that, under field conditions, the total crop Zn uptake and grain Zn concentration were positively correlated with colonization by AMF, due to enhanced Zn uptake after anthesis of the wheat plant.

26.6 Inoculation Techniques

Compared with pot experiments, less work has been done under field conditions. Field responses to mycorrhizal inoculation were often disappointing, especially in high-input agricultural systems. It has been concluded by many researchers that mycorrhizae have little practical importance in agriculture, since each agro-ecosystem has its own ecological conditions, such as nutrient status of the soil, mycorrhizal dependency, inoculum potential of the indigenous mycorrhizal fungi, crop rotation and fallow systems. Agriculturists should appreciate the distribution of mycorrhizae within their systems and understand the impact of their management decisions on mycorrhizal functioning (Ortaş et al. 2002). Inoculum potential can be adversely affected by management practices such as fertilizer application, pesticide use, crop rotation, fallowing, tillage and topsoil removal. Field experiments have shown that most agricultural plants are colonized by mycorrhizal fungi, which have a substantial impact, both positive and negative, on crop productivity (Johnson 1993). Under field conditions, native inoculum potential is generally low and sometimes ineffective (Ortaş 2003). It is very important to know the mycorrhizal inoculation potential before using mycorrhizal inoculum.

Since AM fungi cannot be grown on laboratory media, the production of a large quantity of inoculum is difficult, as is the inoculation of soil under field conditions. Also since most of the commercially important crops are mainly horticultural plants and are raised under nursery conditions before being transplanted to the main field, the inoculation of soil in the nursery would not only result in a saving of the cost of production of the inoculum but would also help in the better establishment of the transplanted horticultural seedling. Horticulture plants which are grown as seedlings give a high response to mycorrhizae. Mycorrhizal seedlings are more reliable then non-mycorrhiza-inoculated ones. It is sound to produce mycorrhiza-inoculated seedlings before transplanting to the field conditions. Our early results showed that mycorrhiza-infected seedlings are highly resistant to environmental stress factors. Under field conditions, the effect of mycorrhizal inoculation on mortality of seedling was tested. Ortaş et al. (2004) observed that non-mycorrhizal seedlings had a high mortality but mycorrhizal seedlings had less mortality (Table 26.5).

Inoculum strategies are very important. Ortaş et al. (2004) tested several techniques to develop suitable inoculum strategies. Very recently biotechnological

Table 26.5 Mycorrhizal and non-mycorrhizal dead and surviving seedlings after transplanting to field conditions (mean of three replicates; Ortaş, unpublished data)

Plant species	Number of seedlings transplanted to the plot	Number of dead seedlings	% dead seedlings	% surviving seedlings
–Mycorrhizae				
Tomato	48	1	2	98
Pepper	60	4	8	92
Eggplant	48	3	7	93
Bell pepper	60	4	8	92
Marrow	24	3	13	87
Cucumber	24	3	13	87
Melon	24	2	9	91
Watermelon	24	7	29	71
+Mycorrhizae				
Tomato	48	0	0	100
Pepper	60	1	2	98
Eggplant	48	0	0	100
Bell pepper	60	0	0	100
Marrow	24	0	0	100
Cucumber	24	1	4	96
Melon	24	0	0	100
Watermelon	24	2	8	92

techniques were applied before transplanting mycorrhiza-inoculated seedlings to the field conditions, which is a most feasible technique.

Very recently a new technique was tested for better seedling performance under field conditions. Mycorrhiza-inoculated and uninoculated pepper seedlings were transplanted to the field with and without seedlings treated with a liquid solution containing mycorrhizal inoculum. The results showed that mycorrhiza-inoculated seedling production is very important (Fig. 26.6). Also, before transplant to the field conditions, seedlings can be treated with a mycorrhizal liquid solution and are highly responsive to such inoculation. The technique is easy and practical. A liquid solution is prepared using a large quantity of spores (soil, roots, hyphae), mixed with water 1:1, v/v. Seedling roots are dipped into the solution before being transplanted to the nursery hole.

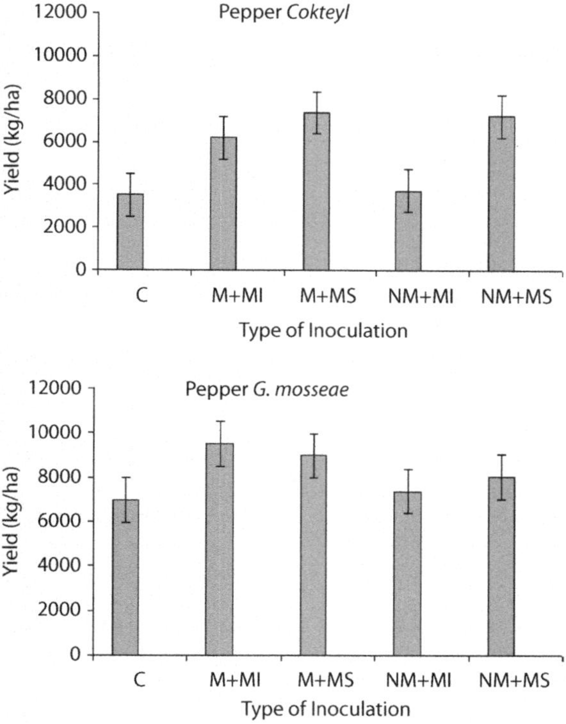

Fig. 26.6 Effect of different inoculation types on paper plant growth under field conditions (Ortaş, unpublished data)

26.7 Conclusion

A large number of studies have expanded our understanding of the potential contribution of mycorrhizae to nutrient uptake under field conditions. Since plant species can give different effects upon mycorrhizal inoculation, these can be related to other beneficial effects of mycorrhizal infection. Also, the response depends on several factors, such as genetic variation and environmental factors.

Recently it was reported that mycorrhizae have several benefits to plants other than nutrient uptake, such as resistance to water deficiency (Bowen and Rovira 1999; Drüge and Schönbeck 1992; Goicoechea et al. 1996). So it is very important to manage the indigenous mycorrhizae when soil nutrients, especially P, are limited under field conditions. For sustainable P management, soil and crop management can help to get maximum benefit from indigenous mycorrhizae (Ortaş 2003). Bowen and Rovira (1999) suggested that a good managed rhizosphere would increase soil and plant quality. Rizosphere management can increase the useful micro-organisms in the plant–soil system. For this reason, the management of mycorrhizae is very important for agricultural sustainability.

References

Abbott LK, Robson AD (1978) Growth of subterranean clover in relation to the formation of endomycorrhizas by introduced and indigenous fungi in a field soil. New Phytol 81:575–587

Abbott LK, Robson AD (1982) The role of vesicular-arbuscular mycorrhizal fungi in agriculture and selection of fungi for inoculation. Aust J Agric Res 33:389–408

Al-Karaki GN, Clark RB (1998) Growth, mineral acquisition and water use by mycorrhizal wheat grown under water stress. J Plant Nutr 21:263–276

Allen MF, Sexton JC, Moore TS Jr, Christensen M (1981) Influence of phosphate source on vesicular-arbuscular mycorrhizae of *Bouteloua gracilis*. New Phytol 87:687–694

Baon JB, Smith SE, Alston AM (1993) Mycorrhizal responses of barley cultivars differing in P efficiency. Plant Soil 157:97–105

Bowen GD, Rovira AD (1999) The rhizosphere and its management to improve plant growth. Adv Agron 66:1–102

Charron G, Furlan V, Bernier-Cordou M, Doyon G (2001) Response of onion plants to arbuscular mycorrhizae. 1. Effects of inoculation method and phosphorus fertilization on biomass and bulb firmness. Mycorrhiza 11:187–197

Drüge Š, Schönbeck F (1992) Effect of vesicular-arbuscular mycorrhizal infection on transpiration, photosynthesis and growth of flax (*Linus usitabissimum* L.) in relation to cytokinin levels. J Plant Physiol 140:40–48

Ellis JR, Watson DMH, Varvel GE, Jawson MD (1995) Methyl bromide soil fumigation alters plant elements concentrations. Soil Sci Soc Am J 59:848–852

Enkhtuya B, Rydlova J, Vosatka M (2000) Effectiveness of indigenous and non-indigenous isolates of arbuscular mycorrhizal fungi in soils from degraded ecosystems and man-made habitats. Appl Soil Ecol 14:201–211

Fries LLM, Pacovsky RS, Safir GR, Kaminski J (1998) Phosphorus effect on phosphatase activity in endomycorrhizal maize. Physiol Plant 103:162–171

Goicoechea N, Antolin MC, Strnad M, Sachez-Diaz M (1996) Root cytokines, acid phosphates and nodule activity in drought-stressed mycorrhizal or nitrogen-fixing alfalfa plants. J Exp Bot 47:683–686

Haas JH, Bar-Yosef B, Krikun J, Barak R, Markovitz T, Kramer S (1987) Vesicular–arbuscular mycorrhizal fungus infestation and phosphorus fertigation to overcome pepper stunting after methyl bromide fumigation. Agron J 79:905–910

Hetrick ABD, Kitt DG, Wilson T (1986) The influence of phosphorus fertilization, drought, fungal species, and non-sterile soil on mycorrhizal growth response in tall grass praiser plants. Can J Bot 64:1199–1203

Hetrick BAD, Wilson GWT, Todd TC (1996) Mycorrhizal response in wheat cultivars: relationship to phosphorus. Can J Bot 74:19–25

Johnson NC (1993) Can fertilization of soil select less mutualistic mycorrhizae. Ecol Appl 3:749–757

Kothari SK, Marschner H, Romheld V (1990) Effect of VA mycorrhizal fungi and rhizosphere microorganisms on root and shoot morphology, growth and water relationship in maize. New Phytol 116:303–311

Li XL, George E, Marschner H (1991) Extension of the phosphorus depletion zone in VA-mycorrhizal white clover in a calcareous soil. Plant Soil 135:41–48

Liu A, Hamel C, Hamilton RI, Smith DL (2000) Mycorrhizae formation and nutrient uptake of new corn (*Zea Mays* L.) hybrids with extreme canopy and leaf architecture as influenced by soil N and P levels. Plant Soil 22:157–166

Medeiros CAB, Clark RB, Ellis JR (1994) Growth and nutrient uptake of sorghum cultivated with vesicular-arbuscular mycorrhiza isolates at varying pH. Mycorrhiza 4:185–191

Olsen JK, Schaefer JT, Edwards DG, Hunter MN, Galea VJ, Muller LM (1999) Effects of mycorrhizae, established from an existing intact hyphal network, on the growth response of capsicum (*Capsicum annuum* L.) and tomato (*Lycopersicon esculentum* Mill.) to five rates of applied phosphorus. Aust J Agric Res 50:223–237

Ortaş I, Harris PJ (1996) The effect of partial soil sterilization and seasonal change on soil degradation (N-mineralization and soil chemical properties). In: Adana (ed) International conference on land degradation. Adana, Istambul, pp 204–217

Ortaş I, Harris PJ, Rowell DL (1996) Enhanced uptake of phosphorus by mycorrhizal sorghum plants as influenced by forms of nitrogen. Plant Soil 184:255–264

Ortaş I, Ortakcı D, Kaya Z (2002) Various mycorrhizal fungi propagated on different hosts have different effect on citrus growth and nutrient uptake. Commun Soil Sci Plant 33:259–272

Ortaş I (2003) Effect of selected mycorrhizal inoculation on phosphorus sustainability in sterile and non-sterile soils in the harran plain in south Anatolia. J Plant Nutr 26:1–17

Ortaş I, Sari N (2003) Enhanced yield and nutrient content of sweet corn with mycorrhizal inoculation under field conditions. Agric Med 133:188–195

Ortaş I, Sari N, Akpınar Ç (2003) Effects of mycorrhizal inoculation and soil fumigation on the yield and nutrient uptake of some solanaceas crops (tomato, eggplant and pepper) under field conditions. Agric Med 133:249–258

Ortaş I, Akpinar A, Coskan A, Demirbas A (2004) Producing mycorrhizal seedling for sustainable agriculture (short presentation at 8th meeting of COST Action 3.83). In: COST (ed) Exploring and exploiting the natural AM fungus diversity in stressed soil–plant system (from molecular techniques to biotechnological tools). COST, Grenada, pp 61

Peterson RL, Massicotte HB, Melville LH (2004) Mycorrhizas: anatomy and cell biology. CABI, Ottawa

Posta K, Fuleky G (1997) The effect of phosphorus on the mycorrhizal colonization of maize by *Glomus Mosseae* (Nicol. Gerd). Novenytermeles 46:573–582

Ryan MH, Angus JF (2003) Arbuscular mycorrhizae in wheat and field pea crops on a low P soil: increased Zn-uptake but no increase in P-uptake or yield. Plant Soil 250:225–239

Sanders FE, Tinker PB (1973) Phosphate inflow into mycorrhizal roots. Pest Sci 4:385–395

Smith SE, Read DJ (1997) Mycorrhizal symbiosis, 2nd edn. Academic, London

Stribley DP, Tinker PB, Rayne, JHB (1980) Relation of internal phosphorus concentration and plant weight in plants infected by vesicular arbuscular mycorrhizas. New Phytol 78:263–268

Tarafdar JC, Marschner H (1994) Phosphatase activity in the rhizosphere and VA mycorrhizal wheat supplied with inorganic and organic phosphorus. Soil Biol Biochem 26:387–395

Tarkalson DD, Jolley VD, Robbins CW, Terry RE (1998) Mycorrhizal colonization and nutrition of wheat and sweet corn grown in manure-treated and untreated topsoil and subsoil. J Plant Nutr 21:1985–1999

Tinker PB (1975) The chemistry of phosphorus and mycorrhizal effects on plant growth. In: Sender FS, Mosse B, Tinker PB (eds) Endomycorrhizas. Academic, London, pp 353–371

Subject Index